Advances in Thin Film Fabrication by Magnetron Sputtering

Advances in Thin Film Fabrication by Magnetron Sputtering

Editor

Rafal Chodun

Basel • Beijing • Wuhan • Barcelona • Belgrade • Novi Sad • Cluj • Manchester

Editor
Rafal Chodun
Warsaw University of
Technology
Warsaw
Poland

Editorial Office
MDPI
St. Alban-Anlage 66
4052 Basel, Switzerland

This is a reprint of articles from the Special Issue published online in the open access journal *Coatings* (ISSN 2079-6412) (available at: https://www.mdpi.com/journal/coatings/special_issues/Magnetron_Sputtering).

For citation purposes, cite each article independently as indicated on the article page online and as indicated below:

Lastname, A.A.; Lastname, B.B. Article Title. *Journal Name* **Year**, *Volume Number*, Page Range.

ISBN 978-3-7258-0607-2 (Hbk)
ISBN 978-3-7258-0608-9 (PDF)
doi.org/10.3390/books978-3-7258-0608-9

Cover image courtesy of Rafal Chodun

© 2024 by the authors. Articles in this book are Open Access and distributed under the Creative Commons Attribution (CC BY) license. The book as a whole is distributed by MDPI under the terms and conditions of the Creative Commons Attribution-NonCommercial-NoDerivs (CC BY-NC-ND) license.

Contents

About the Editor vii

Preface ix

Rafal Chodun, Bartosz Wicher, Katarzyna Nowakowska-Langier, Roman Minikayev, Marlena Dypa-Uminska and Krzysztof Zdunek
On the Control of Hot Nickel Target Magnetron Sputtering by Distribution of Power Pulses
Reprinted from: *Coatings* **2022**, *12*, 1022, doi:10.3390/coatings12071022 1

Omar Gaspar Ramírez, Manuel García Méndez, Ricardo Iván Álvarez Tamayo and Patricia Prieto Cortés
Synthesis of Magnetron-Sputtered TiN Thin-Films on Fiber Structures for Pulsed-Laser Emission and Refractive-Index Sensing Applications at 1550 nm
Reprinted from: *Coatings* **2023**, *13*, 95, doi:10.3390/coatings13010095 15

Young Hyun Jo, Cheongbin Cheon, Heesung Park and Hae June Lee
Particle-in-Cell Simulations for the Improvement of the Target Erosion Uniformity by the Permanent Magnet Configuration of DC Magnetron Sputtering Systems
Reprinted from: *Coatings* **2023**, *13*, 749, doi:10.3390/coatings13040749 35

Patrycja Pokora, Damian Wojcieszak, Piotr Mazur, Małgorzata Kalisz and Malwina Sikora
Influence of Co-Content on the Optical and Structural Properties of TiO_x Thin Films Prepared by Gas Impulse Magnetron Sputtering
Reprinted from: *Coatings* **2023**, *13*, 955, doi:10.3390/coatings13050955 50

Giji Skaria, Ashwin Kumar Saikumar, Akshaya D. Shivprasad and Kalpathy B. Sundaram
Annealing Studies of Copper Indium Oxide ($Cu_2In_2O_5$) Thin Films Prepared by RF Magnetron Sputtering
Reprinted from: *Coatings* **2021**, *11*, 1290, doi:10.3390/coatings11111290 71

Ewa Mańkowska, Michał Mazur, Jarosław Domaradzki and Damian Wojcieszak
P-type (CuTi)Ox Thin Films Deposited by Magnetron Co-Sputtering and Their Electronic and Hydrogen Sensing Properties
Reprinted from: *Coatings* **2023**, *13*, 220, doi:10.3390/coatings13020220 79

Nattakorn Borwornpornmetee, Peerasil Charoenyuenyao, Rawiwan Chaleawpong, Boonchoat Paosawatyanyong, Rungrueang Phatthanakun, Phongsaphak Sittimart, et al.
Physical Properties of Fe_3Si Films Coated through Facing Targets Sputtering after Microwave Plasma Treatment
Reprinted from: *Coatings* **2021**, *11*, 923, doi:10.3390/coatings11080923 97

Malwina Sikora, Damian Wojcieszak, Aleksandra Chudzyńska and Aneta Zięba
Improved Methodology of Cross-Sectional SEM Analysis of Thin-Film Multilayers Prepared by Magnetron Sputtering
Reprinted from: *Coatings* **2023**, *13*, 316, doi:10.3390/coatings13020316 112

Jacob M. Wall and Feng Yan
Sputtering Process of $Sc_xAl_{1-x}N$ Thin Films for Ferroelectric Applications
Reprinted from: *Coatings* **2023**, *13*, 54, doi:10.3390/coatings13010054 125

Ashwin Kumar Saikumar, Sreeram Sundaresh, Shraddha Dhanraj Nehate and Kalpathy B. Sundaram
Properties of RF Magnetron-Sputtered Copper Gallium Oxide ($CuGa_2O_4$) Thin Films
Reprinted from: *Coatings* **2021**, *11*, 921, doi:10.3390/coatings11080921 143

Zhao Wang, Nan Lan, Yong Zhang and Wanrong Deng
Microstructure and Properties of MAO-Cu/Cu-(HEA)N Composite Coatings on Titanium Alloy
Reprinted from: *Coatings* **2022**, *12*, 1877, doi:10.3390/coatings12121877 . . . . **154**

Ping-Hang Chen, Wen-Jauh Chen and Jiun-Yi Tseng
Thermal Stability of the Copper and the AZO Layer on Textured Silicon
Reprinted from: *Coatings* **2021**, *11*, 1546, doi:10.3390/coatings11121546 . . . . **167**

About the Editor

Rafal Chodun

Rafal Chodun works at the Faculty of Materials Science and Engineering of the Warsaw University of Technology as an Assistant Professor. His scientific interests are related to the unconventional synthesis of materials using plasma. The notion of unconventional synthesis means using an approach that extends the conventional field of thermodynamic parameters of state (T, P and C) to unconventional parameters, e.g., electric field and magnetic field. He deals with modifying plasma surface engineering methods, so his work aims to use technological solutions to support the plasma reactions and plasma interaction with the surface of the solids. His research concerns the characterization of the plasma environment and the characterization of the coating material. The utilitarian purpose of Guest Editor research is the development of deposition technology methods with specific structural, chemical, and phase structure features that are unachievable by mainstream plasma engineering techniques (conventional). He is the author of 54 publications, co-author of 2 patents, contributor to the conception and development of the Gas Injection Magnetron Sputtering method, and contributor to industrial applications of magnetron techniques. He is a member of the International Association of Advanced Materials, the Society of Vacuum Coaters, and the Polish Vacuum Society.

Preface

Magnetron sputtering is one of the fastest-emerging plasma surface engineering technologies. As a result, this reprint is aimed at those involved in developing magnetron sputtering technology in its broadest sense. It presents cutting-edge developments with ideas on using novel instruments to improve the film fabrication efficiency of various functional materials: simulation tools, construction modifications, material synthesis boosters, optimized film material parameters, etc.

Rafal Chodun
Editor

Article

On the Control of Hot Nickel Target Magnetron Sputtering by Distribution of Power Pulses

Rafal Chodun [1,*], Bartosz Wicher [1], Katarzyna Nowakowska-Langier [2], Roman Minikayev [3], Marlena Dypa-Uminska [1] and Krzysztof Zdunek [1]

[1] Faculty of Materials Science and Engineering, Warsaw University of Technology, Woloska 141, 02-507 Warsaw, Poland; bartosz.wicher.dokt@pw.edu.pl (B.W.); marlena.dypa.stud@pw.edu.pl (M.D.-U.); krzysztof.zdunek@pw.edu.pl (K.Z.)
[2] National Centre for Nuclear Research, Andrzeja Soltana 7, 05-400 Otwock-Swierk, Poland; katarzyna.nowakowska-langier@ncbj.gov.pl
[3] Institute of Physics of the Polish Academy of Sciences, al. Lotnikow 32/46, 02-668 Warsaw, Poland; minik@ifpan.edu.pl
* Correspondence: rafal.chodun@pw.edu.pl; Tel.: +48-22-234-8704

Abstract: This paper presents the experimental results of high-temperature sputtering of nickel targets by the Gas Injection Magnetron Sputtering (GIMS) technique. The GIMS technique is a pulsed magnetron sputtering technique that involves the generation of plasma pulses by injecting small doses of gas into the zone of the magnetron target surface. Using a target with a dedicated construction to limit heat dissipation and the proper use of injection parameters and electrical power density, the temperature of the target during sputtering can be precisely controlled. This feature of the GIMS technique was used in an experiment with sputtering nickel targets of varying thicknesses and temperatures. Plasma emission spectra and current-voltage waveforms were studied to characterize the plasma process. The thickness, structure, phase composition, and crystallite size of the nickel layers produced on silicon substrates were investigated. Our experiment showed that although the most significant increase in growth kinetics was observed for high temperatures, the low sputtering temperature range may be the most interesting from a practical perspective. The excited plasma has the highest energy in the sputtering temperature range, just above the Curie temperature.

Keywords: ferromagnetic target sputtering; nickel sputtering; hot sputtering; hot target; nickel coatings; gas injection magnetron sputtering; GIMS; magnetron sputtering

1. Introduction

One branch of magnetron sputtering technology has attracted strong interest recently. This branch is the so-called Hot Target Magnetron Sputtering (HTMS) technique [1,2]. HTMS aims to run a cathode sputtering process with an increased temperature. It is possible by modifying the construction of the magnetron target and involves the creation of a break in the contact area with the cooled surface. The flux of dissipated heat is reduced, and the target itself accumulates a significant heat load in its volume, raising its temperature. The primary purpose of this procedure is to increase the target material's vapor pressure and lead to its sublimation. The stream of particles produced during such a process of sputtering and reaching the substrate is significantly enriched by the stream of sublimated species. Densified plasma flux is reflected in a considerable increase in the growth kinetics of the coatings and leads to a significant economization of the technological process [3]. One of the varieties of HTMS processes is the intentional increase in target temperature above the melting point of Liquid Target Magnetron Sputtering [4]. This way, the species flux reaching the substrate is enriched with evaporated particles. The kinetics of growth is one of the critical parameters determining the applicability of the magnetron sputtering technique and seems to be of particular importance in its variants where low growth kinetics is a genetic feature of the technique. This group of methods includes pulsed methods such as High Power Magnetron Sputtering (HiPIMS). A variation in HTMS has also found positive effects in HiPIMS technology [5–9].

One practical purpose of using HTMS is to sputter target ferromagnetic materials effectively. Sputtering of magnetic materials is highly difficult in standard magnetron sputtering techniques as ferromagnetic materials confine magnetic field lines within their volume. The magnetic field strength above the surface of the target is attenuated, resulting in difficulty initiating and sustaining a glow discharge [10]. Magnetron sputtering technology employs various approaches to overcome these obstacles, such as using additional sources of plasma excitation [11–13], thin targets [14], and particular magnetic field configuration [15,16]. Although the mentioned methods have undoubted advantages, they are not ideal, and new solutions are still being sought. One of the more exciting and newer ideas is the HTMS technique to conduct the sputtering process at the target temperature above the Curie temperature T_C. This way, sputtering of ferromagnetic materials is efficient. The HTMS technique has already been used to sputter nickel targets [17,18]. So far, the authors are not aware of any published work showing the application of HTMS in the cases of iron and cobalt. This is due to the shielding effect of the magnetic field and the difficulty in controlling the temperature. These metals have much higher T_C than nickel, 770 and 1121 °C, respectively. The magnitude of the difficulty is mainly indicated by the approach of the T_C to the melting point; in these cases, they are, respectively: 1538 and 1495 °C. "Temperature window" of sputtering processes is thus narrower and at a higher temperature. The most promising development track of the HTMS technique is the effective control of the target temperature.

Target temperature control is a critical issue in HTMS technology. Without it, the magnetron target can be plastically deformed due to creeping or whole magnetron construction can be seriously damaged. The permanent magnet system is particularly vulnerable. Another risk is the possibility of melting the target material. Another factor to consider is the effect of heat radiation on the plasma and the substrate–growing coating system. These issues have not yet been widely discussed in the literature, and the reason can be speculated to be the difficulty in achieving arbitrary sputtering temperature determination. The factor that is standardly used to establish the temperature of the target, the electrical power density, is probably insufficient in HTMS technological practice. Recently, our team has used Gas Injection Magnetron Sputtering (GIMS) for HTMS technology [19]. At that stage, we thought that by using GIMS, we would introduce a crucial temperature control parameter. The GIMS technique is a pulsed technique that uses the generation of discrete plasma streams (plasma pulses) in an environment of dynamically varying working gas concentration, controlled by high-speed pulsed gas valves [20,21]. Typically, in GIMS, layers are produced due to the generation of $10–10^2$ ms plasma pulses at frequencies of $10^{-1}–10^1$ s. The pulse power density is of the order of 10^2 W/cm^2. Controlling the pulsing parameter, i.e., frequency and single pulse lifetime, allowed us to set the sputtering temperature reasonably precisely.

In this experiment, we aimed to exploit the advantages of using the GIMS technique in HTMS technology and see if it is effective in sputtering nickel targets. The second objective of our experiment was to characterize the HTMS process of nickel targets with varying thicknesses (magnetic field shielding strength) at various temperatures.

2. Materials and Methods

2.1. Apparatus

HTMS processes were carried out in a vacuum apparatus presented in Figure 1a. The vacuum chamber was equipped with a vacuum pump system consisting of turbomolecular, roots, and rotary pumps. The vacuum pump unit could reach a base pressure of 5×10^{-3} Pa. An extremely unbalanced, according to Gencoa criterion [22], circular magnetron was used as the plasma source. The 50 mm diameter nickel targets (Grade 1 purity) were installed in the magnetron.

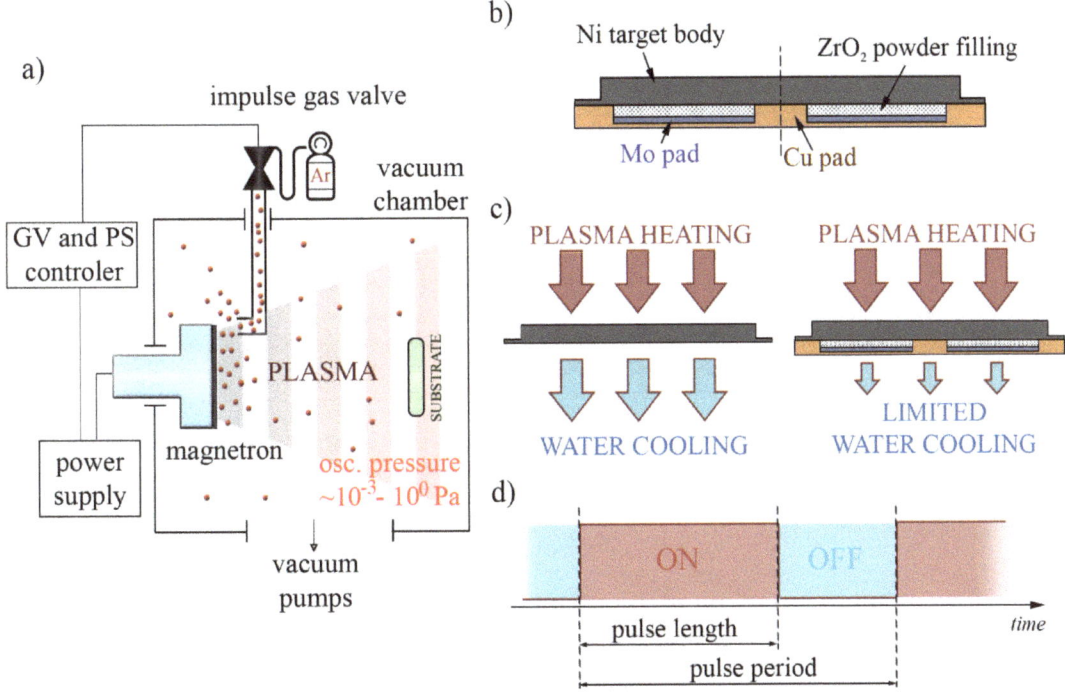

Figure 1. Diagram illustrating: (**a**) the apparatus configuration, (**b**) construction of the hot target system, (**c**) the heat transfer through the standard (left) and hot (right) target system, (**d**) plasma pulse distribution over the time.

Figure 1b presents the sandwich construction of the used targets. The target sandwich consisted of a nickel target, ZrO_2 powder filling, molybdenum foil, and a copper pad. The 1.5, 2, 2.5, and 3 mm thick nickel targets were used for the experiment. A 1.5-mm deep channel was fabricated in the copper pad. The channel was filled with ZrO_2 powder to provide low thermal conductivity at the high-temperature zone of the target system. Additionally, a zirconia filling prevented material deformation due to high temperature. During the coating deposition, the bottom surface of the copper pad was cooled with water. The qualitative difference between hot and standard sputtering is presented in Figure 1c.

2.2. Coatings Deposition Process

Coating deposition processes were carried out using argon as sputtering gas (N5.0 purity). The working atmosphere was created by injections from a fast pulse valve directly at the target surface. The pressure at the valve inlet was set at 1.5×10^5 Pa. The opening time of the Ar valve was fixed at 4 ms. The applied gas settings resulted in a pressure oscillation with a peak of about 10^{-1} Pa. A precise determination of the oscillation was impossible due to the high time inertia of the used vacuum gauge. Pawlak in [23] studied a more accurate study of the pressure characteristic in GIMS.

The plasma 500 ms pulse length and period were the parameters controlling the target temperature in the experiment. Individual power pulses are electronically coupled to the opening moment of the pulsed gas valve. The electronic coupling of the power supply to the valve operation allows the precise determination of the glow discharge at a fixed time from the gas injection. In this experiment, we established the discharge when opening the valve and sustaining it for 500 ms. The interaction of plasma energetic species with the target surface induces thermal energy in the target material. The energy distribution has a pulsed manner in GIMS. Therefore, relatively frequent events of thermal excitation (plasma pulses) increase the temperature of the target material. During the interval between pulses, the accumulated heat is partially dissipated. The heat dissipation is effective when the impulse interval is longer [19]. The pulse length and pulse periods are described in detail in Table 1 with the other process parameters. Figure 1d schematically presents the power pulses time distribution in the GIMS process. The nature of the thermal excitation was pulsed, and the target was continuously heated up during the plasma phase until a specific target temperature resulting from the system's ability to dissipate heat was established. Each discharge is followed by the period of the dissipation of the accumulated heat. The temperature characteristics during the process of sputtering can be considered quasi-stationary. The temperature of the targets was measured using an MM2MH Raytek pyrometer and high-speed Optris Xi400 thermal camera. The center of the target was the spot with the highest temperature. The temperature data in Table 1 shows the pyrometer measurement results at the target's center. As the sensitivity of the pyrometer starts at 450 °C, the process labeled "<450" in the table was performed at low-temperature conditions, below the T_C.

Table 1. Parameters of hot nickel target sputtering in the experiment.

Target Thickness (mm)	Temperature	Period (ms)	Deposition Time (s)
1.5	<450	2000	720
	450	1300	468
	600	1150	414
	800	900	324
	1000	650	234
2	<450	2000	720
	450	1500	540
	600	1100	396
	800	900	324
	1000	650	234
2.5	<450	2000	720
	450	1100	396
	600	730	263
	800	580	209
	1000	550	198
3	<450	2000	720
	450	1500	540
	600	1000	360
	800	650	234
	1000	550	198

The n-type 100 silicon substrates were prepared by ultrasonic cleaning in acetone and drying. After that, they were located perpendicularly to the magnetron Z-axis at 10 cm from its surface. The substrate stage was electrically grounded during coating deposition with the magnetron anode and vacuum chamber. The magnetron system was powered using a Dora Power Systems pulsed DC power supply with a carrier frequency of 125 kHz [24,25].

The power waveforms were calculated from the current and voltage characteristics, measured using the 1 A: 0.1 V Rogowski coil and 1 kV: 1 V voltage, and registered by a Rigol oscilloscope. The pulse energy was calculated using the integral over time. All the calculations were made using the Origin Lab software. The distribution of normal magnetic field $B_\perp$ at the surface of the nickel targets used in the experiments was investigated by SMS-102 Asonik Hall effect meter.

2.3. Plasma Characterization

The optical emission spectroscopy (OES) measurements were obtained using an Optel energy-dispersive optical spectrometer at a wavelength of 200–900 nm. The signals were collected through a quartz window and an optical fiber. The exposition time was set at a full single plasma pulse. During these experiments, the optical collimator was perpendicular to the magnetron Z-axis and placed 55 mm away from the cathode surface.

2.4. Coatings Characterization

The deposited coatings were further characterized by scanning electron microscopy (SEM). The cross-section of nickel coatings fabricated at silicon substrates was observed at a 45° to the surface plane using a ZEISS Ultra Plus device. The cross-sections of the samples were obtained by brittle fracture just after cooling them in liquid nitrogen.

The crystal structure of the coatings was analyzed by X-ray diffraction (XRD) using Cu-$K\alpha$ radiation. The XRD profiles of the layers were measured at a 2θ range of 15–120°. The Debye–Scherrer equation was used to calculate the size of crystallites. Crystalline size D, is expressed:

$$D = \frac{K\lambda}{\beta \cos \theta} \tag{1}$$

where D is the size of the grain, K is known as the Scherer's constant (K = ~0.9), λ is the X-ray wavelength (for Cu 1.54178Å), β is full width at half maximum (FWHM) of the diffraction peak, and θ is the angle of diffraction.

3. Results and Discussion

3.1. Plasma Characterization

The first step in the experiment was to characterize the sputtering process of the nickel target during its progressive heating. For this purpose, current and voltage waveforms were measured at distinct stages of heating a nickel target 1.5 mm thick. Power waveforms were determined based on the indications of the electric power supply integrating the current and voltage waveforms in situ during its operation. The results are shown in Figure 2. The figure shows the fitting of the discharge power curve during the heating of the target. At the initial stage, when the target material's volume is below the Curie temperature, the glow discharge is characterized by relatively low power. The power increases when an increased volume of the target material exhibits a transition from a ferromagnetic state to a non-magnetic state. The nature of the transition is relatively smooth and is associated with a gradual increase in the volume of the non-magnetic fraction of the nickel target. In the final part of the curve, the power value stabilizes. Power stabilization is associated with forming a constant volume of non-magnetic material of the target. Images taken with a thermal camera at distinct stages of heating the target are attached to the figure. When considering the temperature of the target in the HTMS process, it is necessary to keep in mind its irregular distribution in the volume. The images clearly show that its hottest parts are primarily in the center. As the sputtering temperature increases, the pulse power waveforms also change significantly. The figure presents examples of the curve's power waveforms measured in characteristic parts. The power pulses were integrated, and the energy of a single pulse was determined—also shown in the figure. The nature of the changes in the energy values is related to the amplitude of the power waveform, which is the result of changes in the discharge current waveform.

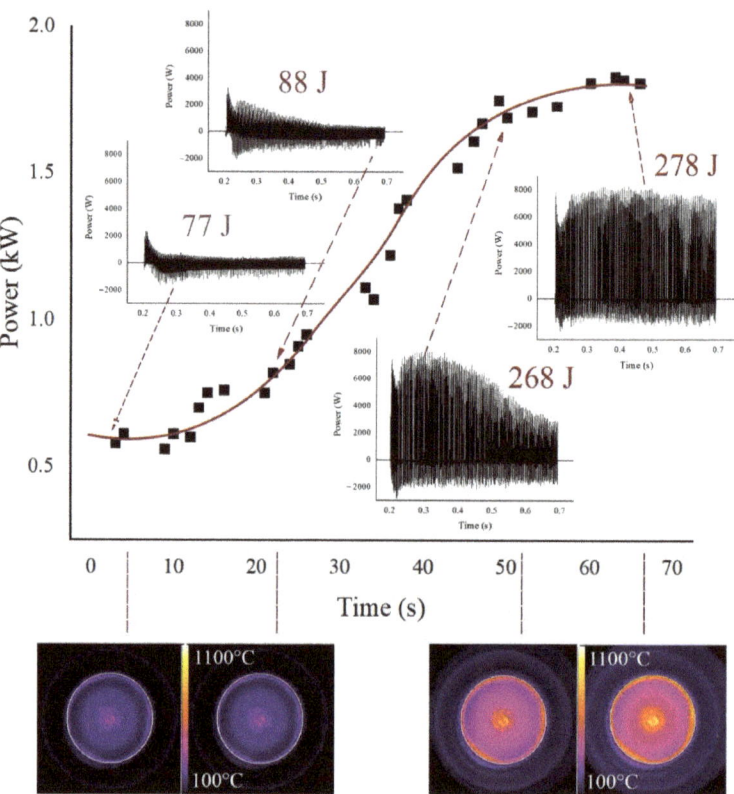

Figure 2. Diagram presenting the evolution of glow discharge power during the sputtering of hot nickel target and images of temperature distribution on the target surface.

The next stage of our experiment was to study the effect of temperature and thickness of the nickel target on the energy of the glow discharge. Sputtering processes of nickel targets were carried out in the temperature range up to 1000 °C. The temperature was controlled by adjusting the period of triggering the plasma pulse according to Table 1. Targets with thicknesses of 1.5, 2, 2.5, and 3 mm were sputtered. Using targets of different thicknesses, we could characterize the discharge under various magnetic field energy supports. The effect of target thickness on $B\perp$ is shown in Figure 3. The magnetron used in the initial state has a maximum intensity of 90 mT. Using a cathode system with a 1.5 mm nickel target reduced the $B\perp$ maximum to ~16 mT. $B\perp$ of ~1.5 mT characterized the 3 mm target. The magnetic field intensity distribution over the target itself does not change much. Figure 3 also shows the energy characteristics as a function of sputtering temperature. The thickness of the target, that is, the $B\perp$ value, significantly affects the pulse energy. The characteristics themselves are interesting. In the range of temperatures, in which we conducted measurements, the maximum energy was noted at temperatures ~100 °C above T_C. In the range of about 500–900 °C, we observed a decrease in the pulse's energy. At temperatures above 900 °C, the energy value increased again.

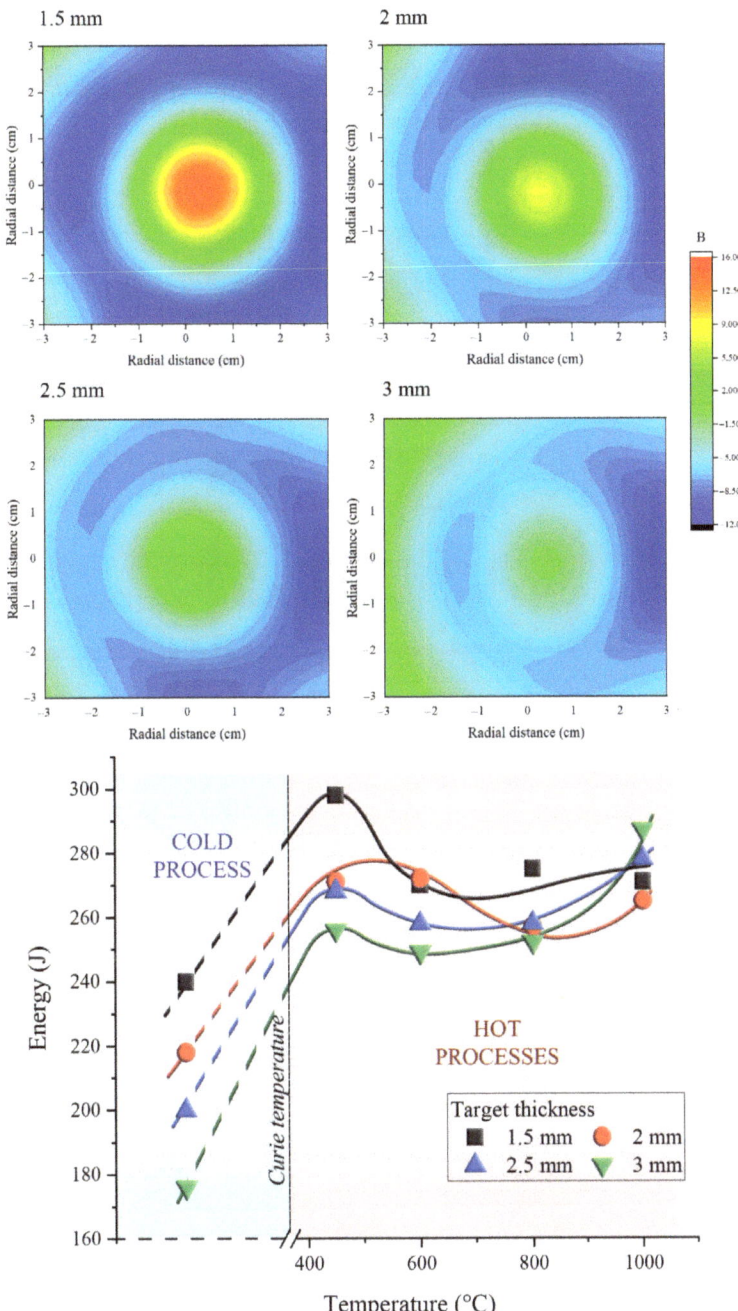

Figure 3. Distribution of $B\perp$ at the surface of nickel targets with various thickness and characteristics of plasma pulse energy during the sputtering process.

The parameter controlling the pulse energy value is the character of the power waveform, which in turn is determined by the discharge current characteristics. Figure 4 shows the discharge current waveforms measured while sputtering a 1.5 mm nickel target at different temperatures. Temperatures at which the pulse energy was relatively low are characterized by current decay at the end of the discharge duration.

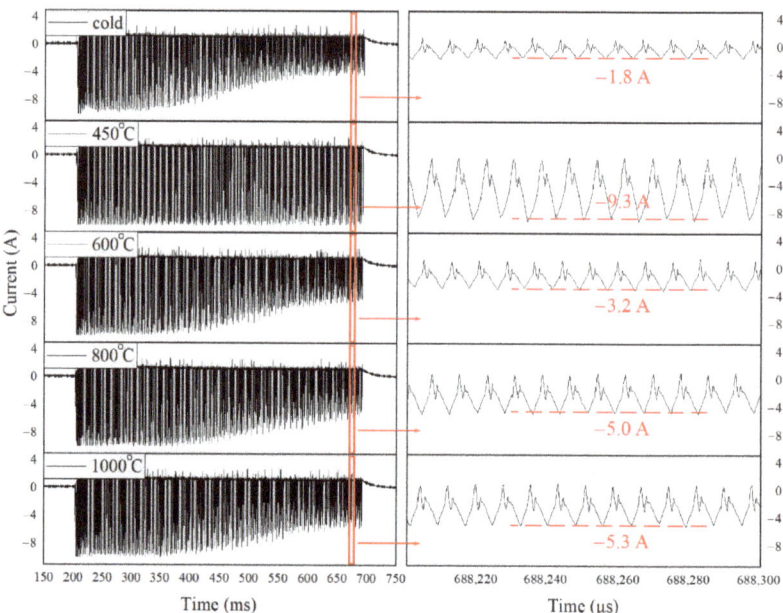

Figure 4. Current waveforms measured during the sputtering of 1.5 mm nickel target at various temperatures.

The observed decay is not an effect related to the characteristics of the GIMS process, i.e., dissipation of the gas doses in the volume of the vacuum chamber. The gas dosage settings we use are established by years of practice, and so far, we have not observed this effect. In our opinion, the disappearance of the discharge current in the temperature range of 500–900 °C can be caused by the so-called gas rarefaction phenomenon [26–28]. The rarefaction phenomenon occurs when the gas is overheated due to high-power sources and additional heat sources for synthesis [29]. The effect of this phenomenon is the thinning of the gas atmosphere and, thus, the disappearance of current electric carriers, which is reflected in the measured current characteristics. This effect is evident at the end of the discharge since there is a temperature peak, even though the target temperature is quasistationary. At very high temperatures, we observed an increase in the current amplitude value. It seems that the increase in the population in the gaseous atmosphere of sublimating molecules is responsible for this state. We can expect sublimation to occur as early as <900 °C. However, its effects are not so apparent since the flux of particles produced during sublimation depends on the emitting surface, and the temperature distribution on the surface of the target is not homogeneous.

OES studies were performed to determine how the sputtering temperature affects the nickel particle population in the plasma and whether the measured spectra are reflected in previous studies. Although the intensities of the optical emission lines do not directly indicate the species population in the plasma, the changes in the line intensity ratios between each species group can provide valuable information on the plasma content with the change of process parameters [30]. Figure 5 shows exemplary sections of the plasma emission spectrum created during the sputtering of a 1.5 mm nickel target. For presentation purposes, the spectral range in which the intensity of excited nickel N I particles was the highest was selected for analysis. The presented range included the lines with the characteristics listed in Table 2. The rest of the figure shows the spectral intensity for different target thicknesses and temperatures.

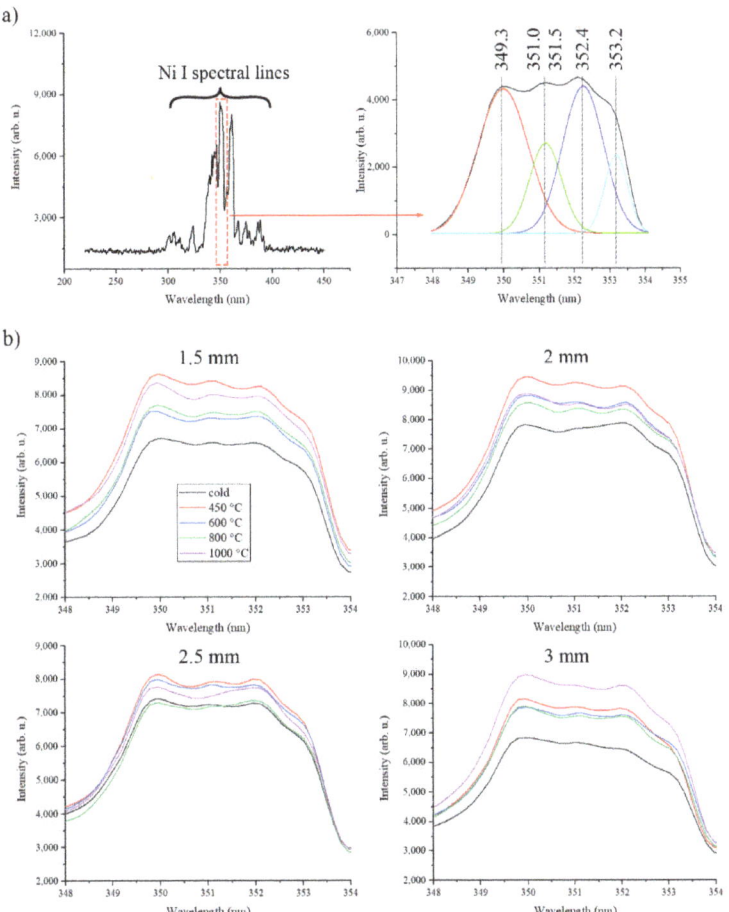

Figure 5. OES spectrum of plasma registered and fitted with basic components during the sputtering of nickel target (**a**) and collection of OES spectra registered at various temperatures (**b**).

Table 2. OES lines attributed to the measured spectra [31].

Specie	Wavelength (nm)	Rel. Intensity	Lower Level	Upper Level
Ni I	349.3	5500	$3d^9(^2D)4s$	$3d^9(^2D)4p$
Ni I	351.0	2600	$3d^9(^2D)4s$	$3d^9(^2D)4p$
Ni I	351.5	6600	$3d^9(^2D)4s$	$3d^9(^2D)4p$
Ni I	352.4	8200	$3d^9(^2D)4s$	$3d^9(^2D)4p$
Ni I	353.2	1100	$3d^8(^3F)4s^2$	$3d^9(^2D)4p$

Comparison of the individual spectra leads to the conclusion that these results are consistent with calculated energy pulses (Figure 3). The intensity of the lines is correlated with the pulse energy and indicates the sputtering effectiveness of the nickel target surface. The sputtering process's spectrum at $T < T_C$ shows the lowest intensity. The highest intensity was registered for a temperature of 450 °C. The spectra collected in the range of 500–900 °C are moderate intensities.

3.2. Coatings Characterization

The SEM images of the nickel coatings' structure was shown in Figure 6. SEM investigations show the structural morphology of the deposited coatings and their thickness. The structure of the coatings is columnar, typical for the magnetron sputtering method [32,33]. In some cases, it is difficult to observe the columnar structure, probably due to the plastic deformation during the cracking, despite prior freezing by liquid nitrogen. Thickness results read from the images are shown in Table 3.

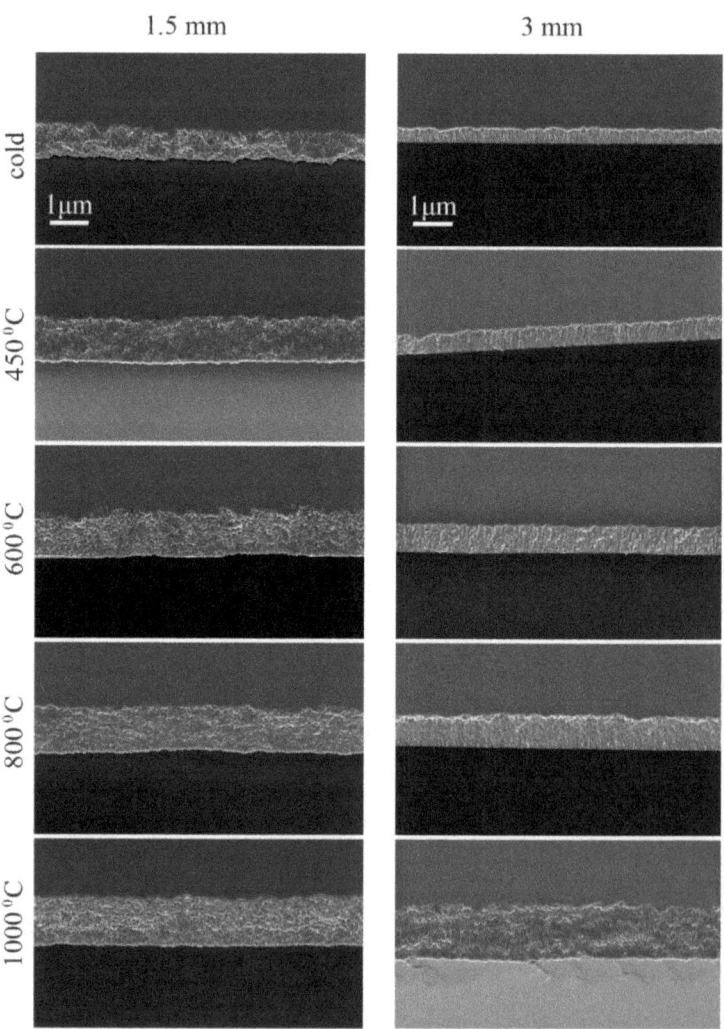

Figure 6. SEM images of nickel coatings structures deposited during the sputtering at various temperatures.

Table 3. Nickel coatings thickness deposited during the HTMS processes.

1.5 mm Thick Target	3 mm Thick Target	Temperature (°C)
900 nm	380 nm	<450
950 nm	500 nm	450
1070 nm	640 nm	600
1080 nm	680 nm	800
1100 nm	1140 nm	1000

The conclusion that comes immediately to mind is that the coatings deposited by sputtering a 3 mm nickel target are characterized by a smaller thickness than those obtained with a thinner target. As expected, the thin nickel target sputters more efficiently than the thick target. The effect of the transformation of ferromagnetic to non-magnetic material is mainly evident in the case of the thick target. The growth kinetics in its case increased by 200%, while the thin target increased by only 22%. There is a noticeable increase in the thickness of coatings from a temperature of 450 °C in both cases. Previous results indicated that the best thickness results could be expected at 450 °C. However, it seems that the effects of high discharge energies and the presence of energetic particles in the plasma cannot be combined with the density of the particle flux emitted from the targets. The jet of sputtered particles is enriched with sublimated particles. These, in turn, are low-energy and do not participate in the processes of ionization and internal excitation, which is why previous studies have not considered their contribution to the population.

XRD technique was used to examine the structure of the coatings in more detail. Figure 7 shows a comparison of the diffraction patterns of the fabricated coatings. We have limited the image to the coatings deposited by sputtering 1.5 and 3 mm nickel targets for presentation purposes. Diffraction patterns are very similar to each other. Reflections from the same set of crystallographic planes were identified in all coatings. XRD peaks were used to calculate the size of the crystallites, according to equation 1. The averaged values of the crystallite size are shown in Table 4.

Table 4. Averaged crystallites' size of nickel coatings deposited in experiment.

1.5 mm Thick Target	3 mm Thick Target	Temperature (°C)
7.8 nm	18.2 nm	<450
7.3 nm	10.2 nm	450
7.2 nm	13.1 nm	600
7.7 nm	15.2 nm	800
12.6 nm	15.0 nm	1000

A certain consistency can be seen in the results, in which the smallest crystallites presented layers fabricated at mid-range temperature. Previous studies have suggested that plasma species may have the highest energy under such conditions. This is reflected in the structure and size of the crystallites. It is known that the coating tends to defect and defragment the crystallites during growth under energetic particles bombardment [34]. Above this temperature, an increase in size is observed. This may be due to the increased fraction of low-energy particles originating from sublimation.

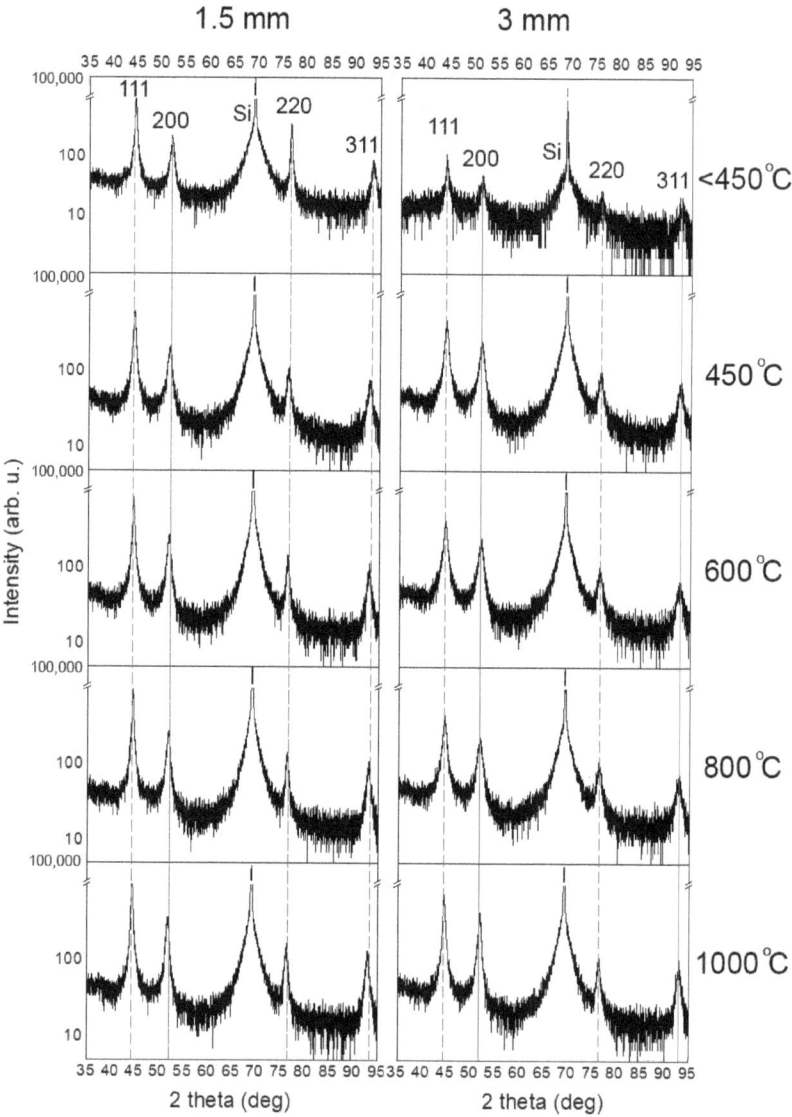

Figure 7. Diffraction patterns of nickel coatings deposited in the experiment.

4. Conclusions

In this article, we have attempted to actively employ the technological parameters of the GIMS technique to control the temperature in the HTMS process of nickel targets. The period of plasma pulses distribution was a parameter by which we could precisely establish the sputtering temperature. As part of the experiment, we attempted to characterize the sputtering process of a temperature-controlled target, and as we expected, the most significant increase in growth kinetics was observed for the highest temperatures. Our experiment also showed that the sputtering model of ferromagnetic targets at temperatures slightly above T_C is exciting and worthy of closer study. From our preliminary findings, one may be inclined to the thesis that this temperature range may be the most favorable for the generation of plasmas whose particles have the highest energies. In this temperature range, we recorded the highest-energy discharges, and the OES spectrum of the plasma excited under these conditions was most enriched with excited particles of sputtered nickel. The preservation of high energy of plasma particles during film fabrication is one of the most raised issues in the literature. Acquiring control over the execution of HTMS processes is a crucial skill in mastering this important parameter of the plasma state. The GIMS technique, at its current stage of development, seems to enable this satisfactorily. As a result, we can use this tool in our future research plans.

Author Contributions: Conceptualization, R.C. and B.W.; methodology, R.C. and B.W.; formal analysis, R.C.; investigation, R.C., K.N.-L., R.M., and M.D.-U.; resources, B.W.; data curation, R.C. and M.D.-U.; writing—original draft preparation, R.C.; writing—review and editing, R.C.; visualization, R.C.; supervision, R.C.; project administration, K.Z.; funding acquisition, K.Z. All authors have read and agreed to the published version of the manuscript.

Funding: This research was funded by the National Science Centre (grant no. 2018/31/B/ST8/00635).

Institutional Review Board Statement: Not applicable.

Informed Consent Statement: Not applicable.

Data Availability Statement: Data is contained within the article.

Conflicts of Interest: The authors declare no conflict of interest.

References

1. Wolfe, J.C.; He, W.-S.; Licon, D.L.; Chau, R.Y. Ion Assisted Deposition Process with Reactive Source Gassification. U.S. Patent 5,415,756, 16 May 1995.
2. Chau, R.Y.; He, W.-S.; Wolfe, J.C.; Licon, D.L. Effect of target temperature on the reactive d.c.-sputtering silicon and niobium oxides. *Thin Solid Films* **1996**, *287*, 57–64. [CrossRef]
3. Bleykher, G.A.; Borduleva, A.O.; Krivobokov, V.P.; Sidelev, D.V. Evaporation factor in productivity increase of hot target magnetron sputtering systems. *Vacuum* **2016**, *132*, 62–69. [CrossRef]
4. Sidelev, D.V.; Bleykher, G.A.; Bestetti, M.; Krivobokov, V.P.; Vicenzo, A.; Franz, S.; Brunella, M.F. A comparative study on the properties of chromium coatings deposited by magnetron sputtering with hot and cooled target. *Vacuum* **2017**, *143*, 479–485. [CrossRef]
5. Karzin, V.V.; Komlev, A.E.; Karapets, K.I.; Lebedev, N.K. Simulation of heating of the target during high-power impulse magnetron sputtering. *Surf. Coat. Technol.* **2018**, *334*, 269–273. [CrossRef]
6. Tesař, J.; Rezek, J.M. On surface temperatures during high power pulsed magnetron sputtering using a hot target. *Surf. Coat. Technol.* **2011**, *206*, 1155–1159. [CrossRef]
7. Bleykher, G.A.; Sidelev, D.V.; Grudinin, V.A.; Krivobokov, V.P.; Bestetti, M. Surface erosion of hot Cr target and deposition rates of Cr coatings in high power pulsed magnetron sputtering. *Surf. Coat. Technol.* **2018**, *354*, 161–168. [CrossRef]
8. Grudinin, A.; Bleykher, G.A.; Sidelev, D.V.; Krivobokov, V.P.; Bestetti, M.; Vicenzo, A.; Franz, S. Chromium films deposition by hot target high power pulsed magnetron sputtering: Deposition conditions and film properties. *Surf. Coat. Technol.* **2019**, *375*, 352–362. [CrossRef]
9. Sidelev, D.V.; Bleykher, G.A.; Krivobokov, V.P.; Koishybayeva, Z. High-rate magnetron sputtering with hot target. *Surf. Coat. Technol.* **2016**, *308*, 168–173. [CrossRef]
10. Chodun, R.; Nowakowska-Langier, K.; Zdunek, K.; Okrasa, S. The role of magnetic energy on plasma localization during the glow discharge under reduced pressure. *Nukleonika* **2016**, *61*, 191–194. [CrossRef]
11. Meckel, B.B. Magnetic Target Plate for Use in Magnetron Sputtering of Magnetic Films. U.S. Patent No. 4,299,678, 10 November 1981.
12. Ho, K.K.; Carman, G.P. Sputter deposition of NiTi thin film shape memory alloy using a heated target. *Thin Solid Films* **2000**, *370*, 18–29. [CrossRef]
13. Loch, D.A.; Gonzalvo, Y.A.; Ehiasarian, A.P. Nickel coatings by inductively coupled impulse sputtering (ICIS). *Surf. Coat. Technol.* **2015**, *267*, 98–104. [CrossRef]
14. Brewer, J.A.; Migliuolo, M.; Belan, R.M. Magnetron sputter deposition of magnetic materials from thick targets. *Proc. Annu. Tech. Conf. Soc. Vac. Coaters.* **1990**, *33*, 37–42.
15. Chang, S.A.; Skolnik, M.B.; Altman, C. High rate sputtering deposition of nickel using dc magnetron mode. *J. Vac. Sci. Technol. A* **1986**, *4*, 413–416. [CrossRef]
16. Wegmann, U. A Cathode Arrangement for the Erosion of Material of a Target in an Apparatus for Cathodic Sputter Coating. G.B. Patent No. 2,090,872, 10 November 1983.
17. Sidelev, D.V.; Bleykher, G.A.; Grudinin, V.A.; Krivobokov, V.P.; Bestetti, M.; Syrtanov, M.S.; Erofeev, E.V. Hot target magnetron sputtering for ferromagnetic films deposition. *Surf. Coat. Technol.* **2018**, *334*, 61–70. [CrossRef]

18. Caillard, A.; El'Mokh, M.; Lecas, T.; Thomann, A.L. Effect of the target temperature during magnetron sputtering of nickel. *Vacuum* **2018**, *147*, 82–91. [CrossRef]
19. Chodun, R.; Dypa, M.; Wicher, B.; Nowakowska-Langier, K.; Okrasa, S.; Minikayev, R.; Zdunek, K. The sputtering of titanium magnetron target with increased temperature in reactive atmosphere by gas injection magnetron sputtering technique. *Appl. Surf. Sci.* **2022**, *574*, 151597. [CrossRef]
20. Chodun, R.; Nowakowska-Langier, K.; Wicher, B.; Okrasa, S.; Kwiatkowski, R.; Zaloga, D.; Dypa, M.; Zdunek, K. The state of coating–substrate interfacial region formed during TiO_2 coating deposition by Gas Injection Magnetron Sputtering technique. *Surf. Coat. Technol.* **2020**, *398*, 126092. [CrossRef]
21. Chodun, R.; Nowakowska-Langier, K.; Wicher, B.; Okrasa, S.; Minikayev, R.; Dypa, M.; Zdunek, K. TiO_2 coating fabrication using gas injection magnetron sputtering technique by independently controlling the gas and power pulses. *Thin Solid Films* **2021**, *728*, 138695. [CrossRef]
22. Gencoa: Balanced and Unbalanced. Available online: https://www.gencoa.com/balance-unbalance (accessed on 26 June 2022).
23. Pawlak, W.; Jakubowska, M.; Sobczyk-Guzenda, A.; Makówka, M.; Szymanowski, H.; Wendler, B.; Gazicki-Lipman, M. Photo activated performance of titanium oxide coatings deposited by reactive gas impulse magnetron sputtering. *Surf. Coat. Technol.* **2018**, *349*, 647–654. [CrossRef]
24. Posadowski, W.M.; Wiatrowski, A.; Dora, J.; Radzimski, Z.J. Magnetron sputtering process control by medium-frequency power supply parameter. *Thin Solid Films* **2008**, *516*, 4478–4482. [CrossRef]
25. Dora, J. Resonant Power Supply. Polish Patent 313150, 6 March 1996.
26. Palmero, A.; Rudolph, H.; Habraken, F.H. Study of the gas rarefaction phenomenon in a magnetron sputtering system. *Thin Solid Films* **2006**, *515*, 631–635. [CrossRef]
27. Huo, C.; Raadu, M.A.; Lundin, D.; Gudmundsson, J.T.; Anders, A.; Brenning, N. Gas rarefaction and the time evolution of long high-power impulse magnetron sputtering pulses. *Plasma Sources Sci. Technol.* **2012**, *21*, 045004. [CrossRef]
28. Horwat, S.; Anders, A. Compression and strong rarefaction in high power impulse magnetron sputtering discharges. *J. Appl. Phys.* **2010**, *108*, 123306. [CrossRef]
29. Palmero, A.; Rudolph, H.; Habraken, F.H. Gas heating in plasma-assisted sputter deposition. *Appl. Phys. Lett.* **2005**, *87*, 071501. [CrossRef]
30. Ganesan, R.; Murdoch, B.J.; Treverrow, B.; Ross, A.E.; Falconer, I.S.; Kondyurin, A.; McCulloch, D.G.; Partridge, J.G.; McKenzie, D.R.; Bilek, M. The role of pulse length in target poisoning during reactive HiPIMS: Application to amorphous HfO_2. *Plasma Sources Sci. Technol.* **2015**, *24*, 035015. [CrossRef]
31. Meggers, W.F.; Corliss, C.H.; Scribner, B.F. Tables of Spectral-Line Intensities, Part I – Arranged by Elements, Part II—Arranged by Wavelengths. *Nat. Bur. Stand. U.S.* **1975**, *600*. [CrossRef]
32. Thornton, J. Influence of apparatus geometry and deposition conditions on the structure and topography of thick sputtered coatings. *J. Vac. Sci. Technol.* **1974**, *11*, 666. [CrossRef]
33. Messier, R.; Giri, A.P.; Roy, R.A. Revised structure zone model for thin films physical structure. *J. Vac. Sci. Technol. A* **1984**, *2*, 500. [CrossRef]
34. Anders, A. A structure zone diagram including plasma-based deposition and ion etching. *Thin Solid Films* **2010**, *518*, 4087. [CrossRef]

Article

Synthesis of Magnetron-Sputtered TiN Thin-Films on Fiber Structures for Pulsed-Laser Emission and Refractive-Index Sensing Applications at 1550 nm

Omar Gaspar Ramírez [1], Manuel García Méndez [2,*], Ricardo Iván Álvarez Tamayo [3] and Patricia Prieto Cortés [4]

[1] Postgraduate Program, Faculty of Physics and Mathematics, Universidad Autónoma de Nuevo León, San Nicolás de los Garza 66455, Mexico
[2] Faculty of Physics and Mathematics, Universidad Autónoma de Nuevo León, San Nicolás de los Garza 66455, Mexico
[3] Faculty of Mechatronics, Bionics and Aerospace, Universidad Popular Autónoma del Estado de Puebla, Puebla 72410, Mexico
[4] Mechatronics Division, Universidad Tecnológica de Puebla, Puebla 72300, Mexico
* Correspondence: manuel.garciamnd@uanl.edu.mx

Abstract: In this work, a set of titanium nitrides thin-films was synthesized with the technique of reactive RF and DC magnetron-sputtering. To demonstrate the versatility and effectiveness of the deposition technique, thin films were deposited onto different fiber structures varying the deposition parameters for optical applications as saturable absorbers in passively q-switched fiber lasers and as lossy mode resonance fiber refractometers. After deposition, optical and electronical properties of samples were characterized by UV–Vis and XPS spectroscopies, respectively. Samples presented coexisting phases of Ti nitride and oxide, where the nitride phase was non-stoichiometric metallic-rich, with a band gap in the range of E_g = 3.4–3.7 eV. For all samples, glass substrates were used as templates, and on top of them, optical fibers were mounted to be covered with their respective titanium compounds.

Keywords: titanium nitride; DC/RF magnetron-sputtering; saturable absorber materials; Q-switched fiber lasers; fiber micro-ball lens; no-core fiber; lossy mode resonance; fiber refractometers

1. Introduction

In recent years, thin film manufacturing has increased greatly due to its multiple applications such as semiconductor processing, insulating materials, tool strengthening or improving the efficiency of the surface properties of various materials. For a given material, in order to obtain an improvement in its surface properties, physical or chemical surface treatment techniques are required. A very popular technique for surface treatment is thin film deposition, which consists of applying a nanometric thin layer of a material on an object. The object is covered with a material such as a thin film, which can then be used for the development of specific applications such as sensors, solar cells, optical devices, etc. Thus, it is necessary to use efficient techniques that ensure quality and reproducibility during the film-growth process, such as chemical vapor deposition (CVD), sol–gel deposition, chemical bath deposition and reactive magnetron sputtering [1–4].

Among them, an ideal and minimally invasive method of thin film coating would be sputtering, a technique that consists of bombarding a target made of the material to be deposited with highly energetic particles, usually of an inert gas such as argon, capable of breaking the atomic bonds on the surface of the material; after that, the atoms removed from the target will be deposited onto the object where the thin film is desired (flat substrate, optical fibers). The whole process takes place inside a high vacuum chamber. To supply the full amount of energy, the magnetron is connected to a power source, which can be either a radio frequency (RF) or direct current (DC) power source. The magnetron will generate a

Citation: Ramírez, O.G.; Méndez, M.G.; Tamayo, R.I.Á.; Cortés, P.P. Synthesis of Magnetron-Sputtered TiN Thin-Films on Fiber Structures for Pulsed-Laser Emission and Refractive-Index Sensing Applications at 1550 nm. *Coatings* 2023, 13, 95. https://doi.org/10.3390/coatings13010095

Academic Editor: Rafal Chodun

Received: 2 December 2022
Revised: 21 December 2022
Accepted: 30 December 2022
Published: 4 January 2023

Copyright: © 2023 by the authors. Licensee MDPI, Basel, Switzerland. This article is an open access article distributed under the terms and conditions of the Creative Commons Attribution (CC BY) license (https://creativecommons.org/licenses/by/4.0/).

magnetic field where the gas ions will be confined, causing an increase in the shocks of the deposit and a decrease in the working pressure. When a reactive gas is used, together with Ar, for example oxygen or nitrogen, the technique is denominated as reactive magnetron sputtering. With the control of experimental parameters, such as base pressure, working pressure, flow of reactive gas, power source, it is possible to tailor the film properties [5,6].

Moreover, the advantages of optical fibers include low cost, small size, and immunity to electromagnetic interference; thus, the use of optical fibers has been very attractive in a variety of research areas such as data transmission, optical instrumentation, optical sensing, fiber lasers, among others. The geometry of the fibers also plays a very important role in the efficiency of their results; as a result, different special fiber structures such as no-core fibers (NCF) and microball lenses (MBL) have been studied in order to take advantage of particular optical properties to interact with the surrounding environment. In this regard, improved with a material coating, the aforementioned fiber structures exhibit light-material properties suitable for sensing and laser development applications. As a result, in recent years, the study of promising materials deposited onto the fiber surface and the deposition method has been of increasing interest for fiber optical applications and systems. Particularly, transition metal oxides (TMO) and transition metal nitrides (TMN) have demonstrated their reliability as candidate materials for the development of saturable absorbers and for electromagnetic resonance phenomenon. Particularly, titanium nitride with varying concentrations of Ti, N and O has been demonstrated as desirable for its saturable absorption [4] and plasmonic properties [7]. Transition metal nitrides such as TiN are ceramic non-stoichiometric materials. The composition and hence the optical properties depend significantly on the preparation method and conditions. Some of these nitrides possess metallic properties at visible wavelengths because of large free carrier concentrations ($\sim 10^{22}$ cm^{-3}). High interband losses make many of these compounds unattractive for plasmonic applications. Titanium nitride, however, exhibits smaller interband losses in the visible spectrum and a small negative real permittivity. It is therefore a material of significant interest for plasmonic applications in the visible and near-IR ranges, which are denominated as double epsilon-cero materials [7–10]. Additionally, with the use of reactive magnetron sputtering technique (DC/RF), it is possible to grow thin films with tunable properties at the optoelectronical level, as this technique allows for the synthesis of films with precise thickness control and greater deposit uniformity and reproducibility [5,11–14].

In this paper, we experimentally explore the reliability of DC/RF magnetron sputtering for accurate titanium nitride compound deposition onto optical fiber structures for their application as saturable absorbers for passively Q-switching fiber laser pulsed laser generation and for their use as fiber refractometers based on LMR phenomenon. The two applications presented in this work are:

(a) For saturable absorber (SA) application, a fiber microball lens coated with a thin film of titanium nitride by RF is used for generation of short high-energy pulses within a linear fiber laser configuration. The coating acts as an SA, modulating the losses inside the cavity by means of a passive Q-switching technique. Then, short pulses with a repetition rate in the kHz order are obtained depending on the variation of the pumping power of the laser.

(b) As a fiber refractometer, a multimode NCF fiber structure coated with a film of titanium nitride by DC is used to generate electromagnetic resonances at the fiber–material interface from the light energy transferred to a plasma wave along the material surface by lossy mode resonance phenomenon. The NCF fiber coated with the film material is highly sensitive to refractive index variations on the medium surrounding the fiber structure.

2. Deposition of Films by RF for Saturable Absorber Application

2.1. Fiber Ball Lens Deposition

Fiber microball lens (MBL) consist of a microsphere at one end of a silica optical fiber. By using a fusion splicer with special fiber processing characteristics (Fujikura ArcMaster

FSM-100M, Tokyo, Japan) and the spherical lens fusion program of its software (AFL Fiber Processing Software FPS, ver. 1.2b, Tokyo, Japan), MBL were constructed on SMF-28 single-mode fiber segments. With the options of the FPS program, MBLs with diameters of 300, 350 and 400 µm were obtained.

Prior to deposition process, optical fibers were mounted on top of glass substrates. Fibers were fixed with a special vacuum type. The synthesis equipment consists of a glass bell chamber (deposition chamber) connected to a vacuum system, made up of mechanical and turbomolecular pumps. The chamber is fed with external tank gases (Ar, N and O high purity, 99.99%), connected with valves and controlled with electronic mass flow meters. Into the chamber, a magnetron is connected to a RF power-source. The magnetron is designed to place targets of 1″ in diameter and 1/8″ thick. The substrate holder is placed in front of the magnetron, 5 cm away. A mechanical shutter is placed between the target and substrate, which prevents deposit of the material when not desired. A schematic diagram of the RF equipment used for the synthesis of this film is included in the Supplementary Materials.

In this work, two batch of samples were prepared: the first batch of three samples (labeled F01, F02 and F03), and the second batch of four samples (labeled F04, F05, F06 and F07).

For the first batch, depositions were performed using a ceramic circular target (TiN target, by Plasmaterials, (Livermore, CA, USA). The depositions were made inside a glass bell chamber, with a base pressure of 7.5×10^{-5} Torr. A supply of Ar was used in the reaction, with a flow rate of 20 sccm. Plasma was generated by applying 200-watt RF power and a working pressure of 70 mTorr. With the shutter activated, the target was first pre-sputtered for 5 min to remove any contamination. The deposition rate was controlled with a quartz crystal sensor connected to an external computer. The deposition time was 15, 20 and 25 min for samples labeled as: F01 (fiber of 300 µm), F02 (fiber of 350 µm) and F03 (fiber of 400 µm), respectively.

For the second batch, depositions were performed using:

(a) Metallic circular target (Ti target, by Kurt-Lesker, Jefferson Hills, PA, USA)—The depositions were made inside a glass bell chamber, with a base pressure of 7.5×10^{-5} Torr. A mixture of high purity gases, Ar and N, was used in the reaction with a flow rate of 20 and 5 sccm, respectively. The gas flow is monitored with mass flow meters. Plasma was generated by applying 300-watt RF power and a working pressure of 70 mTorr. The target was pre-sputtered for 5 min to remove any contamination. The deposition rate was controlled with a quartz crystal sensor connected to an external computer. The deposition time was 15 min. In this case, in a single substrate, two optical fibers were placed. The fibers were labeled F04 (Fiber of 300 µm) and F05 (Fiber of 350 µm). The substrate used as template was then characterized with XPS and UV–Vis spectroscopy and labeled as F0405.

(b) Ceramic circular target (TiN target, by Plasmaterials)—The depositions were made inside a glass bell chamber, with a base pressure of 7.5×10^{-5} Torr. A supply of Ar was used in the reaction, with a flow rate of 20 sccm. Plasma was generated by applying 300-watt RF power and a working pressure of 70 mTorr. With the shutter activated, the target was first pre-sputtered for 5 min to remove any contamination. The deposition rate was controlled with a quartz crystal sensor connected to an external computer. The deposition time was 15 min. In this case, in a single substrate, two optical fibers were placed. Fibers were labeled F06 (fiber of 300 µm) and F07 (fiber of 350 µm). The substrate used as a template was then characterized with XPS and UV–Vis spectroscopy and labeled as F0607.

After deposition, areas of substrate surrounding the fibers were characterized with XPS (K-ALPHA, Thermo Fisher Scientific, Waltham, MA, USA) and UV–Vis-NIR spectrophotometer (V-770, Jasco, Easton, MD, USA) spectroscopies. In addition, Filmetrics equipment was employed to measure the optical thickness of the films.

In this case, samples of first batch were analyzed with XPS and UV–Vis spectroscopies. As the second batch is a preliminary study, intended to focus on saturable absorption for Ti nitrides, obtained from two different targets, only the UV–Vis results will be presented.

The composition of the films was analyzed by XPS using a monochromatic source of AlKα of 1486.68 eV. The XPS technique can extract the atomic concentration of the elements and their respective chemical states. During measurements, samples were first bombarded with Ar^+ ions for a few seconds to remove any contamination adsorbed at their surface. The C1s peak position at 284.8 eV is used as a binding energy reference. The procedure consisted of making measurements in survey mode (from 1486.68 to 0 eV; 1 eV/step), followed by windows of high resolution for the C1s, O1s, N1s and Ti2p windows (0.2 eV/step). Spectra were measured before and after sputtering. From the position in binding energy (BE), the chemical state was determined. From the sensitivity factors and the area below the curve for each transition, the atomic concentration (atomic %) of Ti and N was obtained. From the Ti2p and N1s signals, spectra were deconvoluted using lorentzian curves.

The optical properties were analyzed using the UV-Vis–NIR spectrophotometer, in the wavelength range of 300 to 2500 nm. Curves of transmittance, reflectance and absorbance were obtained. With the thickness and transmittance spectra, Tauc curves were constructed. From them, optical parameters: band gap (E_g), absorption coefficient (α), refractive index (n), extinction coefficient (k), dielectric (ε_∞) and static (ε_0) constants), were extracted.

2.2. Optoelectronical Characterization

Figure 1 displays the Ti2p and N1s transitions for F01, F02, and F03 before sputtering. Figure 2 displays the Ti2p and N1s transitions for F01, F02, and F03 after sputtering. In Figure 1, Ti appears at the surface in an oxidized state, which is expected at this level of analysis. Some traces of N can be detected. Since all samples are thin, it is expected that even during surface-level analysis, it is possible to detect signals from the entire thickness of the sample.

In Figure 2, a convolution of oxide/nitride states, for the Ti2p transition, can be observed. In the figure, deconvoluted curves for both states are included. Oxide and nitride states are clearly differentiable, both in binding energy and in the shape of the transition. The shape of the Ti in the nitride state is modified because of the generation of satellites (caused by the extra conduction electrons) to the left of each transition, which produce asymmetry [11,14]. In Figure 1, a more symmetric shape in the Ti2p transition can be noted. Again, in Figure 2, the BE of Ti-oxide appears at 457–458 eV, while BE of Ti-nitride is about 1 eV lower, at 454–456 eV. This is expected, as the oxidized state tends to be more stable. The values of BE detected in our samples agree with the ones reported in the literature for oxide/nitride [12,13,15–17]. No oxy-nitride signal was detected in the Ti2p window, which is very distinguishable between the Ti-oxide and Ti-nitride and which appears at the middle in energy value [17,18]. In addition, metallic states of Ti were discarded. The assignation of the chemical states of samples was also supported by additional XPS analysis of the Ti and TiN targets, which are included in the Supplementary Materials.

From the area below the curve of the Ti2p3/2-nitride and N1s, and from using their respective sensitivity factors, the empirical formula for the nitride phase was obtained. Table 1 shows a summary of data obtained from the XPS analysis.

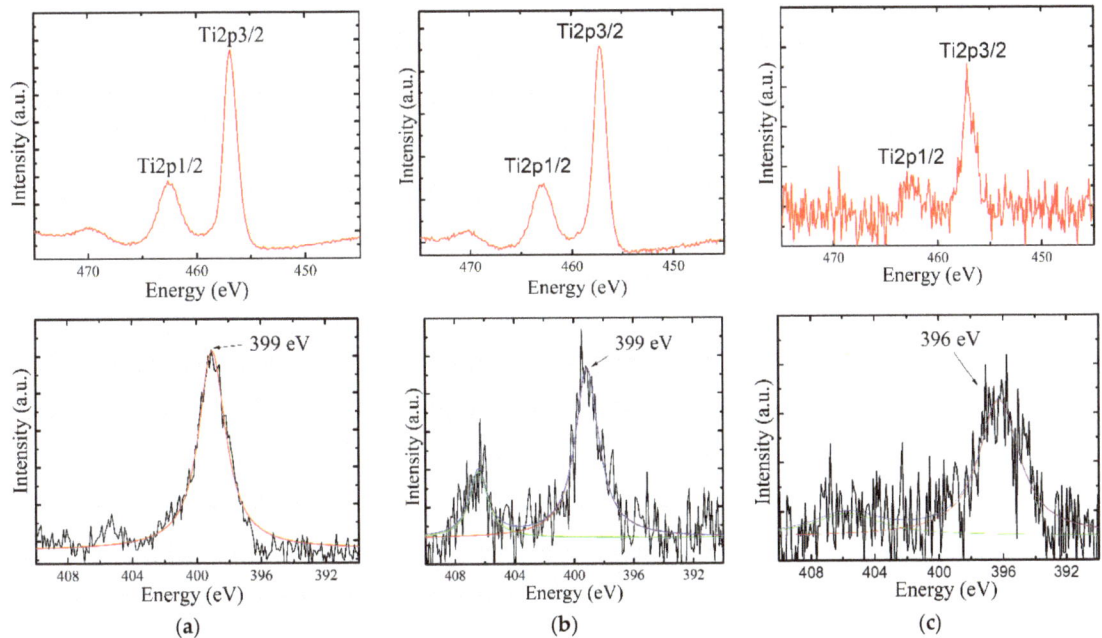

Figure 1. High-resolution spectra for the Ti2p (**top**) and N1s (**bottom**) windows for samples (**a**) F01, (**b**) F02 and (**c**) F03, before sputtering.

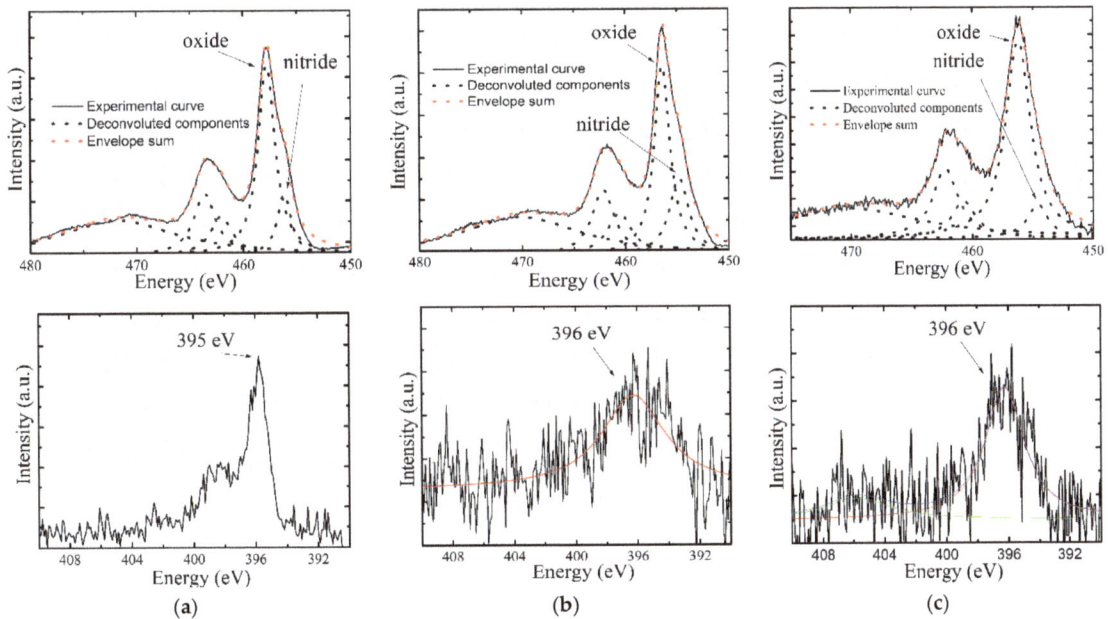

Figure 2. High-resolution spectra for the Ti2p (**top**) and N1s (**bottom**) windows for samples (**a**) F01, (**b**) F02 and (**c**) F03 after sputtering.

Table 1. Summary of data obtained from XPS analysis. Second column is an account of the BE for each chemical state (eV). Third column is an account of the percentage of each chemical state. Last column is the obtained empirical formula. The analysis of F01, F02 and F03 corresponds to the substrates used as a template on which optical fibers were mounted.

Sample	Binding Energy (eV)		Ti Percentage (%)		Empirical Formula
	Ti-N	Ti-O	TiN	Ti-O	
F01	456.19	457.7	20	80	$Ti_{0.6}N_{0.4}$
F02	454.8	456.3	33	67	$Ti_{0.8}N_{0.2}$
F03	454.52	456.19	12	88	$Ti_{0.4}N_{0.6}$

From the XPS analysis, it is possible to conclude that the samples presented a mixture of coexisting phases of titanium nitride and titanium oxide. The most prevalent is the oxidized one. The nitride phase for samples F01 and F02 looks more metallic than for F03. It is possible that the higher deposition time favors the generation of more oxidation events.

The transmittance curves obtained from the UV–Vis measurements are included in Figure 3 for samples F01, F02 and F03, together with the Tauc curves at inset, and the n, k vs. λ of all samples in one graph. Figure 4 includes the transmittance curves for samples F0405 and F0607.

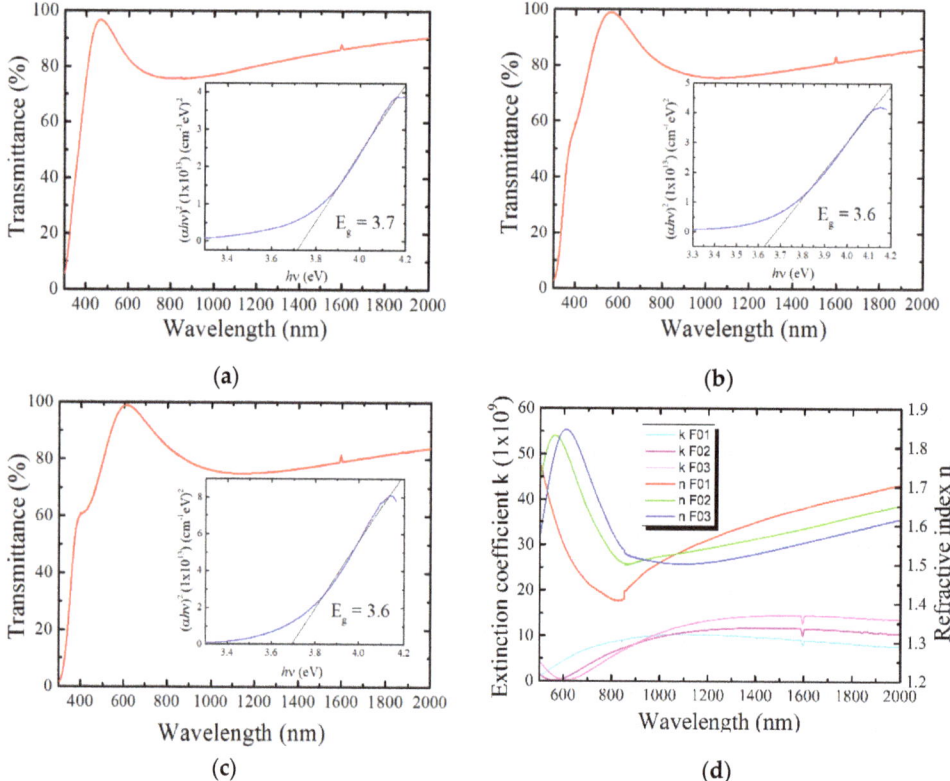

Figure 3. Transmittance spectra for samples (**a**) F01, (**b**) F02 and (**c**) F03. Tauc curves are included in inset; (**d**) n and k vs. wavelength curves for the three samples.

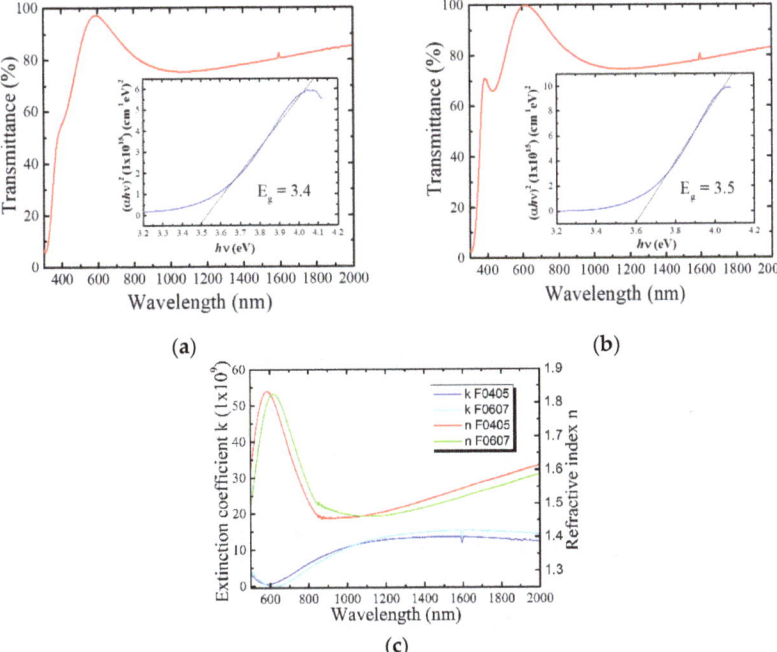

Figure 4. Transmittance spectra for samples (**a**) F0405 and (**b**) F0607. Tauc curves are included in inset; (**c**) n and k vs. wavelength curves for the two samples.

The optical parameters were obtained from the transmittance spectra. Then, by the use of the Drude–Lorentz model, E_g, α, n and k were extracted. Applying the single-oscillator model developed by Wemple and Domenico, ε_∞ (dielectric constant, real part, at high-frequency values, for electronic transitions) and ε_0 (static constant, real part, at low-frequency values, related to the lattice contribution) were extracted [19]. A detailed procedure about how to use these models has been published by us elsewhere [20].

The TiN compound is formed from the hybridization of the orbitals Ti: $3d^24s^2$ and N: $2p^3$. Thus, electronic levels are distributed in energy to conform the sp^3d^2 hybrid-orbital. Previous to hybridization, there are four and three valence electrons for Ti and N, respectively. When hybridization takes place, there is room for six electrons, and one remains free as a conductor electron. This final configuration is very similar to gold: $5d^{10}6s^1$. For this reason, TiN contains a high concentration of conduction electrons, $\approx 10^{22}$ cm^{-2}, proportional to the conduction concentration of metallic elements, in this case to gold [7,10,15]. This metallic-like behavior for TiN can be observed in the transmittance spectra. The wavelength located at the onset, the region when transmittance starts to experiment a sudden drop, is defined as the screened plasma energy, λ_{ps}, or the energy region where the real part of the dielectric constant $\varepsilon_\infty = 0$. This wavelength, for metals such as gold and Ti nitrides, can be located within the visible spectral range and is affected by both intraband and interband characteristics (this is the cause of the yellowish color for both gold and stoichiometric TiN). At wavelengths $\lambda < \lambda_{ps}$, it is also expected to produce a sharp rise in the refractive index, n, because of light attenuation due to the interband absorption, together with a reduction in transmittance and a maximum absorbance [7]. For regions $\lambda > \lambda_{ps}$, an almost constant transmittance or transparent zone is expected. A drop in the transmittance is expected in the IR region at the onset of vibrational modes in the lattice [19,20].

At this region (transparent zone), a condition of high transmittance (low reflectance) is expected for typical semiconductors (SiO_2, TiO_2, sapphire). For metallic-like compounds, the transmittance depends on the density of the free carriers, which is influenced by the deposition conditions and film thickness. Thus, high reflectivity, similar to that of metals, coupled with low transmittance, may be possible.

In all the transmittance curves, the onset of the absorption edge is located at ~500–600 nm. This onset corresponds with wavelengths within the visible part of the spectra, and thus, all our films presented with a yellowish color, which is expected for TiN [21]. In this case, at this border, there are intraband transitions due to the interaction of light with the conduction electrons, which produce a maximum absorption at $\approx \lambda = 600$ nm. Between $\lambda = 600$ nm and $\lambda = 800$ nm, there is a transition zone. At $\lambda = 800$ nm, a transmittance of $\approx 80\%$ is observed. Then, at the onset of the absorption edge, the absorption coefficient (a) can be calculated from the relation [18–20]:

$$\alpha = \frac{1}{d} \ln\left[\frac{100}{T}\right], \quad (1)$$

where T is the measured transmittance, and d is the optical thickness. The optical band gap (E_g) was extracted from Tauc's relation:

$$(\alpha h v)^r = A(h v - E_g), \quad (2)$$

where A is the edge width parameter. The value of E_g is obtained by $(\alpha h v)^r$ vs. $h v$ with $r = 2$ for a direct transition, as reported for TiN [15,22]. In a direct transition, the maxima of density of states (DOS) at the top of valence level are aligned with the maxima of DOS at the bottom of the conduction level at $k = 0$ [20].

From the dispersion curve (n vs. λ), dielectric and static constants are obtained [19,20]. The procedure employed to extract the band gap from the Tauc's curves is included in the Supplementary Materials.

In Table 2, a summary of the optical parameters extracted from the experimental transmittance curves is displayed.

Table 2. Summary of optical parameters extracted from transmittance curves and applying the optical models. The analysis of F01, F02 and F03 corresponds to the substrates used as a template, where optical fibers were mounted. In addition, F0405 is the substrate used as a template, on which F04 and F05 were mounted (use of Ti target). Likewise, F0607 is the substrate used as a template, on which F06 and F07 were mounted (use of TiN target).

Sample	Deposition Time (min)	Thickness (nm)	E_g (eV)	ε_∞	ε_L
F01	15	20	3.7	1.9	2.0
F02	20	22	3.6	2.2	2.3
F03	25	20	3.6	2.1	2.4
F0405	15	20	3.4	1.9	2.1
F0607	15	20	3.5	2.1	2.1

All films are similar in thickness, and the presented values for E_g range from 3.5 to 3.7 eV. These values agree for the titanium nitride films reported on by other authors, from 3.0 to 3.5 eV [15,21]. Values of $E_g > 3.5$ eV can be attributed to more metallic-like states, due to the non-stoichiometry of our films and the higher amount of loaded Ti, as the XPS analysis suggests. These additional electronic stats tend to be redistributed between the valence and conduction levels, giving place to band gap widening. In our case, the lowest value of the band gap at 3.4 eV corresponds to the film synthetized from the Ti target. Regarding the dielectric constants, the values were almost similar for all samples. The ε_∞ has values in the range of $E_g = 1.9$–2.2 eV, where the value for F01 (synthesized from TiN target) is similar to F0405 (synthesized from Ti target). From these last values, we could observe that the optical properties using both synthesis routes would produce Ti nitride films with similar properties.

From Figures 3d and 4c, "*n*" presents somewhat monovaluated stable values in the region of 1000–2000 nm, a transition region with an onset at ≈800 nm, with a sharp increase at 600 nm, which corresponds to the yellow region of the visible spectra, which is very distinguishable in the samples (see Figure S1c; Supplementary Materials).

From the data provided by XPS and UV–Vis characterization of the samples from the first batch and the UV–Vis characterization of samples from the second batch, we can conclude that the thin films of the non-stoichiometric TiN were synthesized. This films turned out to be metallic-rich, coexisting with a phase of titanium oxide.

2.3. Passive Q-Switched Laser Pulse Generation

For pulsing to occur, the SA accumulates a large amount of energy supplied by the pumping source, until saturation overflow is reached, leading to a release of energy in the form of a pulse of the order of microjoules. The experimental setup is a linear cavity laser containing an Er/Yb-doped double-clad fiber (EYDCF) with a length of ~1.2 m (DCF-EY-10/128, CorActive, Québec, QC, Canada), with a core absorption of 85 dB m^{-1} at 1535 nm and inner cladding absorption of 2 dB m^{-1} at 915 nm), which functions as a gain medium. The EYDCF pumps with a high-power 25 W laser at 976 nm, and the light beam passes through of a 2 × 1 + 1 combiner that allows for the passage of wavelengths at 976 and 1550 nm at the same time. On one side of the cavity is a fiber loop mirror (FLM) that consists of a 50/50 coupler with its output ports connected, which is used as a mirror capable of reflecting 100% of the incident light. At the other end is a 90/10 coupler from which the 10% output port was connected to an MBL fiber that is fixed horizontally to the base, with a flat mirror to the front. The 90% output port was the output of the laser signal and where the data measuring the pulsations were detected with the help of a photodetector, of which were read by means of an oscilloscope. Figure 5 shows the laser setup.

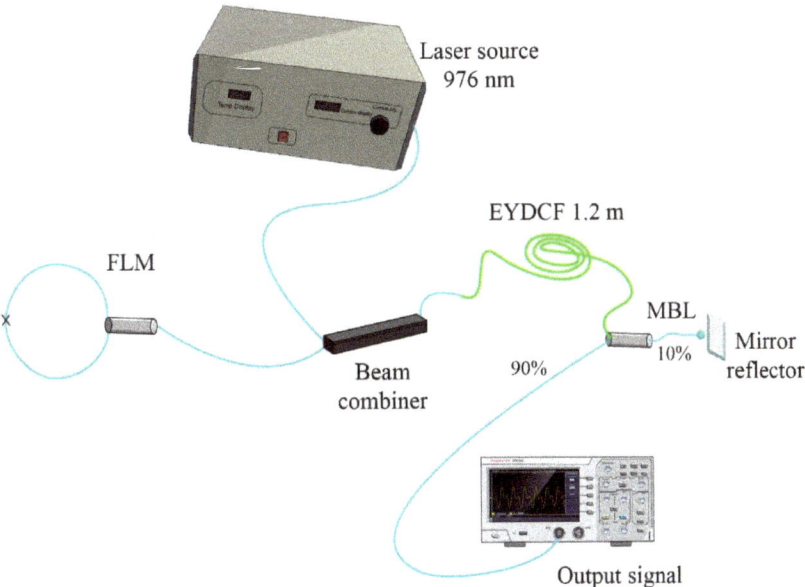

Figure 5. Experimental setup of the PQS laser.

The laser pulses were detected with a photodetector and were recorded by an oscilloscope. Previously, test measurements were made with an MBL without deposited material to detect possible self-Q switches and to rule them out later, ensuring only to include the pulses generated by the nonlinear optical properties of the material. Table 3 is shown below, detailing the different MBLs that were used to perform the measurements.

Table 3. MBL diameters for each sample measured.

Sample	MBL Diameter (μm)
F01	300
F02	350
F03	400
F04	300
F05	350
F06	300
F07	350

The characteristics of the generated PQS laser pulses are shown in Figure 6, where it can be seen that all the samples maintain stability in the pump power range of 0.457 to 0.98 W, except in sample F03, in which unstable pulses were observed, attributed to the lack of available conduction electrons, since this sample had a lower amount of Ti (see Table 1). We think that this factor, and not the diameter of the optical fiber, explains this behavior.

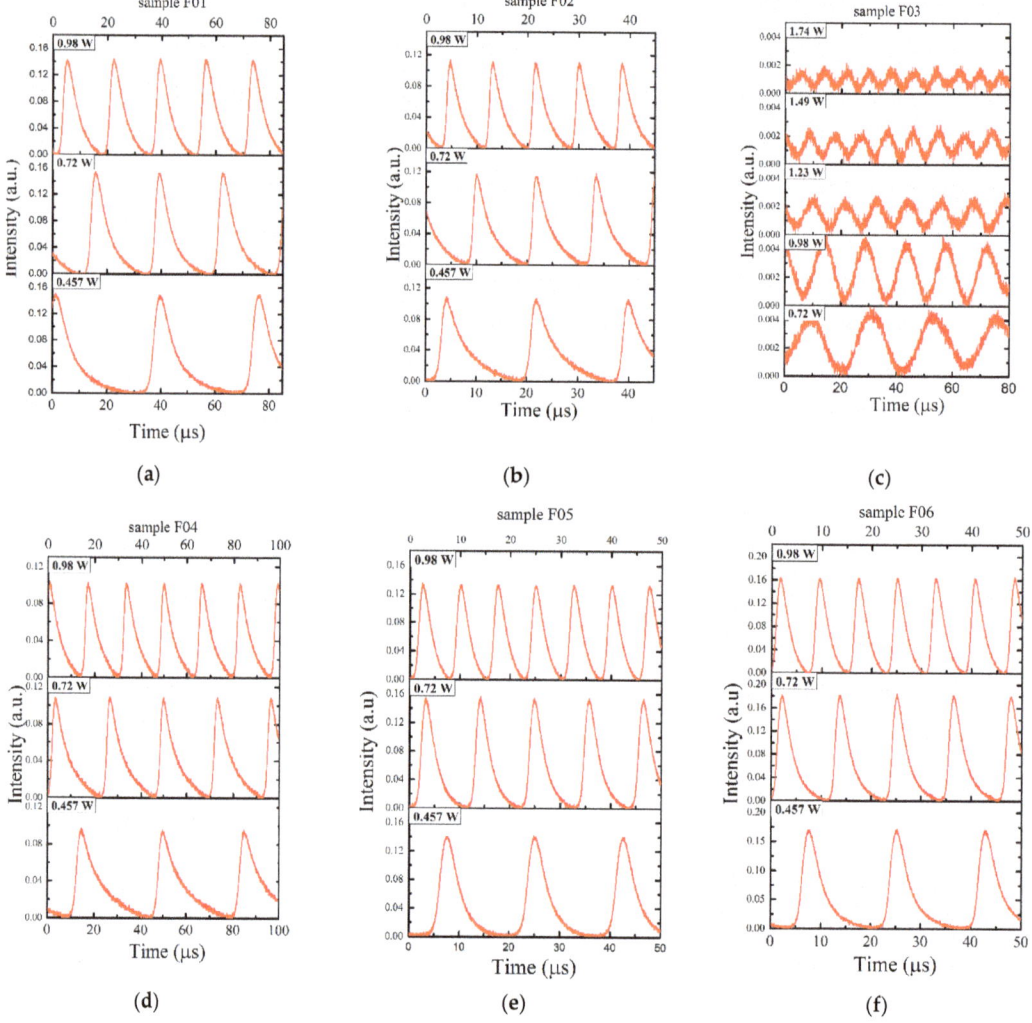

Figure 6. Cont.

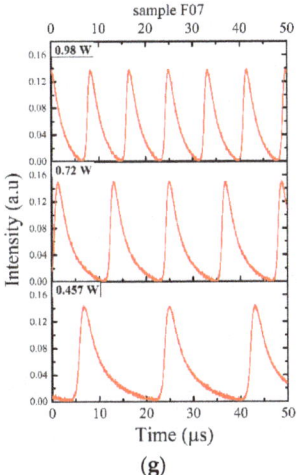

Figure 6. Train of PQS pulses of the laser at different pump powers for the different MBL samples: (**a**) sample F01, (**b**) sample F02, (**c**) sample F03, (**d**) sample F04, (**e**) sample F05, (**f**) sample F06, (**g**) sample F07.

All the characteristics obtained from the information provided by the pulse trains are summarized in Table 4, while Figures 7 and 8 appear in graphic form for each of the samples.

Table 4. Characteristics of the PQS laser pulses generated in each of the samples while the pump power increases in a range of 0.57 to 0.98 W.

Sample	Pump Power (W)	Repetition Rate (kHz)	Pulse Width (µs)	Peak Power (W)	Pulse Energy (µJ)
F01	0.457	26.12	4.19	0.01115	0.04671
	0.72	41.98	2.85	0.02624	0.0748
	0.98	59.24	2.6	0.03363	0.08744
F02	0.457	28.24	4.15	0.01126	0.04674
	0.72	42.45	3.03	0.02433	0.07373
	0.98	59.38	2.55	0.03474	0.08858
F04	0.457	28.24	8.58	0.00582	0.04993
	0.72	43.1	5.94	0.01297	0.07703
	0.98	60.97	4.92	0.0179	0.08808
F05	0.457	28.53	4.02	0.01186	0.04767
	0.72	46.06	2.92	0.02387	0.06969
	0.98	66.69	2.61	0.03091	0.08067
F06	0.457	28.49	3.91	0.01356	0.053
	0.72	43.4	2.97	0.02661	0.07903
	0.98	64.24	2.68	0.03229	0.08655
F07	0.457	27.58	3.78	0.01391	0.05257
	0.72	42.43	2.92	0.02696	0.07872
	0.98	60.97	2.52	0.03619	0.09119

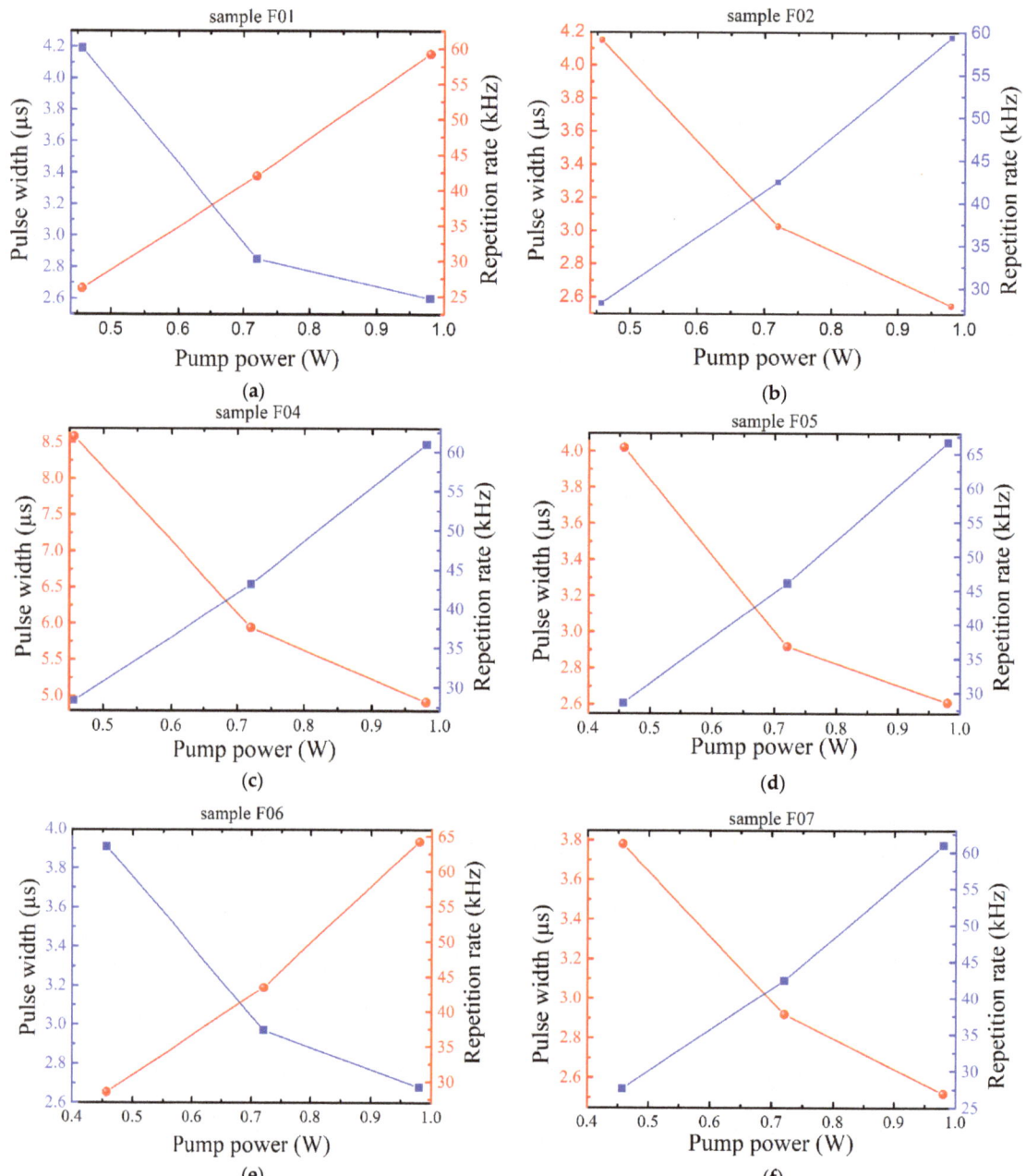

Figure 7. Pulse width and repetition rate of the PQS pulses as a function of the pump power of MBL sample: (**a**) sample F01, (**b**) sample F02, (**c**) sample F04, (**d**) sample F05, (**e**) sample F06, (**f**) sample F07.

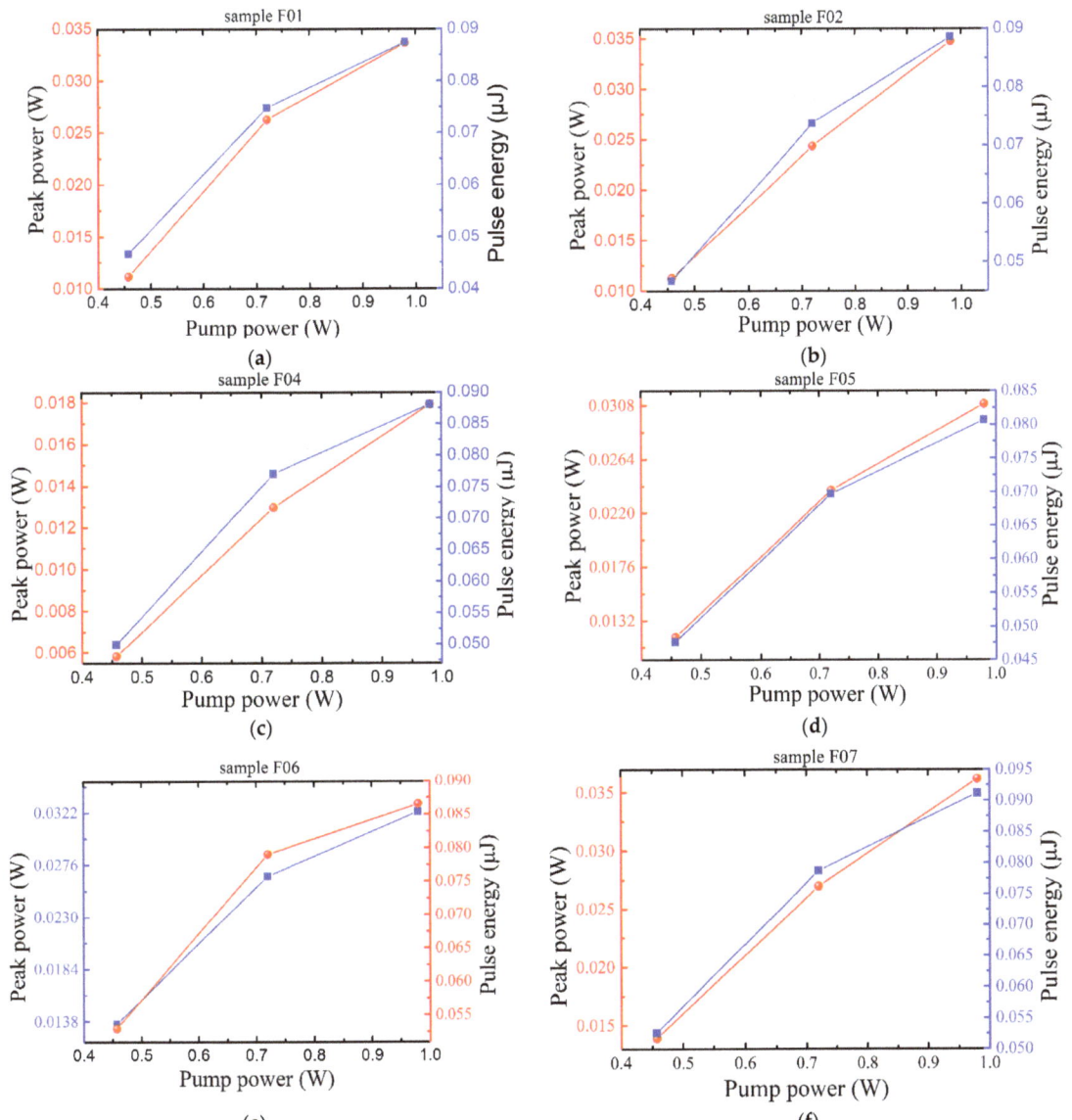

Figure 8. Peak power and pulse energy of the PQS pulses as a function of the pump power of all MBL samples. (**a**) sample F01, (**b**) sample F02, (**c**) sample F04, (**d**) sample F05, (**e**) sample F06, (**f**) sample F07.

With this, it can be observed that the obtained laser pulses exhibit a typical behavior of PQS generation. By increasing the pump power, the repetition rate of the PQS pulses increases while the pulse width decreases.

From the figures, it can be shown that all the samples generate pulses at powers of 0.457, 0.72 and 0.98 W. These pulses are generated due to the material that covers each fiber. However, sample F03, in particular, presents alterations or modifications that are not detected in the others. Based on the data obtained from the analysis of the samples (XPS: see Table 1), it was observed that F03 presents a reduction in the amount of Ti when compared to the others ($Ti_{0.4}N_{0.6}$). It is possible that the reduction of the metal also reduces the amount of carriers that favor the interaction of the laser with the conduction electrons of the deposited material. In this way, it has been reported that the more metallic-rich the TiN phase, the better it is to sustain a pulsed laser operation by using the passive q-switching technique. Regarding the pulses in general, a similar and favorable response is observed in all the samples, regardless of the experimental conditions used in the RF deposition.

3. Deposition of Films by DC for LMR Refractometer Application

3.1. No-Core Fiber Structure and Deposition

The dielectric properties of the thin material film coated onto the optical fiber surface, which interacts with the light propagated through the fiber, are exploited in fiber sensors based on electromagnetic resonance responses, such surface plasmon resonance (SPR) and lossy mode resonance (LMR) phenomena. Instead of using a typical Kretschmann–Raether configuration for SPR and LMR bulky approaches, fiber structures effectively guide multiple modes supported by the fiber coating, with an angle greater than the attenuated total reflectance (ATR) angle, by total internal reflection (TIR). Then, with a significant phase coincidence between the guided light and the material coating, SPR and LMR occur in the optical fiber structures at the fiber-coating interface, where an evanescent wave is then produced. Depending on the wavelength of the light and the thickness of the material, mode coupling takes place when fiber-guided modes over the cut-off condition are guided through the material surface. As a result, energy is transferred from the light modes to a plasma wave generated at the interface surface at the resonance wavelength. Optically, the transferred energy generates a lack of intensity in the light through the fiber, which is detected as an intensity notch at the resonance wavelength in the output spectrum of the fiber transmission expressed by the power distribution of the input light source $p(\theta)$, the TIR critical angle θ_c, and the reflected light at the fiber–material interface $R^{N_r(\theta)}$, as [23]:

$$T(\lambda) = \frac{\int_{\theta_c}^{90°} p(\theta) R^{N_r(\theta)}(\theta,\lambda) d\theta}{\int_{\theta_c}^{90°} p(\theta) d\theta}, \qquad (3)$$

where $N_r(\theta) = L/d \tan\theta$ stands for the number of reflections, depending on the coated fiber length L and the fiber diameter d.

In the case of SPR, electromagnetic resonance occurs at a single resonance wavelength; however, for LMR candidate materials, multiple resonant orders exist along a broadband wavelength range. Here, the exhibited plasmonic effect depends on the permittivity of the material coating and, as a consequence, on the complex refractive index conditions. LMR is obtained from materials on the low imaginary part of the refractive index. In this sense, different transition metal oxides (TMO) and metal nitrides have demonstrated to be good candidates to explore the LMR effect.

Moreover, in order to achieve generation of the LMR effect in optical fibers, it is necessary to ensure the strong interaction between the light guided through the fiber and the material deposited onto the fiber surface. For this purpose, different fiber structures such as chemically etched fibers, tapers, D-shaped fibers, and no-core fibers (NCF) have been studied for their application in fiber refractometers based on the LMR effect. In this case, although higher sensitivity properties on D-shaped fibers have been demonstrated, no-core fibers facilitate the implementation and deposition process, since special fiber preparation is not required. In our approach, NCF were used for the proposed application. The use of a multimode NCF eases the generation of strong evanescent waves from high-order modes excited from the fundamental mode propagated along the fiber surface. Figure 9 shows the fiber structure formed by an NCF segment spliced between two MMF segments. The proposed structure also facilitates the splicing process during the fabrication process due to fiber compatibility.

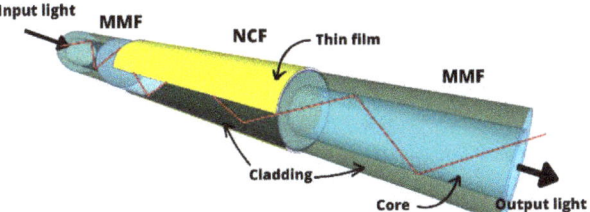

Figure 9. No-core multimode fiber structure.

For the NCF deposition preparation, high vacuum tape and glass substrate were used for mounting and to fix the fiber structure. The film was deposited over the fiber surface as well as in the substrate. The film deposited in the substrate near the NCF was used for XPS characterization and optical thickness measurement. The same target (surface native oxide) was used to supply oxygen. A Filmetrics device in reflectance mode was used to measure the thickness of the films. From simulation analyses and from the results of our research group, it was demonstrated that for Ti compounds, film thickness from 200 to 300 nm allows for the generation of a strong second LMR order around 1500 nm [24]. For this purpose, DC magnetron sputtering from a TiN target was performed with a power of 250 W at 5 cm distance between the substrate and the target for 18 min. Oxygen was supplied from the same target (surface native oxide).

The deposition was made using a gas mixture containing argon (Ar) and nitrogen (N_2), where Ar and N_2 were the working and reactive gases, respectively. High purity (99.999%) Ar and N_2 were employed for the TiO_xN_y deposition. The chamber was pumped down to a base pressure of 6.6×10^{-5} Torr before N_2 and Ar were introduced. The flow rate of both gases was controlled during deposition by using gas flowmeters. The target-substrate distance d was fixed at 5 cm. A source from the DC Sputtering Ion Magnetron (Materials Science Inc., San Diego, CA, USA) was used during the deposition process. The pre-used vacuum evaporation was performed using two pumps: a mechanical JEOL75 G (Agilent, Santa Clara, CA, USA) and RP-250 Turbo Macrotorr turbomolecular V (Agilent, Santa Clara, CA, USA). The gas pressure was established using the flowmeter AERAFC-7800CD. After deposition, the thickness of the resulting film was measured ex situ by the Filmetrics equipment in reflectance mode. The measured film thickness was ~250 nm. A schematic diagram of the DC equipment used for the synthesis of this films is included in the Supplementary Materials.

Elemental characterization of the film was performed by the XPS equipment. A monochromatic AlKα source of 1486.68 eV was used to obtain the spectra. For the analysis, the binding energies were calibrated with the C1s peak at 284.5 eV. Figure 10 shows the XPS high-resolution spectra for Ti2p, O1s and N1s windows before and after sputtering.

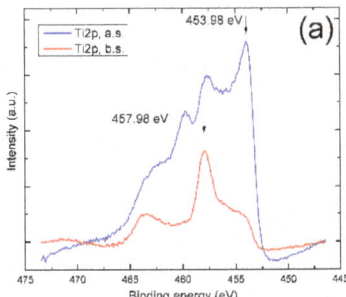 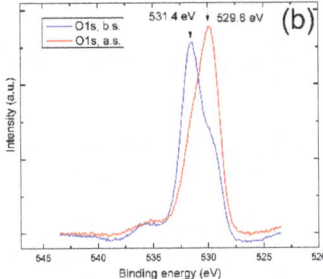 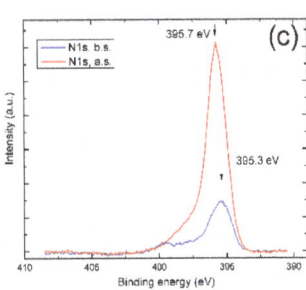

Figure 10. XPS high-resolution spectra for sample 01 film for (**a**) Ti2p, (**b**) O1s and (**c**) N1s windows. The spectra before (b.s.) and after sputtering (a.s.) is presented.

The binding energies (BE) obtained from XPS spectra are presented in Table 5.

Table 5. Binding energy values of the film.

Transition	BE (Before Sputtering)	BE (After Sputtering)
Ti2p$_{3/2}$	457.98 eV	453.98 eV
O1s	531.4 eV	529.8 eV
N1s	395.3 eV	395.7 eV

For the Ti2p window shown in Figure 10a, before sputtering, the BE of Ti2p$_{3/2}$ can be associated with a nonstoichiometric native oxide Ti oxide, as the metallic state is discarded. After sputtering, that Ti2p$_{3/2}$ signal can be associated with an oxynitride phase. In addition, it a tendency of the BE to become lower can be observed, from 457.98 to 453.98 eV when Ti tends to nitridize. For the O1s window of Figure 10b, the BE before sputtering can also be related to native oxide. After sputtering, O1s can be related to an oxynitride phase. For the case of O1s, the BE before sputtering (531.5 eV) closely agrees with the reported value of the native oxide phase [25]. Conversely, this is not the case for the oxynitride phase, due to the wide variety of substoichiometric TiN$_x$O$_y$ phases with different contents of O and N, which can turn into several variations in the chemical state and therefore into BE values. For N1s in Figure 10c, the BE change is not significant before and after sputtering. Thus, those signals can be related to the oxynitride phase. From the calculated atomic concentrations of Ti, N and O, extracted from the XPS spectra after sputtering, the empirical formula of the film was Ti$_{0.37}$N$_{0.41}$O$_{0.21}$.

3.2. LMR Refractometer Results

In order to characterize the sensitivity of the fiber structure to refractive index variations of a liquid medium surrounding the coated fiber, the experimental setup of Figure 11 was used. The fiber structure was fixed between the holders of a mechanical device, and then, the NCF was immersed in liquid solutions from a set with different refractive indices. A broadband NIR LED source with emissions in the range of 1400 to 1700 nm was used as the input light source. The transmitted optical spectrum was measured to determine the characteristics of the output light by using a NIR spectrometer with a wavelength range from 970 to 1700 nm.

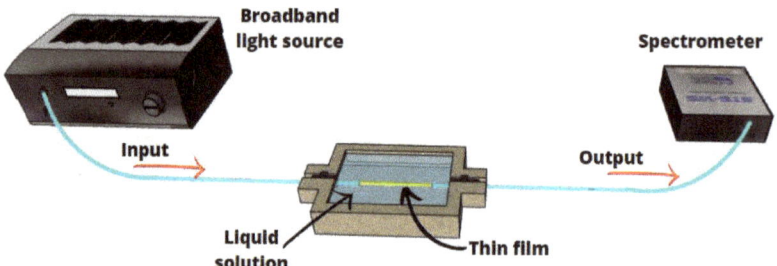

Figure 11. Experimental setup for LMR refractometer.

The measurements of the output transmitted signal and the transmission of the coated fiber are shown in Figure 12. The spectrum of the input light source (black line) was directly obtained with the source connected to the spectrometer. As can be observed, the spectrum of the broadband source spans from ~1400 to ~1650 nm. The spectrum of the transmitted light through the fiber (red line) exhibits a transmission notch at 1512.3 nm. The inset of Figure 12 shows the transmission calculated as the division of the transmitted signal over the input light signal, shown in dBm units. The results were obtained for the NCF structure submerged in a solution with a refractive index of 1.36. The observed transmission notch corresponds to the second LMR order, which exhibits a depth of ~6 dB.

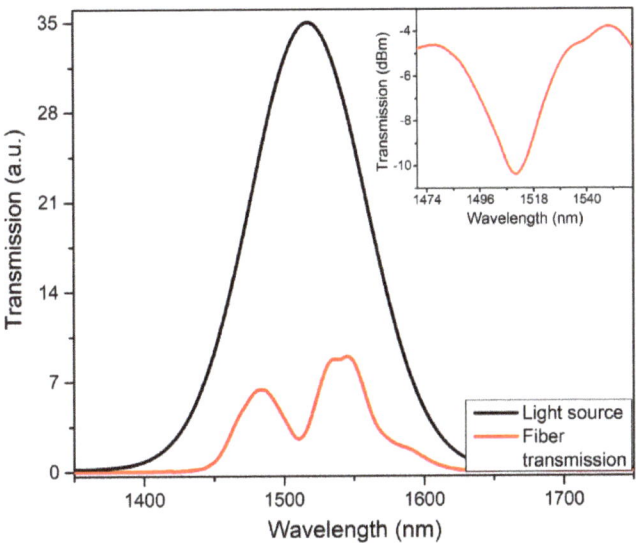

Figure 12. Output spectrum and transmission of the fiber structure.

The characterization of the wavelength displacement of the transmission notch as a function of the variations on the refractive index of the liquid medium surrounding the coated fiber is shown in Figure 13. As can be observed, with the increase in the refractive index of the surrounding medium, the transmission notch displaces toward longer wavelengths as the depth of the notch decreases.

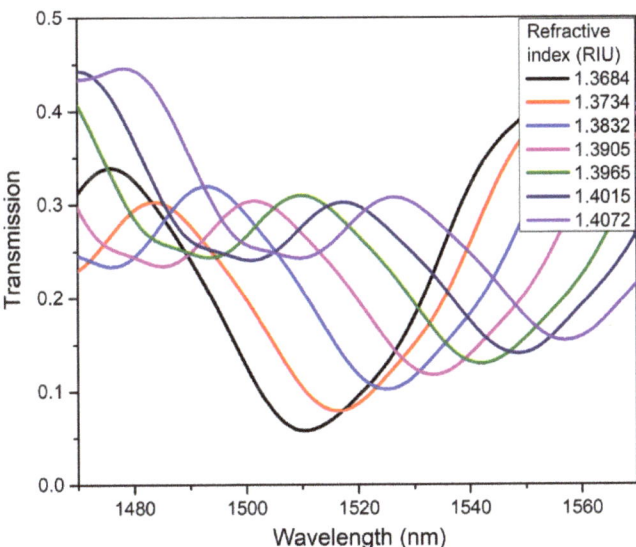

Figure 13. Wavelength displacement of the transmission notch as a function of the refractive index variations.

From the results obtained in Figure 13, the characteristic curve of the refractometer is obtained by wavelength displacement interrogation, as shown in Figure 14. As it can be observed, the results can be fitted to a linear function with R^2 of 0.996. The sensitivity, obtained from the slope, is 1184.91 nm/RIU in a refractive index ranging from 1.3684 to 1.4072.

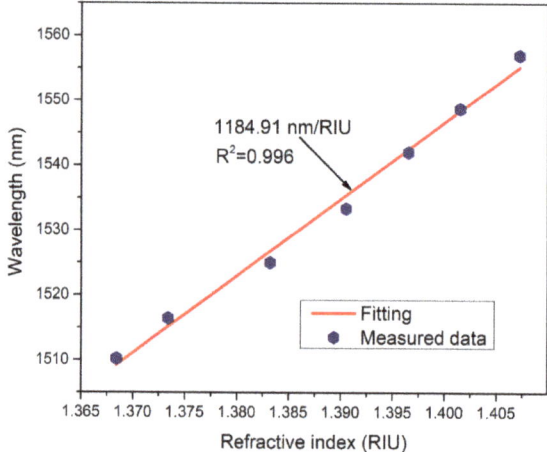

Figure 14. Wavelength displacement of the transmission notch as a function of RI variation wavelength shift sensitivity.

4. Conclusions

Titanium nitride thin films were synthesized by the technique of RF reactive magnetron sputtering. The thin films were deposited on a glass substrate, on which optical fibers were placed. After deposition, the substrates were characterized by means of XPS and UV–Vis spectroscopy. The analysis showed the formation of oxide and nitride phases, where the TiN phase was non-stoichiometric and metal-rich. The forbidden width of the nitride phase presented values in the range of E_g = 3.4–3.7 eV. The coated fibers were then placed in a Q-switch laser optical system, with the purpose of detecting and characterizing the plasmonic response at an energy region of around 1.55 µm. Regular pulses were detected in the region close to the output energy of the laser, which was generated by the interaction of light with the plasmonic modes of the thin film.

Finally, we report the applications as a saturable absorber and LMR fiber refractometer of different fiber structures coated with thin films of titanium nitride compounds deposited by DC/RF magnetron sputtering. The obtained results demonstrate the reliability of the sputtering technique to equalize the deposition parameters in order to obtain specific optical properties for the proposed applications.

Supplementary Materials: The following supporting information can be downloaded at: https://www.mdpi.com/article/10.3390/coatings13010095/s1, Figure S1: (a) Sputtering system for RF deposition (b) Schematic detail of the deposition chamber; Figure S2: Sputtering system for DC deposition; Figure S3: Tauc´s curve with the procedure used to obtain E_g; Figure S4: (a) Ti target (b) TiN target; Figure S5: XPS spectra for the Ti2p and N1s high resolution windows for TiN target; Figure S6: XPS spectra for the Ti2p and N1s high resolution windows for Ti target.

Author Contributions: R.I.Á.T. and M.G.M. conceived and designed the experiments; M.G.M., P.P.C. and O.G.R. contributed to the sputtering deposition and thin-film characterization by XPS; R.I.Á.T. and O.G.R. performed the experiments of the PQS laser; P.P.C. and R.I.Á.T. performed the experiments of the LMR refractometer; M.G.M. and R.I.Á.T. contributed the reagents, materials, and analysis tools; O.G.R., P.P.C., R.I.Á.T. and M.G.M. wrote, reviewed, and edited the paper. All authors have read and agreed to the published version of the manuscript.

Funding: This research was funded in part by Universidad Autónoma de Nuevo León: 245-CE-2022; Universidad Popular Autónoma del Estado de Puebla: Fondo de investigación UPAEP 2022.

Institutional Review Board Statement: Not applicable.

Informed Consent Statement: Not applicable.

Data Availability Statement: Not applicable.

Acknowledgments: This work was supported by the PAICyT UANL project under grant 245-CE-2022. Omar Gaspar Ramírez is thankful for CONACyT Posgraduate scholarship grant 817293.

Conflicts of Interest: The authors declare no conflict of interest.

References

1. Ponja, S.D.; Williamson, B.A.; Sathasivam, S.; Scanlon, D.O.; Parkin, I.P.; Carmalt, C.J. Enhanced electrical properties of antimony doped tin oxide thin films deposited via aerosol assisted chemical vapour deposition. *J. Mater. Chem. C* **2018**, *6*, 7257–7266. [CrossRef]
2. Mohammed, M.A.; Salman, S.R.; Wasna'a, M.A. Structural, optical, electrical and gas sensor properties of ZrO_2 thin films prepared by sol-gel technique. *NeuroQuantology* **2020**, *18*, 22. [CrossRef]
3. Ashok, A.; Regmi, G.; Romero, N.A.; Solis, L.M.; Velumani, S.; Castaneda, H. Comparative studies of CdS thin films by chemical bath deposition techniques as a buffer layer for solar cell applications. *J. Mater. Sci. Mater. Electron.* **2020**, *31*, 7499–7518. [CrossRef]
4. Ricardo, I.Á.T.; Omar, G.R.; Patricia, P.C.; Manuel, G.M.; Antonio, B.P. TiOxNy Thin Film Sputtered on a Fiber Ball Lens as Saturable Absorber for Passive Q-Switched Generation of a Single-Tunable/Dual-Wavelength Er-Yb Double Clad Fiber Laser. *Nanomaterials* **2020**, *10*, 923.
5. Nikhil, K.P.; Daniel, J.R.A.; Erhan, A.; King, P.J.; Srinivas, G.; Kelvin, S.K.K.; Anthony, O.N. Effect of deposition conditions and post deposition anneal on reactively sputtered titanium nitride thin films. *Thin Solid Film* **2015**, *578*, 31–37.
6. Swann, S. Magnetron sputtering. *Phys. Technol.* **1988**, *19*, 67. [CrossRef]
7. Jarosław, J.; Piotr, W.; Paweł, P.M.; Monika, O.; Bartłomiej, W.; Aleksandra, S.; Michał, S.; Cezariusz, J.; Krzysztof, Z. Titanium Nitride as a Plasmonic Material from Near-Ultraviolet to Very-Long-Wavelength Infrared Range. *Materials* **2021**, *14*, 7095.
8. Luca, M.; Tapan, B.; Beatrice, R.B.; Filip, M.; Andrea, L.B.; Stepan, K.; Alberto, N. Controlling the plasmonic properties of titanium nitride thin films by radiofrequency substrate biasing in magnetron sputtering. *Appl. Surf. Sci.* **2021**, *554*, 149543.
9. Chun, C.C.; John, N.; Yang, Z.P.; Wilton, J.M.K.-P.; Willard, R.; Ting, S.L.; Diego, A.R.D.; Abul, K.A.; Hou-Tong, C. Highly Plasmonic Titanium Nitride by Room-Temperature Sputtering. *Sci. Rep.* **2019**, *9*, 15287.
10. Gururaj, V.N.; Jeremy, L.S.; Xingjie, N.; Alexander, V.K.; Timothy, D.S.; Alexandra, B. Titanium nitride as a plasmonic material for visible and near-infrared wavelengths. *Optical Materials Express* **2012**, *2*, 478–489.
11. Jiachang, B.; Ruyi, Z.; Shaoqin, P.; Jie, S.; Xinming, W.; Wei, C.; Liang, W.; Junhua, G.; Hongtao, C.; Yanwei, C. Robust plasmonic properties of epitaxial TiN films on highly lattice-mismatched complex oxides. *Physical Review Materials* **2021**, *5*, 075201.
12. Barhai, P.K.; Neelam, K.; Banerjee, I.; Pabi, S.K.; Mahapatra, S.K. Study of the effect of plasma current density on the formation of titanium nitride and titanium oxynitride thin films prepared by reactive DC magnetron sputtering. *Vacuum* **2010**, *84*, 896–901. [CrossRef]
13. White, N.; Campbell, A.L.; Grant, J.T.; Pachter, R.; Eyink, K.; Jakubiak, R.; Martinez, G.; Ramana, C.V. Surface/interface analysis and optical properties of RF sputter-deposited nanocrystalline titanium nitride thin films. *Appl. Surf. Sci.* **2014**, *292*, 74–85. [CrossRef]
14. Available online: https://www.thermofisher.com/mx/es/home/materials-science/learning-center/periodic-table.html (accessed on 13 November 2022).
15. Solovana, M.N.; Brusa, V.V.; Maistruka, E.V.; Maryanchuk, P.D. Electrical and Optical Properties of TiN Thin Films. *Inorg. Mater.* **2014**, *50*, 40–45. [CrossRef]
16. Van Bui, H.; Groenland, A.W.; Aarnink, A.A.I.; Wolters, R.A.M.; Schmitz, J.; Kovalgin, A.Y. Growth kinetics and oxidation mechanism of ALD TiN thin films monitored by in situ spectroscopic ellipsometry. *J. Electrochem. Soc.* **2011**, *158*, H214–H220. [CrossRef]
17. Saha, N.C.; Tompkins, H.G. Titanium nitride oxidation chemistry: An x-ray photoelectron spectroscopy study. *J. Appl. Phys.* **1992**, *72*, 3072–3079. [CrossRef]
18. Solis Pomar, F.; Nápoles, O.; Vázquez Rbaina, O.; Gutierrez Lazos, C.; Fundora, A.; Colin, A.; Pérez Tijerina, E.; Melendrez, M.F. Preparation and characterization of nanostructured titanium nitride thin films at room temperature. *Ceram. Int.* **2016**, *42*, 7571–7575. [CrossRef]
19. Wemple, S.H.; DiDomenico Jr, M. Behavior of the electronic dielectric constant in covalent and ionic materials. *Phys. Rev. B* **1971**, *3*, 1338–1351. [CrossRef]
20. Manuel, G.M.; Alvaro, B.C.; Ricardo, R.S.; Víctor, C. Investigation of the annealing effects on the structural and opto-electronical properties of RF-sputtered ZnO films studied by the Drude-Lorentz model. *Appl. Phys. A* **2015**, *120*, 1375–1382.
21. Arnaud, V.; Maria Alejandra, U.H.; Gaylord, G.; Nicolas, C.M.; Damien, J.; Marion, H.; Jean, Y.M.; Stéphanie, R.; Francis, V.; Carmen, J.; et al. Optical, electrical and mechanical properties of TiN thin film obtained from a TiO_2 sol-gel coating and rapid thermal nitridation. *Surf. Coat. Technol.* **2021**, *413*, 127089.
22. Hengyong, W.; Mingming, W.; Zhanliang, D.; Ying, C.; Jinglong, B.; Jian, L.; Yun, Y.; Yingna, W.; Yi, C.R.W. Composition, microstructure and SERS properties of titanium nitride thin film prepared via nitridation of sol–gel derived titania thin films. *J. Or Raman Spectrosc.* **2017**, *48*, 578–585.
23. Dwivedi, Y.S.; Sharma, A.K.; Gupta, B.D. Influence of skew rays on the sensitivity and signal-to-noise ratio of a fiber-optic surface-plasmon-resonance sensor: A theoretical study. *Appl. Opt.* **2007**, *46*, 4563–4569. [CrossRef] [PubMed]

24. Ricardo Iván, Á.T.; Patricia, P.C.; Manuel, G.M.; Abel, F.C. Lossy mode resonance refractometer operating in the 1.55 μm waveband based on TiOxNy thin films deposited onto no-core multimode fiber by DC magnetron sputtering. *Opt. Fiber Technol.* **2022**, *71*, 102929.
25. Lubinda, L. Debate over Mapostori use of open spaces. *Harare News* **2015**, *16*, 10–17.

Disclaimer/Publisher's Note: The statements, opinions and data contained in all publications are solely those of the individual author(s) and contributor(s) and not of MDPI and/or the editor(s). MDPI and/or the editor(s) disclaim responsibility for any injury to people or property resulting from any ideas, methods, instructions or products referred to in the content.

Article

Particle-in-Cell Simulations for the Improvement of the Target Erosion Uniformity by the Permanent Magnet Configuration of DC Magnetron Sputtering Systems

Young Hyun Jo [1,2], Cheongbin Cheon [1], Heesung Park [1] and Hae June Lee [1,*]

[1] Department of Electrical Engineering, Pusan National University, Busan 46241, Republic of Korea
[2] Mechatronics Research, Samsung Electronics Company, Ltd., Hwaseong 18448, Republic of Korea
* Correspondence: haejune@pusan.ac.kr

Abstract: Improving the target erosion uniformity in a commercial direct current (DC) magnetron sputtering system is a crucial issue in terms of process management as well as enhancing the properties of the deposited film. Especially, nonuniform target erosion was reported when the magnetic flux density gradient existed. A two-dimensional (2D) and a three-dimensional (3D) parallelized particle-in-cell (PIC) simulation were performed to investigate relationships between magnetic fields and the target erosion profile. The 2D PIC simulation presents the correlation between the heating mechanism and the spatial density profiles under various magnet conditions. In addition, the 3D PIC simulation shows the different plasma characteristics depending on the azimuthal asymmetry of the magnets and the mechanism of the mutual competition of the E × B drift and the grad-B drift for the change in the electron density uniformity.

Keywords: DC magnetron sputtering; particle-in-cell simulation

1. Introduction

Direct current magnetron sputtering (DCMS) is a standard physical vapor deposition (PVD) technology for metal deposition in a semiconductor process. The DCMS system applies DC electric power to a negatively biased metal target, whose atoms are sputtered by the ion bombardment. The atoms move through the chamber before they are deposited on a substrate on the opposite side. DCMS systems are usually operated at pressures lower than a few mTorr to minimize disturbance to diffusion and deposition of sputtered atoms. When the gas pressure is high, the sputtered atoms collide with neutrals, making the deposition inefficient. On the contrary, it is difficult to discharge in a low-pressure condition because the breakdown voltage rapidly increases on the left-hand side of Paschen's breakdown curve for a low-pressure regime. In order to keep a stable discharge under low-pressure conditions, a static magnetic field is applied to reduce electron transport in the perpendicular direction. In the parallel direction to the magnetic field, the magnetic mirror effect enhances the electron confinement, and finally, the plasma density is high at a particular location above the target. However, the magnetic field should not be too high to reduce the ion bombardment on the target. That is to say, the magnetic flux density should be in the range to magnetize not only electrons but also ions. Therefore, the design of the magnetic field profile under a given gas pressure and device structure is a critical issue. Many studies have been dedicated to revealing the physics of DCMS plasmas [1–6].

A standard DCMS system contains a cathode with a metal sputtering target and permanent magnets. The magnetic fields of permanent magnets make charged particles rotate and confine them to the circular motion commonly known as gyromotion. Magnetic confinement is the reason why DCMS systems sustain high-density plasmas at such low pressures. Plasmas consist of magnetically confined electrons, called magnetized electrons, and unmagnetized ions. They are generated near the sputtering target in DCMS systems.

Then, energetic ions accelerated in the sheath region of the plasmas, bombarding and sputtering atoms of the target materials. The sputtered atoms are deposited on a substrate. They can also be ionized while passing through plasma. The source of the deposited particles in a DCMS system is the ion bombardment on the target, which is controlled by the density profiles of the magnetized electrons. Therefore, it is crucial to understand the mechanism of plasma formation in the DCMS system.

We primarily need to design or modify components related to DCMS plasmas, such as the system length between a target and a substrate, target radius, magnetic configurations, and so on. The magnetic configuration is one of the most critical parameters to control the characteristics of the process results, including the target erosion uniformity. The conventional balanced magnetrons confine plasmas near the target well. In contrast, the unbalanced magnetrons can cause plasmas to escape toward the substrate since the magnetic field lines from the stronger magnets are not fully closed with the weaker magnets [1,4,5]. However, the specific characteristics and spatial distributions of plasmas depending on the magnet design are not so easy to anticipate, even though they are necessary for costly and closed processes such as semiconductor processing.

The magnetic field intensity of a DCMS system is typically a few hundred Gauss. The radius of the gyromotion, which is called the gyro-radius or Larmor radius, is $mv_\perp/|q|B$, where m is the mass of the charged particle, $v_\perp$ is the velocity perpendicular to the direction of the magnetic field, q is the electric charge of the charged particle, B is the intensity of the magnetic field. In a typical DCMS system, the gyro-radius of an ion is tens of centimeters, while the gyro-radius of an electron is an order of a micrometer. It means that most of the ions in DCMS plasmas are not magnetized since the system length of a standard DCMS system is between a few centimeters and tens of centimeters, while electrons are magnetized and confined in the system. Therefore, the magnetized electrons directly affect the ionization region and the spatial distribution of DCMS plasmas. DCMS plasmas were simulated with fluid approaches [7–10], kinetic approaches [11–24], or hybrid methods [6,7,25–29]. However, the validity of fluid models is limited by gas pressure, which should be high enough to apply the fluid assumptions. That is why a kinetic approach is necessary to simulate low-pressure DCMS plasmas. Some previous studies used only the trajectories of electrons to reduce computational load without coupling the charged particle dynamics with the self-consistent field solver [30–34].

Most of the previous research has chosen the particle-in-cell (PIC) Monte Carlo collisions (MCC) method to simulate low-pressure DCMS plasmas. A PIC-MCC is a computational methodology to simulate plasma discharges [35]. The interactions between charged particles and the electromagnetic field are simulated by repeatedly calculating the Newton-Lorentz equation for the charged particle motion and the Maxwell equation for the electromagnetic field. The Poisson equation can be used instead of the Maxwell equation to obtain the electric field in the electrostatic system. MCC is an efficient method to calculate collisions with a preprocess of computing particles to collide in the PIC simulation, proposed by Vahedi et al. [36]. A full PIC-MCC simulation includes no assumptions about energy distributions in plasma dynamics and is thus considered the most accurate way of running a plasma simulation to date. Highly accelerated PIC-MCC simulations by graphics processing unit (GPU)-based parallelization have recently been reported [37–39].

As DCMS is a well-established technique, most fundamental physics are well understood, except for the effect of 3D magnetic field structure, which requires a three-dimensional (3D) PIC simulation. However, a 3D PIC simulation is still challenging and time-consuming because of the realistic size of the DCMS device. In this study, we investigate the effect of the curvature variation in 3D magnetic fields. Using the PIC-MCC method, this paper focuses on the fundamental relationship between the physics of DCMS plasmas and the magnetic field configurations of permanent magnets in the DCMS system. First, the influences of several actual magnetic field configurations, including asymmetric magnets for DCMS plasmas, are investigated with a two-dimensional (2D) PIC-MCC simulation as a preliminary study in Section 2. Then, the effects of azimuthal asymmetry and asymmetry of permanent magnets on DCMS plasmas are discussed in more detail with a 3D PIC-MCC simulation in Section 3. Finally, the conclusion is followed in Section 4.

2. Two-Dimensional Simulation for the Variation of Magnet Configurations

This section elaborates on the effects of magnetic flux density with various configurations of permanent magnets using a 2D electrostatic PIC-MCC simulation (version 2023.02.01). The computational details are presented in Section 2.1. The effects of magnetic field magnitude on DCMS plasmas are discussed as a fundamental study in Section 2.2. The influence of the thickness of the magnetic yoke is discussed in Section 2.3. Finally, the effects of asymmetric magnets are investigated in Section 2.4.

2.1. Computational Details

An in-house 2D electrostatic PIC-MCC simulation code is utilized here [37]. The code is parallelized using CUDA and performed with an NVIDIA GeForce RTX 3090 (Gigabyte Technology Co., Ltd., New Taipei, Taiwan). This code calculates the charged particle motions for loops based on the cell number. It means that the simulation needs the dynamic load balance to improve the calculation speed for spatially nonuniform DCMS plasmas. However, the computation speed is fast enough to obtain each 2D result shown in this study within 24 h. A schematic diagram of the simulation domain is depicted in Figure 1. It represents a typical DCMS system with a copper sputtering target, which is also a cathode in this system. The applied electric power of the cathode is set to 100 W. The other conductors at the boundaries are all grounded and disconnected from the cathode, with a tiny gap between them. The upper cathode in the domain is covered with the substrate. The gas pressure is 1 mTorr of argon feed gas, which is considered uniform in the background of the discharge region for simplicity. Only argon plasmas are simulated without consideration of both sputtering processes and metal plasmas generated by sputtered atoms to focus on the influence of magnetic field configurations on the plasma profile. The ion-induced secondary electron emission coefficient of the copper target is assumed to be 0.2. The number of cells is 512×180, and the time step Δt is 10^{-11} s in the simulation. The applied magnetic field is calculated using a freeware program called Finite Element Method Magnetics (FEMM) (version 4.2, 2019).

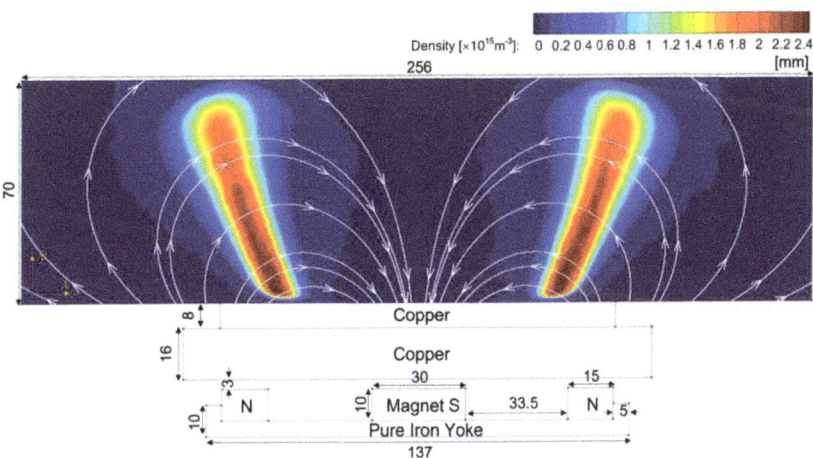

Figure 1. A schematic domain of the two-dimensional simulation with the plasma density and magnetic field lines. The length scale is in mm.

2.2. Influences of Magnetic Field Intensity

At first, two cases were selected to investigate the role of the magnitude of the magnetic flux density. Here, the configuration of the permanent magnets is the same, but only the magnitude is different. The profiles of magnetic fields are depicted in Figure 2, where the maximum intensity of magnetic fields is 350 G in Figure 2a and 500 G in Figure 2b. They have the same magnetic field profiles, but the color bar scales and maxima differ for both figures. The height of the yoke is 10 mm, and the widths of the magnets are 30 mm for the south pole and 15 mm for the north pole. The gap distance from the north pole to the south pole is 33.5 mm and symmetric.

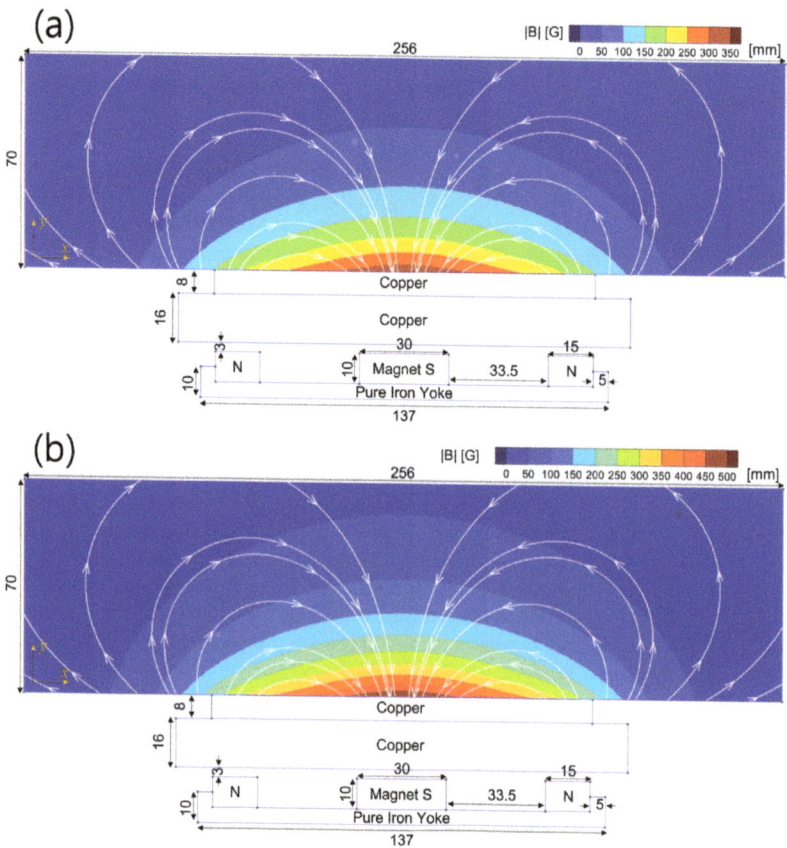

Figure 2. Magnetic fields with the maximum magnetic flux density of (**a**) 350 G and (**b**) 500 G.

The results of the PIC-MCC simulation are shown in Figure 3 for the plasma density and potential profiles. Only half of the domain is depicted here because the profiles are symmetric. The plasma density of the 350 G case is higher than that of the 500 G case. It is an unexpected result, as it is usually thought that stronger magnetic fields would confine plasmas better. Comparing Figure 3c,d, the sheath thickness in the 500 G case is broader than in the 350 G case since the plasma density is lower with the same target voltage of −470 V.

Figure 4 shows the Ohmic electron heating, $J \cdot E$, and the electron temperature T_e, where J is the electron current density, and E is the electric field intensity. Figure 4a shows enhanced electron heating in a broader space for the 350 G case compared with Figure 4b for the 500 G case. The electron temperature is higher near the substrate than the target, as shown in Figure 4c,d. Most ionizations occur in the high electron temperature region near the substrate. The source of the energetic electrons is the ion-induced secondary electron emission (SEE) from the target. They are accelerated by the electric field inside the sheath in front of the target to ionize the neutral gas or pass through the magnetized region until they meet the sheath in front of the substrate. These electrons make the high-temperature region near the substrate. The ions generated there are accelerated directly toward the target to induce SEE without collisions at very low pressure. That is why the plasma density of the 500 G case is lower even though it has stronger magnetic fields and higher electron temperatures near the target than that of the 350 G case. It indicates that more ionization near the substrate is the key to enhancing the DCMS discharge at very low pressure. Another noticeable difference is that heated electrons in the 350 G case are the primary reason for the striation. The fundamentals of this phenomenon are not fully analyzed here, but an additional electron heating mechanism exists in addition to the γ-mode in typical DC or DCMS plasmas, unlike the 500 G case. The striation of DCMS plasmas is barely found at pressures higher than a few mTorr, as shown in the previous study with a similar magnetic condition [20]. The striation phenomenon may have a sudden transition regime depending on the magnetic field intensity, which will be investigated more in future work.

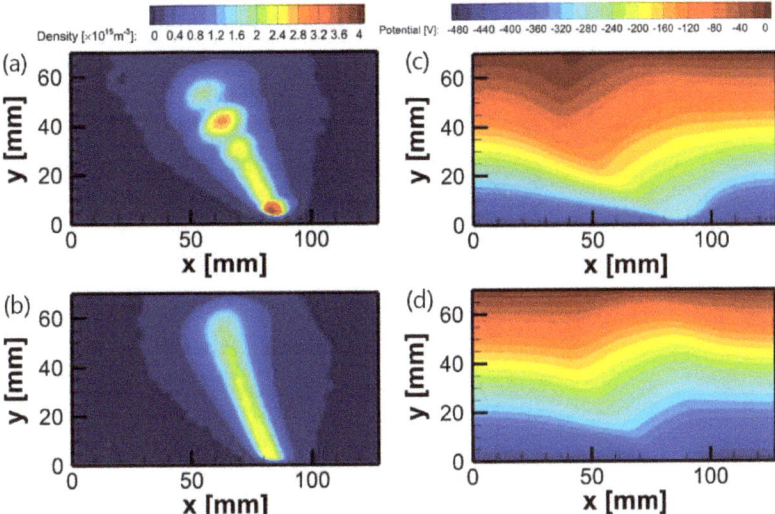

Figure 3. (**Left**) Plasma densities with the maximum magnetic flux densities of (**a**) 350 G and (**b**) 500 G. (**Right**) Electric potentials with the maximum magnetic flux densities of (**c**) 350 G and (**d**) 500 G.

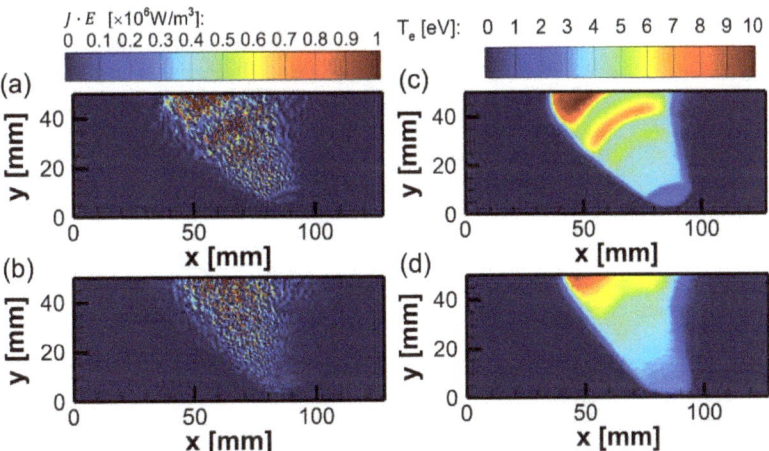

Figure 4. Profiles of $J \cdot E$ with the maximum magnetic flux densities of (**a**) 350 G and (**b**) 500 G. Profiles of electron temperature with the maximum magnetic flux densities of (**c**) 350 G and (**d**) 500 G.

2.3. Influences of the Yoke Thickness

Magnetic yokes commonly consist of ferromagnetic materials with high permeability and focus magnetic fluxes in the desired direction or position. In a DCMS system, magnetic yokes make magnetic fluxes from permanent magnets concentrate in the direction toward plasmas. The concentrated magnetic fluxes enhance the magnetic fields and the density of plasmas. The effects of various yoke configurations, including a wide yoke that moves the region of target erosion, were previously reported [38]. It shows that it is possible to control plasmas and target erosion profiles using the yoke magnets. In this section, the effects of the yoke thickness are discussed. The magnetic field lines and the magnitude of the magnetic flux density, with a thickness of the magnetic yoke of 15 cm, are depicted in Figure 5. The thicker yoke pulls the magnetic field slightly more toward the target, enhancing the magnetic field near the target.

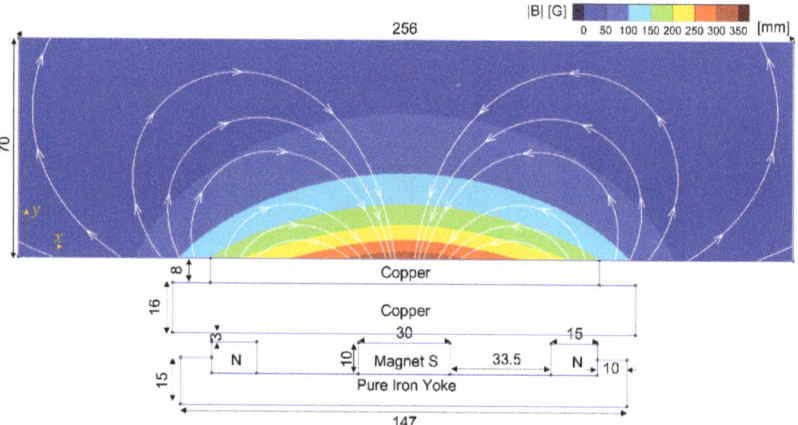

Figure 5. Magnetic field lines and the magnitude of the magnetic flux density of 350 G case with a magnetic yoke with a thickness of 15 mm.

The effects of the yoke thickness on DCMS plasmas are presented in Figure 6 for the plasma density and potential profiles and in Figure 7 for electron heating and the electron temperature. Compared with the change in the magnitude of the magnetic flux density, the plasma density near the target seems unaffected by the yoke thickness. However, the density near the top substrate is higher with the thicker yoke, as shown in Figure 6b. On the other hand, the electron temperature increases with the overall thickness of the yoke. The enhanced discharge with the thicker yoke can be explained by the increasing electron temperature observed in the middle region of the domain, as shown in Figure 7d. The striation patterns are related to the spatial variations of electron temperature caused by electron heating, as shown in Figure 7b. The stronger magnetic fields created by the thicker yoke trap more electrons, even far from the target. As a result, more ionization occurs with enhanced electron heating. Therefore, the yoke thickness can be one of the parameters to control the electron temperature and the density in the region a little bit away from the target. However, the strong magnetic field still limits the perpendicular transport of electrons near the target, and thus the density does not change significantly near the target for $0 < y < 10$ mm.

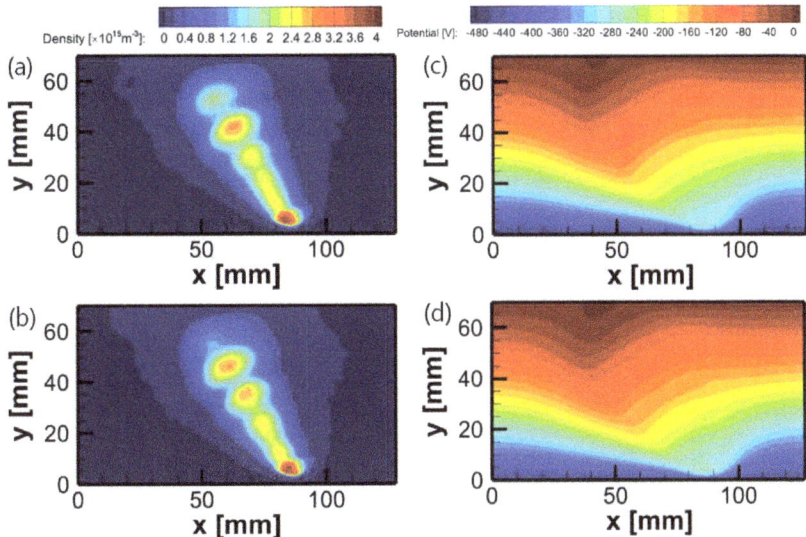

Figure 6. Plasma densities with the yoke thickness of (**a**) 10 mm and (**b**) 15 mm. Electric potentials with the yoke thickness of (**c**) 10 mm and (**d**) 15 mm.

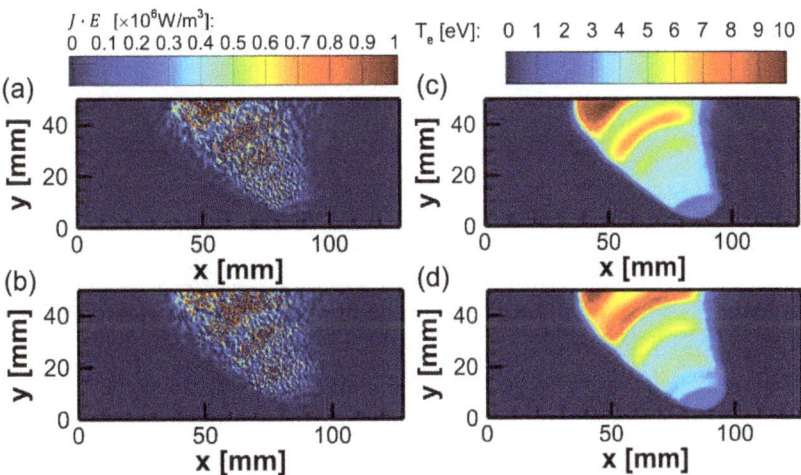

Figure 7. Profiles of $J \cdot E$ with the yoke thickness of (**a**) 10 mm and (**b**) 15 mm. Profiles of electron temperatures with the yoke thickness of (**c**) 10 mm and (**d**) 15 mm.

2.4. Effects of Asymmetric Magnets

A case of asymmetric permanent magnets, where the magnet on the right side is more intensified, is investigated in this section. The magnetic fields in this case are shown in Figure 8. In this case, the magnetic field on the right side with a wide magnet and short gap is stronger than the one on the left side with a narrow magnet and a long gap. The maximum magnitude of the magnetic flux density is set to be 350 G in this case.

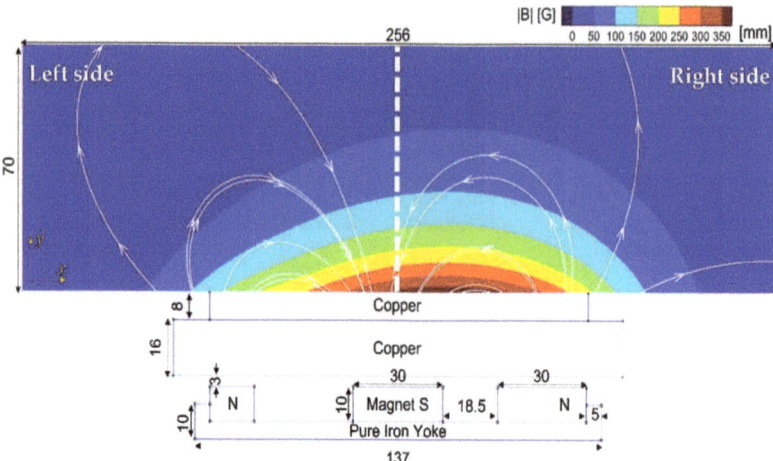

Figure 8. Magnetic field lines and the magnitude of the magnetic flux density for asymmetric magnets where the widths of the magnets are 15 mm on the left side and 30 mm on the right side.

The effects of asymmetric magnets on DCMS plasmas are presented in Figure 9 for the plasma density and electric potential profiles and in Figure 10 for electron heating and the electron temperature. The stronger magnetic fields on the right side confine plasmas in a similar shape to the 500 G case, while the weaker magnetic fields on the left side confine plasmas in a similar shape to the 350 G case. The lower density in the region where the magnetic fields are stronger can be explained in the same way as discussed in Section 2.2, although the striation is slightly less visible on the left side. Even though the electron confinement is better on the right side, the plasma density is high on the left side because the electron heating is enhanced there, as shown in Figure 10a.

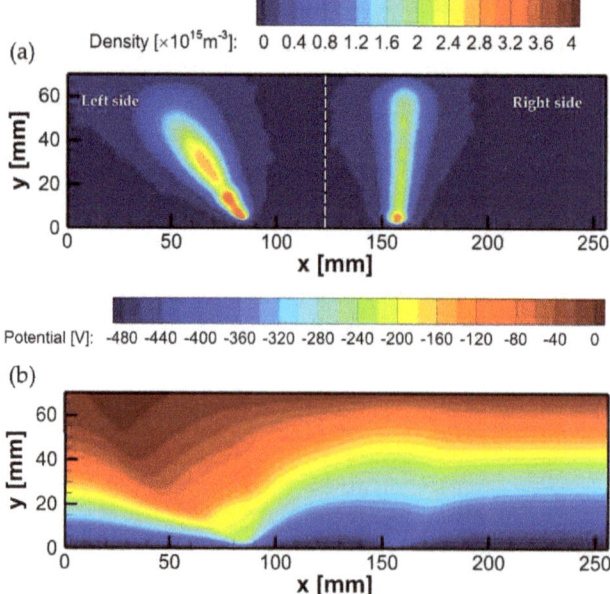

Figure 9. (a) Plasma density and (b) electric potential of the asymmetric magnetic flux density shown in Figure 8.

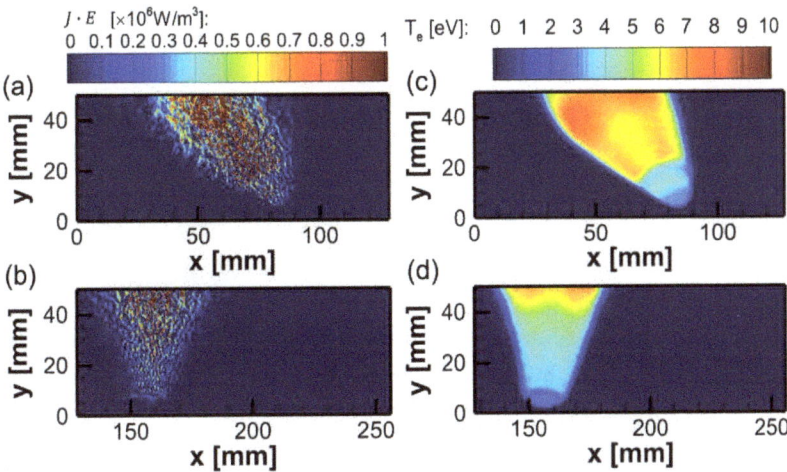

Figure 10. (**Left**) Profiles of $J \cdot E$ of the (**a**) left side and (**b**) right side of the domain for the asymmetric case. (**Right**) Profiles of electron temperatures of (**c**) the left side and (**d**) the right side of the domain in the asymmetric case.

The spatial distribution of plasma density on the right side seems more vertical to the target than on the left side. It is because magnetic field lines cause the difference in the spatial profiles of the plasma density. Plasmas are placed along the direction of $-\nabla B$, where B is the magnitude of the magnetic flux density. The profiles of electron heating and electron temperatures shown in Figure 10 also show the same tendency toward less electron heating (Figure 10b) with more intensified magnetic fields. This result indicates that not only the magnitude of the magnetic flux density but also the magnetic field profile determine the transition of the striation regime. In addition, this kind of asymmetric configuration of magnets can be an option to control target erosion profiles since there will be different sputtering characteristics depending on the sputtering region on the target.

3. Three-Dimensional Simulation for Azimuthal Symmetry of Magnets

In Section 2, the 2D simulation was performed without consideration of the drift motion of electrons. As shown in Figure 11, the magnetic and electric fields normal to the target surface in the z-direction are crossed and generate an azimuthal E × B drift in the 3D space. In addition, $-\nabla B$ is also in the z-direction to cause a grad-B drift in the azimuthal direction. Finally, the density gradient in the z-direction also induces diamagnetic drift. These three types of drift motion play an essential role in transport in the azimuthal direction. Therefore, it is mandatory to include the effect of the drift motion under the variation of the curvature of the magnetic field lines.

The design of permanent magnets can vary depending on either the desired erosion region of the target plate or the characteristics of deposited thin films. For example, azimuthally asymmetric magnets make magnetic fields, and generated plasmas focus on a specific region to be sputtered, which can be used to make even more uniform erosion of the sputtering target [38,39]. This section discusses the effects of magnets' azimuthal symmetry on DCMS plasmas with three-dimensional electrostatic PIC-MCC simulation results. Detailed information, including the numerical method and the performance of the simulation code, was reported in the previous study [22].

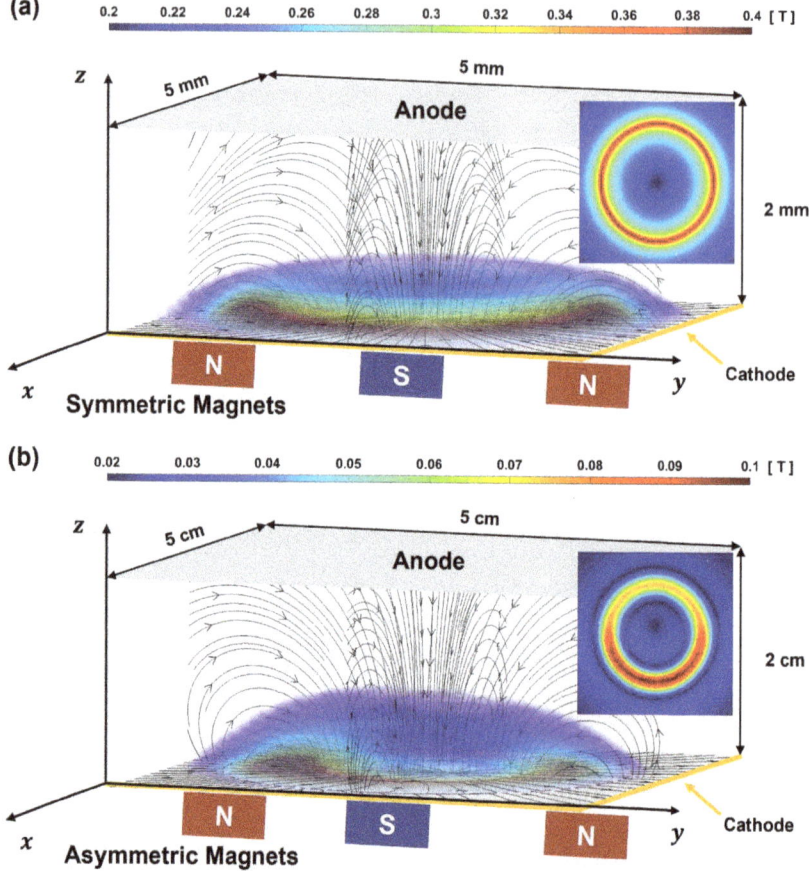

Figure 11. Schematic domains and profiles of magnetic fields in the three-dimensional simulation with (**a**) azimuthally symmetric magnets and (**b**) azimuthally asymmetric magnets.

Schematic diagrams of the 3D simulation domain with both azimuthally symmetric and asymmetric magnetic fields are depicted in Figure 11a,b, respectively. The simulation domains for both cases indicate simplified DCMS systems with boundaries that consist of one cathode surface and five grounded conductors. A fixed electric current is applied to the cathode in this 3D simulation instead of the constant electric power to obtain simulation results more quickly. Note that the simulation conditions differ since the results are obtained from different studies. For the symmetric case, the domain size is 5 mm × 5 mm × 2 mm, and the number of cells is 80 × 80 × 32. The applied current is 2 mA. The time step Δt is 2×10^{-11} s. The gas pressure is set to 150 mTorr. Argon gas is used and assumed to be uniform in the discharge region. For the asymmetric case, the domain size is 50 mm × 50 mm × 20 mm, and the number of cells is 64 × 64 × 32. The applied current is 20 mA. The time step Δt is 1×10^{-11} s. The gas pressure is set at 50 mTorr. It also considers uniform argon feed gas in the discharge region. The data were extracted at $r = 1.6$ mm and $r = 1.6$ cm, respectively, to investigate the change in the azimuth direction over time in the symmetric magnet and asymmetric magnet cases. These radii were set to follow the line of the maximum electron density when extracting data in the azimuth direction. The axial positions of the symmetric and asymmetric magnets were set to 0.75 mm and 3.125 mm, respectively, because the steady state results showed that the electron densities here had peak points on the two-dimensional xy plane.

Figure 12a presents the change in the simulated density of DCMS plasmas under the symmetric magnetic field shown in Figure 11a. Initially, the plasma density profile shows azimuthal symmetry at 0.75 ms. However, after 1 ms, it shows the evolution of the $m = 1$ mode. Finally, after 2.5 ms, the plasma density shows an $m = 2$ mode. Figure 12b shows the time evolution of the plasma density in the azimuthal region at the center, which clearly shows the growth of the m = 1 mode from 1 to 2 ms and the $m = 2$ mode after 2.5 ms. The symmetric magnets cause rotating structures of plasmas, which are well known as a phenomenon of rotating spoke instabilities [40–44]. In this case, the dominant mode is $m = 2$ in the steady state. The direction of rotation is clockwise, and the rotation velocity is 79.4 km/s. At the steady state, the $m = 2$ mode has an oscillation frequency of 3.03 MHz.

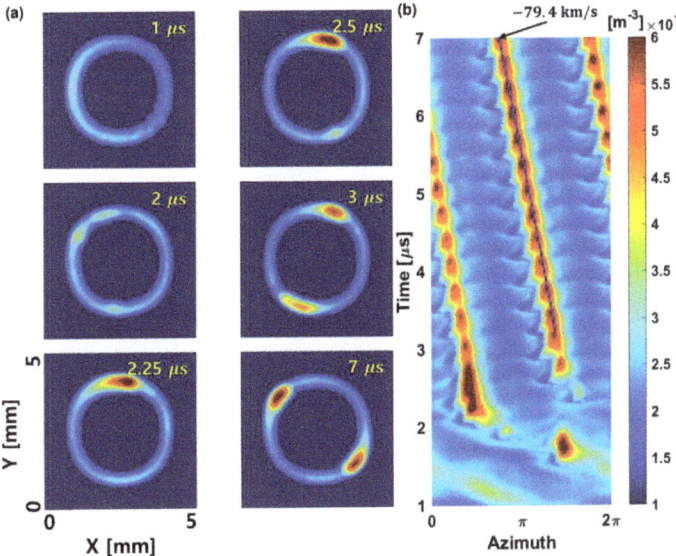

Figure 12. Time evolution of (**a**) spatial distributions of plasma density in xy plane with the symmetric magnet and (**b**) azimuthal distributions of the plasma density.

On the other hand, Figure 13 shows the plasma density profiles for the asymmetric magnetic fields shown in Figure 11b. The asymmetric magnetic field profile generates a quasi-stationary density structure with an azimuthal $m = 1$ mode. In the beginning, there is a transient time that shows a slight rotation of the density profile until $t < 4$ ms. However, the density pattern is almost static after 4 ms, following the change in the magnitude of the magnetic flux density. The high-density region is the same as where the azimuthal direction's drift velocity decreases.

Figure 14 shows the 3 types of azimuthal drift velocities measured at z = 3.125 mm and r = 1.6 cm for the steady state of Figure 13. The grad-B drift is the most dominant, while the E × B drift is relatively small and uniform. In addition, the direction of the grad-B drift is counter-clockwise, but that of the E × B drift is clockwise. Therefore, the most dominant factor triggering the instability is the gradient of the magnetic field profiles.

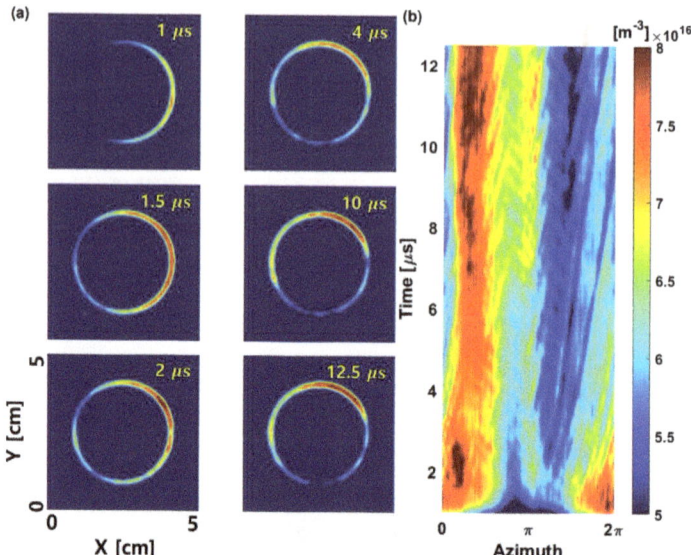

Figure 13. Time evolution of (**a**) spatial distributions of plasma density in xy plane with the asymmetric magnet and (**b**) azimuthal distributions of the plasma density.

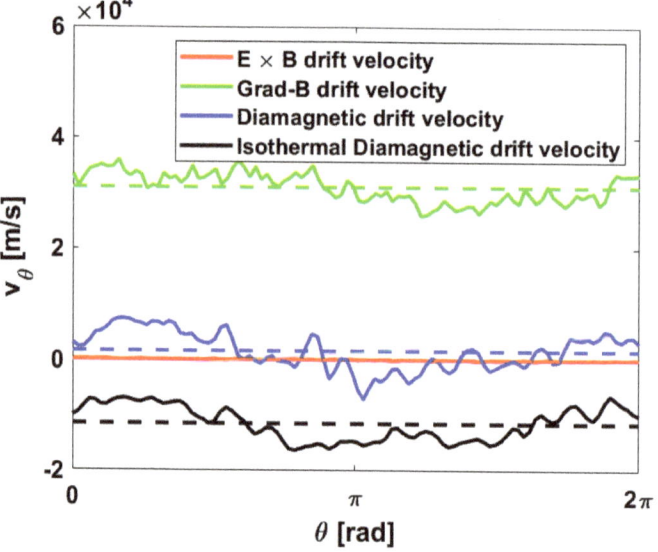

Figure 14. Azimuthal drift velocities were measured at z = 3.125 mm and r = 1.6 cm for E × B (red), grad-B (green), and diamagnetic (blue) drift, respectively. The black line is for the diamagnetic drift under the assumption of an isothermal plasma without a temperature gradient. Dashed lines indicate the mean values of each drift velocity.

The different states of each case can be analyzed with field energy. The electrostatic potential energy is given by $q\phi$ where ϕ is the electric potential. Figure 15 shows the time evolution of field energies for the two cases. The field energy of the symmetric case on a small scale oscillates since there are constant interactions between waves and charged particles, while that in the asymmetric case on a large scale saturates without oscillation. The reasons for the difference can be various since the simulation conditions are also different. However, the fundamental mechanism of the stationary structure with asymmetric magnets is not fully understood yet.

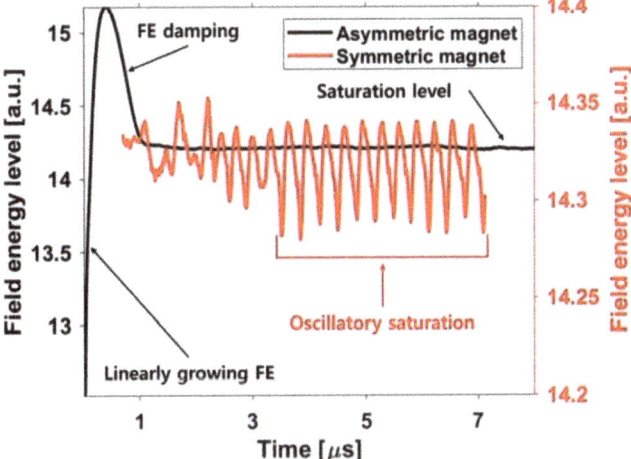

Figure 15. Saturation levels are shown in logarithmic scale for the normalized potential energy with an asymmetric (black) and a symmetric (red) magnet.

4. Conclusions

The effects of various configurations of permanent magnets on DCMS plasmas have been investigated with a 2D and a 3D PIC-MCC simulation. In the 2D PIC-MCC simulation, higher plasma densities with striations were obtained with a magnetic flux density of 350 G on the magnet surface than with 500 G. The increase in plasma density is caused by the enhanced electron heating under the weak magnetic flux density, although the electron confinement is much better under the strong magnetic flux density. The electron heating mechanism is related to wave-like transports along the magnetic field lines.

With the increase in the yoke thickness, the magnetic flux density increases slightly in the plasma region, and the geometry of the field lines changes. The plasma density near the target surface does not change much with the yoke thickness change. However, the density profile far from the target surface changes significantly. It indicates that more heated electrons propagate following the far magnetic field line but not near the target surface, where the strong magnetic field still limits the perpendicular transport of electrons. It shows the importance of the structure of magnetic field lines. Another example to show the importance of the magnetic field profile is the case with an asymmetric magnet structure. Even though the electron confinement is better where the gap from the north to the south pole is shorter, the plasma density is high in the weak-magnetic field region because the electron heating is enhanced.

Another interesting phenomenon of a stationary structure with azimuthally asymmetric magnets is investigated with a 3D PIC-MCC simulation. The difference between cases with azimuthally symmetric and asymmetric magnets is whether the azimuthal drift velocity is uniform. It was discussed along with the time evolution of the plasma density and the field energy. With the symmetric magnet on a small scale, oscillation patterns in the $m = 2$ mode are generated at an oscillation frequency of 3.03 MHz. However, with the asymmetric magnet on a larger scale, a static $m = 1$ mode lasts long without oscillation. We found that the grad-B drift is the most dominant from the measurement of the three drift velocities. Therefore, the instability generating the spoke shown in Figure 12 could be caused by the strong shear of the significant grad-B drift for a high curvature magnet. It is a topic for future work.

Although the number of specific simulation cases is limited in this paper, newly found fundamental influences of magnetic configurations on DCMS plasmas were investigated with kinetic approaches. However, the structures of magnets in a DCMS system can be much more complex depending on the desired characteristics of target sputtering or deposition. For example, additional permanent magnets or electromagnets can be applied to modify plasma distribution or improve the uniformity of the deposition profile. Therefore, more fundamental studies are necessary to determine the specific magnet design and configuration.

Author Contributions: Conceptualization, Y.H.J., C.C. and H.J.L.; Methodology, Y.H.J. and C.C.; Software, C.C. and H.P.; Validation, C.C. and H.J.L.; Formal analysis, Y.H.J., C.C. and H.P.; Investigation, Y.H.J., C.C. and H.P.; Data curation, Y.H.J., C.C. and H.P.; Writing—original draft, Y.H.J.; Writing—review and editing, H.J.L.; Visualization, C.C.; Supervision, H.J.L.; Project administration, H.J.L.; Funding acquisition, H.J.L. All authors have read and agreed to the published version of the manuscript.

Funding: This work was supported by the National Research Council of Science and Technology (NST) grant by the Korean government (MSIT) (No. CRC-20-01-NFRI) and by the National R&D Program through the National Research Foundation of Korea (NRF), funded by the Ministry of Education, Science and Technology (Grant No. NRF-2019R1A2C1088518).

Institutional Review Board Statement: Not applicable.

Informed Consent Statement: Not applicable.

Data Availability Statement: Not applicable.

Conflicts of Interest: The authors declare no conflict of interest.

References

1. Kelly, P.J.; Arnell, R.D. Magnetron sputtering: A review of recent developments and applications. *Vacuum* **2000**, *56*, 159. [CrossRef]
2. Gudmundsson, J.T. Physics and technology of magnetron sputtering discharges. *Plasma Sources Sci. Technol.* **2020**, *29*, 113001. [CrossRef]
3. Buyle, G.; Depla, D.; Eufinger, K.; Haemers, J.; De Gryse, R.; De Bosscher, W. Simplified model for calculating the pressure dependence of a direct current planar magnetron discharge. *J. Vac. Sci. Technol. A* **2003**, *21*, 1218. [CrossRef]
4. Savvides, N.; Window, B.J. Unbalanced magnetron ion-assisted deposition and property modification of thin films. *J. Vac. Sci. Technol. A* **1986**, *4*, 504. [CrossRef]
5. Sproul, W.D. High-rate reactive DC magnetron sputtering of oxide and nitride superlattice coatings. *Vacuum* **1998**, *51*, 641. [CrossRef]
6. Bogaerts, A.; Bultinck, E.; Kolev, I.; Schwaederlé, L.; Aeken, K.V.; Buyle, G.; Depla, D. Computer modelling of magnetron discharges. *J. Phys D Appl. Phys.* **2009**, *42*, 194018. [CrossRef]
7. Bogaerts, A.; Bultinck, E.; Eckert, M.; Georgieva, V.; Mao, M.; Neyts, E.; Schwaederlé, L. Computer Modeling of Plasmas and Plasma-Surface Interactions. *Plasma Process. Polym.* **2009**, *6*, 295. [CrossRef]
8. Chen, L.; Cui, S.; Tang, W.; Zhou, L.; Li, T.; Liu, L.; An, X.; Wu, Z.; Ma, Z.; Lin, H.; et al. Modeling and plasma characteristics of high-power direct current discharge. *Plasma Sources Sci. Technol.* **2020**, *29*, 025016. [CrossRef]
9. Costin, C.; Marques, L.; Popa, G.; Gousset, G. Two-dimensional fluid approach to the dc magnetron discharge. *Plasma Sources Sci. Technol.* **2005**, *14*, 168. [CrossRef]
10. Costin, C.; Popa, G.; Gousset, G. On the secondary electron emission in dc magnetron discharge. *J. Optoelectron. Adv. Mater.* **2005**, *7*, 2465.
11. Pflug, A.; Siemers, M.; Schwanke, C.; Kurnia, B.F.; Sittinger, V.; Szyszka, B. Simulation of plasma potential and ion energies in magnetron sputtering. *Mater. Technol.* **2011**, *26*, 10. [CrossRef]
12. Yagisawa, T.; Makabe, T. Modeling of dc magnetron plasma for sputtering: Transport of sputtered copper atoms. *J. Vac. Sci. Technol. A* **2006**, *24*, 908. [CrossRef]
13. Nanbu, K.; Segawa, S.; Kondo, S. Self-consistent particle simulation of three-dimensional dc magnetron discharge. *Vacuum* **1996**, *47*, 1013. [CrossRef]
14. Kondo, S.; Nanbu, K. Axisymmetrical particle-in-cell/Monte Carlo simulation of narrow gap planar magnetron plasmas. I. Direct current-driven discharge. *J. Vac. Sci. Technol. A* **2001**, *19*, 830. [CrossRef]
15. Kolev, I.; Bogaerts, A.; Gijbels, R. Influence of electron recapture by the cathode upon the discharge characteristics in dc planar magnetrons. *Phys. Rev. E* **2005**, *72*, 056402. [CrossRef] [PubMed]
16. Kolev, I.; Bogaerts, A. PIC—MCC Numerical Simulation of a DC Planar Magnetron. *Plasma Process. Polym.* **2006**, *3*, 127. [CrossRef]

17. Zheng, B.; Fu, Y.; Wang, K.; Tran, T.; Schuelke, T.; Fan, Q.H. Comparison of 1D and 2D particle-in-cell simulations for DC magnetron sputtering discharges. *Phys. Plasmas* **2021**, *28*, 014504. [CrossRef]
18. Shon, C.H.; Lee, J.K.; Lee, H.J.; Yang, Y.; Chung, T.H. Velocity Distributions in Magnetron Sputter. *IEEE Trans. Plasma Sci.* **1998**, *26*, 1635. [CrossRef]
19. Hur, M.Y.; Kim, J.S.; Lee, H.J. The Effect of Negative Ions from the Target on Thin Film Deposition in a Direct Current Magnetron Sputtering System. *Thin Solid Film.* **2015**, *587*, 3. [CrossRef]
20. Hur, M.Y.; Oh, S.H.; Kim, H.J.; Lee, H.J. Numerical Analysis of the Incident ion Energy and Angle Distribution in the DC Magnetron Sputtering for the Variation of Gas Pressure. *Appl. Sci. Converg. Technol.* **2018**, *27*, 19. [CrossRef]
21. Jo, Y.H.; Park, H.S.; Hur, M.Y.; Lee, H.J. Curved-boundary particle-in-cell simulation for the investigation of the target erosion effect of DC magnetron sputtering system. *AIP Adv.* **2020**, *10*, 125224. [CrossRef]
22. Jo, Y.H.; Cheon, C.; Park, H.; Hur, M.Y.; Lee, H.J. Multi-dimensional electrostatic plasma simulations using the particle-in-cell method for the low-temperature plasmas for materials processing. *J. Korean Phys. Soc.* **2022**, *80*, 787. [CrossRef]
23. Kageyama, J.; Yoshimoto, M.; Matuda, A.; Akao, Y.; Shidoji, E. Numerical simulation of plasma confinement in DC magnetron sputtering under different magnetic fields and anode structures. *Jpn. J. Appl. Phys.* **2014**, *53*, 088001. [CrossRef]
24. Matyash, K.; Fröhlich, M.; Kersten, H.; Thieme, G.; Schneider, R.; Hannemann, M.; Hippler, R. Rotating dust ring in an RF discharge coupled with a dc-magnetron sputter source. Experiment and simulation. *J. Phys. D Appl. Phys.* **2004**, *37*, 2703. [CrossRef]
25. Shidoji, E.; Ohtake, H.; Nakano, N.; Makabe, T. Two-Dimensional Self-Consistent Simulation of a DC Magnetron Discharge. *Jpn. J. Appl. Phys.* **1999**, *38*, 2131. [CrossRef]
26. Kolev, I.; Bogaerts, A. Numerical Models of the Planar Magnetron Glow Discharges. *Contrib. Plasma Phys.* **2004**, *44*, 582. [CrossRef]
27. Jimenez, F.J.; Dew, S.K.; Field, D.J. Comprehensive computer model for magnetron sputtering. II. Charged particle transport. *J. Vac. Sci. Technol. A* **2014**, *32*, 061301. [CrossRef]
28. Kwon, U.H.; Choi, S.H.; Park, Y.H.; Lee, W.J. Multi-scale simulation of plasma generation and film deposition in a circular type DC magnetron sputtering system. *Thin Solid Film.* **2005**, *475*, 17. [CrossRef]
29. Kwon, U.H.; Lee, W.J. Multiscale Monte Carlo Simulation of Circular DC Magnetron Sputtering: Influence of Magnetron Design on Target Erosion and Film Deposition. *Jpn. J. Appl. Phys.* **2006**, *45*, 8629. [CrossRef]
30. Ido, S.; Suzuki, T.; Kashiwagi, M. Computational Studies on the Erosion Process in a Magnetron Sputtering System with a Ferromagnetic Target. *Jpn. J. Appl. Phys.* **1998**, *37*, 965. [CrossRef]
31. Ido, S.; Suzuki, T.; Kashiwagi, M. Computational Studies of Plasma Generation and Control in a Magnetron Sputtering System. *Jpn. J. Appl. Phys.* **1999**, *38*, 4450.
32. Kadlec, S. Computer simulation of magnetron sputtering—Experience from the industry. *Surf. Coat. Technol.* **2007**, *202*, 895. [CrossRef]
33. Holik, M.; Bradley, J.; Gonzales, V.B.; Monaghan, D. Monte Carlo Simulation of Electrons' and Ions' Trajectories in Magnetron Sputtering Systems. *Plasma Process. Polym.* **2009**, *6*, S789. [CrossRef]
34. Tsygankov, P.A.; Orozco, E.A.; Dugar-Zhabon, V.D.; López, J.E.; Cárdenas, P.A. Simulation of the electron dynamics in a magnetron sputtering device with equipotential and non-equipotential cathode. *J. Phys. Conf. Ser.* **2019**, *1386*, 012127. [CrossRef]
35. Birdsall, C.K.; Langdon, A.B. *Plasma Physics via Computer Simulation*; McGraw-Hill: New York, NY, USA, 1985.
36. Vahedi, V.; Surendra, M. A Monte Carlo collision model for the particle-in-cell method: Applications to argon and oxygen discharges. *Comput. Phys. Commun.* **1995**, *87*, 179. [CrossRef]
37. Hur, M.Y.; Kim, J.S.; Song, I.C.; Verboncoeur, J.P.; Lee, H.J. Model description of a two-dimensional electrostatic particle-in-cell simulation parallelized with a graphics processing unit for plasma discharges. *Plasma Res. Express* **2019**, *1*, 015016. [CrossRef]
38. Iseki, T. Completely flat erosion magnetron sputtering using a rotating asymmetrical yoke magnet. *Vacuum* **2010**, *84*, 1372. [CrossRef]
39. Iseki, T. Flat erosion magnetron sputtering with a moving unbalanced magnet. *Vacuum* **2006**, *80*, 662. [CrossRef]
40. Hecimovic, A.; Keudell, A. Spokes in high power impulse magnetron sputtering plasmas. *J. Phys. D Appl. Phys.* **2018**, *51*, 453001. [CrossRef]
41. Anders, A.; Yang, Y. Plasma studies of a linear magnetron operating in the range from DC to HiPIMS. *J. Appl. Phys* **2018**, *123*, 043302. [CrossRef]
42. Boeuf, J.P. Rotating structures in low temperature magnetized plasmas—Insight from particle simulations. *Front. Phys.* **2014**, *74*, 1.
43. Panjan, M.; Anders, A. Plasma potential of a moving ionization zone in DC magnetron sputtering. *J. Appl. Phys.* **2017**, *121*, 063302. [CrossRef]
44. Panjan, M. Self-organizing plasma behavior in RF magnetron sputtering discharges. *J. Appl. Phys.* **2019**, *125*, 203303. [CrossRef]

Disclaimer/Publisher's Note: The statements, opinions and data contained in all publications are solely those of the individual author(s) and contributor(s) and not of MDPI and/or the editor(s). MDPI and/or the editor(s) disclaim responsibility for any injury to people or property resulting from any ideas, methods, instructions or products referred to in the content.

Article

Influence of Co-Content on the Optical and Structural Properties of TiO$_x$ Thin Films Prepared by Gas Impulse Magnetron Sputtering

Patrycja Pokora [1], Damian Wojcieszak [1,*], Piotr Mazur [2], Małgorzata Kalisz [3] and Malwina Sikora [1,4]

[1] Faculty of Electronics, Photonics and Microsystems, Wroclaw University of Science and Technology, Janiszewskiego 11/17, 50-372 Wroclaw, Poland
[2] Institute of Experimental Physics, University of Wroclaw, Max Born 9, 50-204 Wroclaw, Poland
[3] Faculty of Engineering and Economics, Ignacy Mościcki State Vocational University in Ciechanów, Narutowicza 9, 06-400 Ciechanów, Poland
[4] Nanores Company, Bierutowska 57–59, 51-317 Wroclaw, Poland
* Correspondence: damian.wojcieszak@pwr.edu.pl; Tel.: +48-713202375

Abstract: Nonstoichiometric (Ti,Co)Ox coatings were prepared using gas-impulse magnetron sputtering (GIMS). The properties of coatings with 3 at.%, 19 at.%, 44 at.%, and 60 at.% Co content were compared to those of TiO$_x$ and CoO$_x$ films. Structural studies with the aid of GIXRD indicated the amorphous nature of (Ti,Co)Ox. The fine-columnar, homogeneous microstructure was observed on SEM images, where cracks were identified only for films with a high Co content. On the basis of XPS measurements, TiO$_2$, CoO, and Co$_3$O$_4$ forms were found on their surface. Optical studies showed that these films were semi-transparent (T > 46%), and that the amount of cobalt in the film had a significant impact on the decrease in the transparency level. A shift in the absorption edge position (from 337 to 387 nm) and a decrease in their optical bandgap energy (from 3.02 eV to more than 2.60 eV) were observed. The hardness of the prepared films changed slightly (ca. 6.5 GPa), but only the CoO$_x$ film showed a slightly lower hardness value than the rest of the coatings (4.8 GPa). The described studies allowed partial classification of non-stoichiometric (Ti,Co)Ox thin-film materials according to their functionality.

Keywords: oxide thin films; TiO$_x$; (Ti,Co)Ox; CoO$_x$; cobalt; gas impulse magnetron sputtering; semitransparent; amorphous coatings

1. Introduction

Materials based on mixtures of Ti and Co oxides have recently gained attention because of their wide range of applications in electronics. Today, there are only a few reports on Co-doped TiO$_2$ [1–5], but mainly in the form of nanopowders, nanoparticles, or nanowires, while thin-film work itself is in a minority. One of the possible reasons for this fact is that the properties of such oxide mixtures are strongly related to the preparation method and additional postprocessing, such as high-temperature annealing. Therefore, a direct comparison of their properties is impossible. There exist only a few reports related to complex modification of the structural, surface, optical, or electrical properties of titanium dioxide as a matrix by doping with Co [6–9].

Our previous work is one of the few examples of such studies [10,11]. We have shown how interesting properties and applications can be obtained by combining the advantages of Ti and Co oxides. For example, the gradient coatings that we obtained were transparent and had a high refractive index despite the high content of cobalt [11]. More interesting is the fact that these coatings exhibited a fully recoverable resistive switching effect, which has not been previously reported in the case of such materials (Figure 1a). Their great potential can be used in innovative transparent electronic devices with a memory effect. Furthermore,

we also prepared transparent (~80%) oxide films based on titanium with a homogeneous distribution of cobalt [10]. In their case, a unipolar memristive effect was also observed, but only for coatings with a low cobalt content (Figure 1b). To date, no similar results have been reported. The aim of the present work was to manufacture nonstoichiometric (Ti,Co)Ox materials (with reduced oxygen content) in order to obtain a lower resistivity as compared to the mentioned coatings. A compromise between optical and electrical properties was planned to be achieved. In addition, we decided to prepare films with a higher amount of cobalt because of the small number of publications in this field.

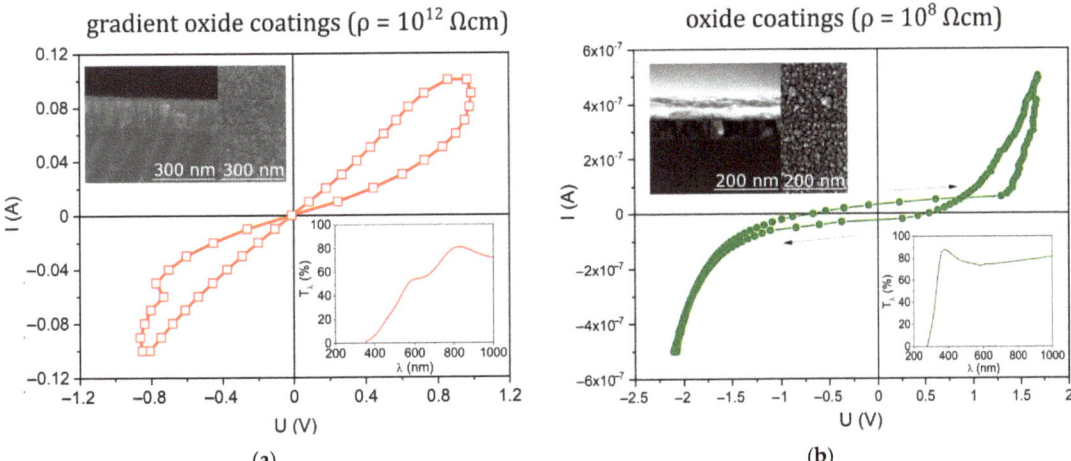

Figure 1. Current to voltage (I–V) of transparent (**a**) gradient [11] and (**b**) oxide [10] coatings (Ti,Co)Ox that exhibit a memristive effect. In the insets, SEM images of surface and cross-section, as well as transmission characteristics, are shown.

The influence of Co on the properties of Ti-based oxides is not yet well understood, which limits the possibility of fully exploiting the potential that the combination of these two materials can offer. Cobalt and its oxides exhibit high stability (especially in the form of CoO and Co_3O_4) [12], as well as high photocatalytic activity [13] and ferromagnetic properties [14]. Co-based oxides and metallic materials can be applied in spintronics [15], magneto-optical devices [16], in the construction of semiconducting sensors [17], electrochromic coatings [18], and heterogeneous catalysts [19]. The properties of oxides based on titanium and cobalt are significantly dependent on their preparation technology. These materials are mostly manufactured in the form of nanotubes [1], nanowires [2], nanorods [3], nanoparticles [4], or various types of thin-film coating (single films or multilayers) [5] (Figure 2). Nanotubes, nanowires, and nanorods make up the largest part of the (Ti,Co)Ox forms reported in publications (Figure 2). The hydrothermal [2] or anodisation [1,3] methods are used mainly for their manufacture. The form of nanotubes is interesting for electrochemical applications (e.g., fuel cells) due to their ability to transport electrons unidirectionally and their large surface area [20,21]. These types of nanomaterials exhibit improved catalytic properties [22,23]. In the case of nanowires, an application in efficient photoelectrochemical (PEC) hydrolysis can be presented [2]. Nanorods, on the other hand, are useful in the oxygen evolution reaction (OER), which is a key process in many technologies, including the conversion of electrochemical energy or the production of zinc–air batteries [3]. Unlike the mentioned forms, these oxides are often manufactured as nanoparticles [4,6,9], which are useful in energy storage or solar energy conversion [4].

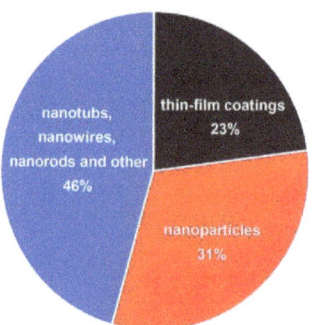

Figure 2. Percentage number of publications on various Ti- and Co-based oxide materials (TiO$_2$:Co, TiO$_2$+CoO$_X$, Ti$_x$Co$_{X-1}$O$_y$) [based on the ScienceDirect database from 1980 to 2022].

Nanoparticles are usually manufactured by sol-gel [9], hydrothermal [8], calcination [24], pyrolysis [25], or plasma thermal treatment [26] methods (Figure 3a). It should be noted that in the case of nanoparticles, typically only crystalline titanium dioxide is identified, and the influence of Co content on their properties has not been well explored [4,26]. Crystallite sizes depend not only on the amount of cobalt but are also determined by the preparation method of given nanoparticles. Therefore, it is difficult to define the relationship between Co content and crystallite size [4,6,8,9,26,27]. Oxide material-based Ti and Co are also prepared as thin-film coatings. They can have the form of single films [5,10,28,29] and multilayers [30–32] (Figure 3b). According to the literature, magnetron sputtering [10,33–35], Metal Organic Chemical Vapor Deposition (MOCVD) [28], epitaxy [5], or sol-gel [7,36] are primarily used for their manufacture. Their (material composition (Co-content) can also vary significantly [29,34,37]. Modification of the sputtering conditions results in the receipt of diversified thin-film materials, which is important in their application area [2,33]. As mentioned above, in the case of oxide nanomaterials based on titanium and cobalt, their material composition is a key factor.

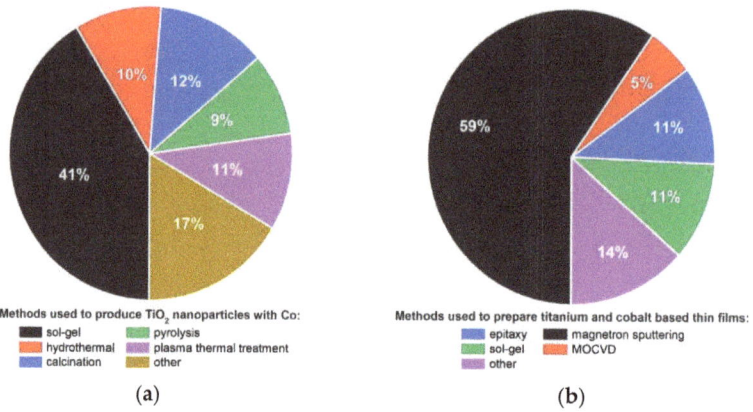

Figure 3. Methods used to prepare TiO$_2$ nanoparticles doped with Co (**a**); preparation methods of thin films based on Ti and Co mixed oxides (**b**) based on the ScienceDirect database from 1980 to 2022.

The current state of the art indicates that in nanoparticles, as well as in thin-film coatings, the amount of cobalt is usually below 15 at.% [5,27,38] (Figure 4), although there are some reports that present results for materials with even 50 at.% or 76 at.% cobalt [26,29]. As can be seen, there is a research gap regarding materials with cobalt content comparable to or greater than titanium. As an analysis of the literature shows, most of these materials are crystalline [5,25,38], while approximately 14% of the total work has been devoted to amorphous materials [31,39]. Therefore, it can be seen that the main focus has been on investigating the influence of cobalt on TiO$_2$ properties only within a narrow range of this element, completely discounting what modifications in (Ti,Co)Ox films would be introduced by a higher amount of cobalt. This fact itself makes conducting research for (Ti,Co)Ox coatings with a higher content of cobalt a novelty in the field of the materials in discussion.

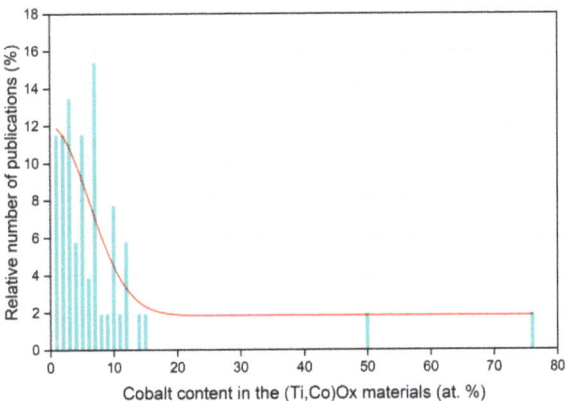

Figure 4. Relative number of publications describing cobalt content in (Ti,Co)Ox films based on the ScienceDirect database from 1980 to 2022.

Among crystalline materials, (Ti,Co)Ox can generally be distinguished by nanoparticles [6,9,37] or also by thin films (but mainly annealed) [28,37,40]. In the case of nanoparticles (often prepared with the sol–gel method), TiO$_2$ with anatase or rutile structure is dominant, while the presence of cobalt as separate phases (oxide or metallic) or as a compound with titanium or its oxides [8] occurs rarely with nanoparticles. However, there exist some works in which crystal forms of their compounds (e.g., CoTiO$_2$ or Co$_2$TiO$_2$) have been identified [26]. Accurate studies on titanium- and cobalt-based thin films are rare, especially in non-stoichiometric (i.e., low oxygen amount) forms. In addition, there is a general strong need to develop multifunctional coatings [10]. This paper investigates the influence of the amount of cobalt on the optical and structural properties of non-stoichiometric thin films based on titanium and cobalt.

2. Materials and Methods

2.1. Preparation of Thin Films

Nonstoichiometric (Ti,Co)Ox thin films with the desired material composition were prepared using the gas impulse magnetron sputtering method (denoted as GIMS). Magnetrons were supplied by an MSS2 2 kW pulsed AC power supply unit (DORA Power System, Wilczyce, Poland) [41–46]. The sputtering system was also equipped with vacuum gauges (Pfeiffer Vacuum, Aßlar, Germany) and a gas flow control system that involves mass flow controllers (MKS Instruments, Andover, MA, USA). In the applied GIMS processes, a gas mixture (Ar:O$_2$) with a low O$_2$ content (10:1) was injected into the working chamber directly on the surface of the metallic Ti, Co, and Ti-Co targets (diameter-30 mm, thickness 3 mm, purity 99.95%) mounted on the magnetron. Ti-Co targets with 2 at.%, 12 at.%, and 50 at.% Co content were used for preparation. Targets were prepared with spark plasma sintering (SPS) using a system provided by FCT GmbH (Rauenstein, Germany) [47–49]. For sintering, Co and Ti nanopowders (99.95%, Kurt Lesker, Dresden, Germany) in the Lukasiewicz Research Network–Institute of Non-Ferrous Metals [49] were used. The targets were sintered at 1200 °C in a graphite matrix. The material composition of the targets was determined using a JXA-8230 X-ray microanalyzer (JEOL) with wave and energy-dispersive spectrometers (WDS and EDS). A detailed description of the Ti-Co target preparation method was described elsewhere [10]. Due to the possibility of sputtering in multimagnetron configuration (Ti,Co)Ox thin films with various Co content (3, 19, 44, and 60 at.%) were obtained. In addition, TiO_x and CoO_x reference films were prepared. The Ar:O_2 gas mixture was obtained due to the use of a gas mixer that includes two individual MKS mass flow controllers. Ar and O_2 flow rates were set at 30 and 3 sccm, respectively. Gas impulses, injected directly into the target, were synchronised with the magnetron supply unit (MSS2 type, Dora Power System), and in each cycle lasted 100 ms. The locally ignited plasma was obtained at $<6 \times 10^{-3}$ mbar, with a supply power of 500 W (500 V, 1 A). The plasma ignition time was 30 ms and the interval between pulses was 70 ms. The sputtering system was equipped with diffusion and rotary pumps. Before the GIMS processes, the vacuum chamber was evacuated to a base pressure of ca. 5×10^{-6} mbar. Thin films were deposited on Si and SiO_2 substrates. The distance between the target and the substrate was 16 cm. Figure 5 shows a schematic layout of thin film preparation.

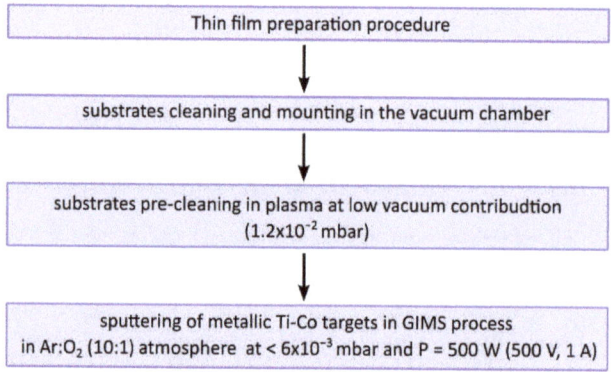

Figure 5. Schematic layout of (Ti,Co)Ox thin-film preparation.

2.2. Methods of Thin Film Characterisation

The surface morphology of the coatings and their chemical composition were investigated using a high-quality SEM/Xe-FIBFEI Helios NanoLab 600i field-emission scanning electron microscope (FEI, Hillsboro, OR, USA) equipped with an energy dispersive X-ray spectrometer (EDS). In addition, high-resolution SEM images of the surface and cross-sections were examined. Analysis of Ti and Co concentrations on the basis of EDS maps of the elemental distribution was also carried out. The structural properties of the thin films were determined based on the results of GIXRD in the incidence mode of grazing (at 3°) mode. For the measurements, an Empyrean X-ray diffractometer (PANalytical, Malvern, UK) with a PIXel3D detector and Cu Kα radiation with a wavelength of 1.5406 Å (40 kV, 30 mA) using Bragg-Brentano reflecting geometry parafocusing optics was used. By comparing the obtained pattern with PDFcards, a phase structure was determined. For data analysis, MDI JADE 5.0 software (ICDD, Newtown Square, PA, USA) was used. For the analysis of the surface state, X-ray photoelectron spectroscopy (XPS) was used. The Specs Phoibos 100 MCD-5 (5 single-channel electron multiplier) hemispherical analyser (SPECS Surface Nanoanalysis GmbH, Berlin, Germany) using a Specs XR-50 X-ray source with Mg Kα (1253.6 eV) beam was used. The XPS spectra were analysed using Casa XPS software. The thicknesses of the manufactured films were verified with the aid of a contactless Taylor Hobson Tally Surf CCI Lite optical profiler (Talysurf CCI Lite, Leicester, UK). The nanoindentation technique was used to determine the hardness of the prepared thin films. The hardness was obtained from experimental load–displacement curves for an indentation experiment using the Oliver and Pharr method [50,51]. Measurements were made with a CSM Instruments (CSM Instruments, Peseux, Switzerland) NHTT 01-03620 nanoindenter model equipped with a Vickers diamond indenter. For optical characterisation, the light transmission method was used. The measurement setup was equipped with an integrated light source DH-2000-BAL (containing a halogen and deuterium lamp) and Ocean Optics QE 65000 and NIR 256-2.1 spectrophotometers (Ocean Optics, Largo, FL, USA). The transmission coefficient was determined from transmittance spectra measured in the wavelength range of 250–2000 nm. The average transmission was evaluated by calculating the integral in visible wavelength range of 300 to 900 nm. On the basis of the measurements, parameters such as the light transmission coefficient (T_λ), position of the optical absorption edge ($\lambda_{cut-off}$), and the width of the optical band gap (E_g^{opt}) were determined.

3. Results and Discussion

3.1. Material Composition of (Ti,Co)Ox Thin Films

In Figure 6b, SEM images of surface topography with EDS maps of elemental distribution of (Ti,Co)Ox thin films are shown. It was found that films with 3 at.%, 19 at.%, 44 at.%, 60 at.%, and 100 at.% of cobalt content in the TiO$_x$ matrix were prepared. EDS maps indicate a homogeneous distribution of Ti, O, and Co, as well as a lack of agglomeration effects. The absence of areas with a clearly higher concentration of cobalt should be emphasised because it proves the high quality of sintered targets and the possibility of manufacturing Ti-based oxide coatings with a homogeneous distribution of Co. Detailed results of the EDS analysis are collected in Table 1.

Table 1. Material composition of thin-film coatings based on Ti and Co and targets used for their deposition by magnetron sputtering.

Target	Co-Content in the Film (at.%)
Ti	-
Ti$_{0.98}$Co$_{0.02}$	3
Ti$_{0.88}$Co$_{0.12}$	19
Ti$_{0.88}$Co$_{0.12}$ + Ti$_{0.50}$Co$_{0.50}$	44
Ti$_{0.50}$Co$_{0.50}$	60
Co	100

3.2. Optical Characterisation of (Ti,Co)Ox Thin Films

The optical properties of the coatings were determined on the basis of transmission characteristics (Figure 7a). In the case of undoped TiO$_x$, the average transmission level of value $T_{\lambda a}$ = 48% (average transmission related to the area under the characteristic) was the highest compared to other films (Figure 7b). The addition of Co resulted in a decrease in the transparency. The (Ti,Co)Ox films with 3 at.% and 19 at.% of cobalt were semitransparent due to the 46% and 29% values of $T_{\lambda a}$, respectively (Figure 7b). The increase in Co content resulted in a significant decrease in the transparency level to <10% for (Ti$_{00.56}$Co$_{0.44}$)Ox and (Ti$_{0.40}$Co$_{0.60}$)Ox, respectively. In the case of CoO$_x$, an opaque film was received ($T_{\lambda a}$ < 2%) (Figure 7b). Except for analysis of transmission level, the position of the optical absorption edge ($\lambda_{cut-off}$) was determined (Figure 7a). It was found that with the increase of cobalt content, the $\lambda_{cut-off}$ position shifts to longer wavelengths ('red shift'). This effect has also been reported in other works [9,27], but it should be noted that the position of the optical absorption edge is also related to the preparation method. Therefore, the 'blue shift' of the $\lambda_{cut-off}$ can also be observed in oxide materials based on Ti and Co [7]. These results are in agreement with other works, e.g., [5,7,36,52–55]. In general, the transmission level of Ti oxides decreases with increasing Co content, but the preparation method and the form of (Ti,Co)Ox material (thin film or nanoparticles) determine these changes [52–54], as can be seen for the thin films, where the light transmission level decreases with increasing Co content [7,36,52].

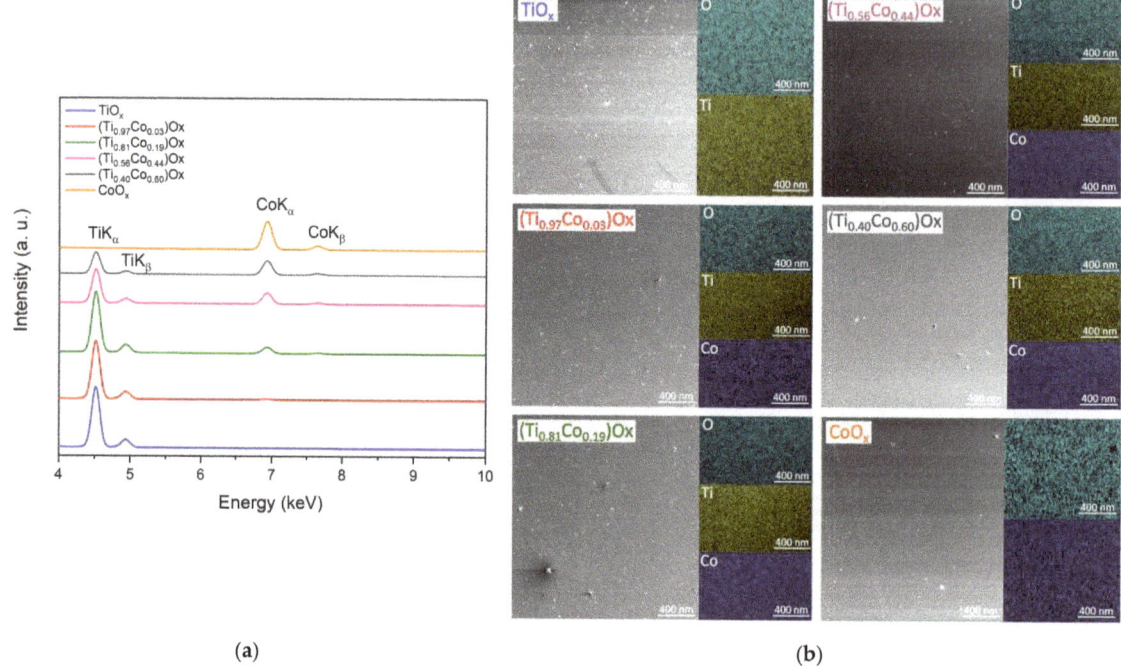

Figure 6. Results of X-ray microanalysis investigations for as-deposited Ti, Co, and O for (Ti,Co)Ox thin films: (**a**) EDS spectra and (**b**) maps of elemental distribution of mixed oxides.

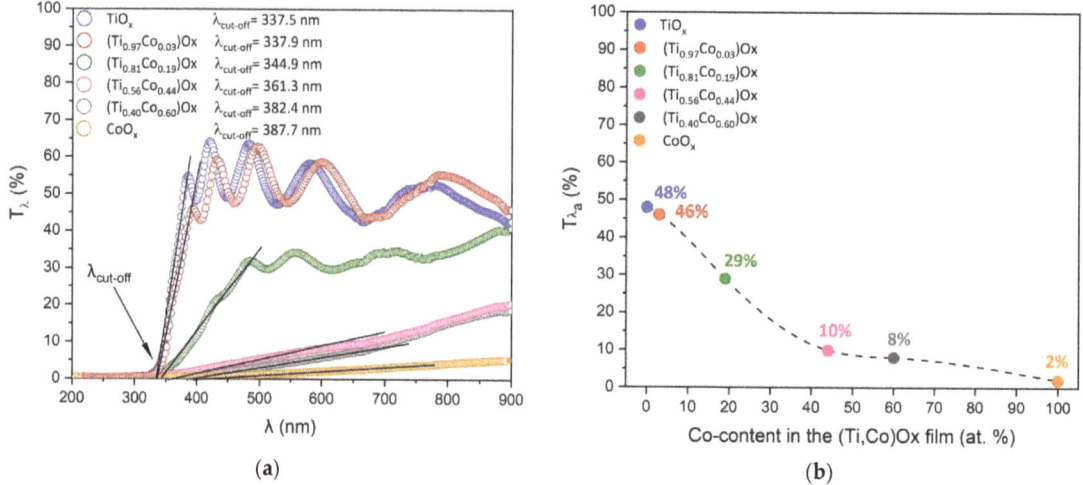

Figure 7. Transmission characteristics of TiO_x (t = 410 nm), $(Ti_{0.97}Co_{0.03})Ox$ (t = 335 nm), $(Ti_{0.81}Co_{0.19})Ox$ (t = 480 nm), $(Ti_{0.56}Co_{0.44})Ox$ (t = 270 nm), $(Ti_{0.40}Co_{0.60})Ox$ (t = 226 nm), and CoO_x (t = 150 nm) thin films (**a**) with average transmission level ($T_{\lambda a}$) as a function of the cobalt content in the film (**b**). Designations: T_λ—transmission, $T_{\lambda a}$—average transmission in the range of 300 to 900 nm, t—thickness of the films.

Films containing a low amount of cobalt are characterised by a level of transparency similar to that of undoped TiO_2 or the mentioned nanoparticles (ca. 80%). Above 10 at.% cobalt results in a significant reduction of transmission (Figure 8a). However, the literature lacks comprehensive analyses of titanium oxide coatings with neither low nor high Co content. A significantly different character can be seen for nanoparticles (Figure 8b) [9,21,25,27]. These nanomaterials have a similar value of the light transmission coefficient (from 60% to 90%). It is difficult to determine the reason for the lack of changes with the increase in Co content. This can be determined by the manufacturing method or may be related to the fact that the core of the nanoparticles is TiO_2 and cobalt is located in its shell [9,25,27]. The solution to this shortcoming may be additional annealing, which will almost certainly cause oxygenation of the film structure.

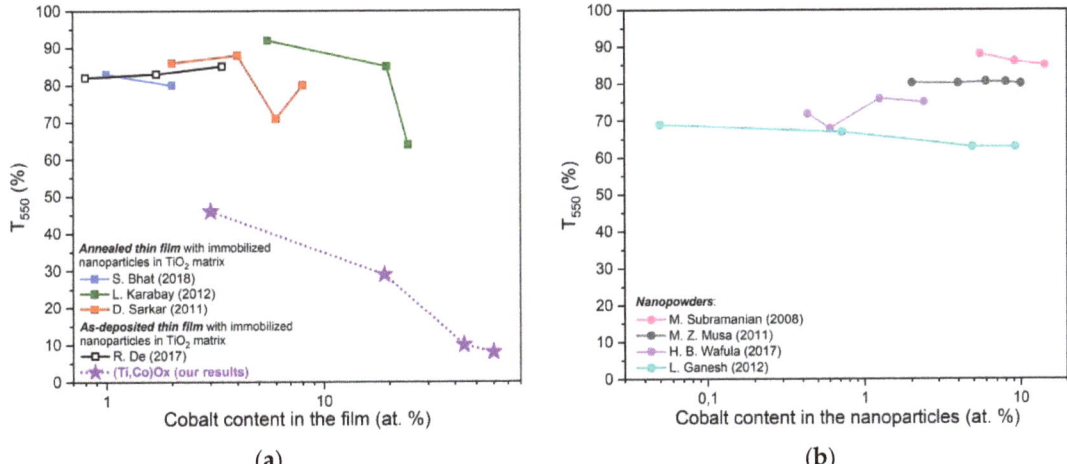

Figure 8. Influence of cobalt content on the optical energy gap coefficient of oxide materials based on Ti: (a) thin films with immobilised nanoparticles in the TiO_x matrix [7,52–54]; (b) nanopowders [9,21,25,27]. On the basis of data from the publication, the cobalt content was converted from wt.% to at.%.

Based on transmission characteristics, the Tauc plots (for indirect transitions) were obtained and the optical band gap was estimated (Figure 9). The E_g^{opt} values for the TiO_x and $(Ti_{0.97}Co_{0.03})Ox$ films were very similar and equal to 3.08 and 3.02 eV, respectively. The increase in the Co content resulted in a decrease in the E_g^{opt} value. In the case of the film with 19 at.% of Co, 2.26 eV was noticed. For films with greater amounts of cobalt, a transmission level that was too low did not allow us to determine the value of the optical band gap. However, it should be noted that the results obtained are consistent with reports in the literature [6,56]. The results available for thin films show the crucial role of their form and method of preparation. There is a general tendency for the value of E_g to decrease with the increase of Co content [7,21,25,27,52–54,57] (Figure 10).

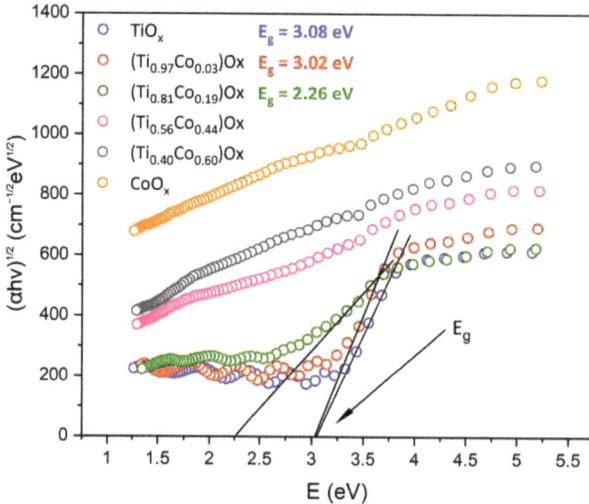

Figure 9. Tauc plots for the thin films of TiO_x, $(Ti,Co)Ox$, and CoO_x with marked areas of determination of the optical band gap (E_g^{opt}).

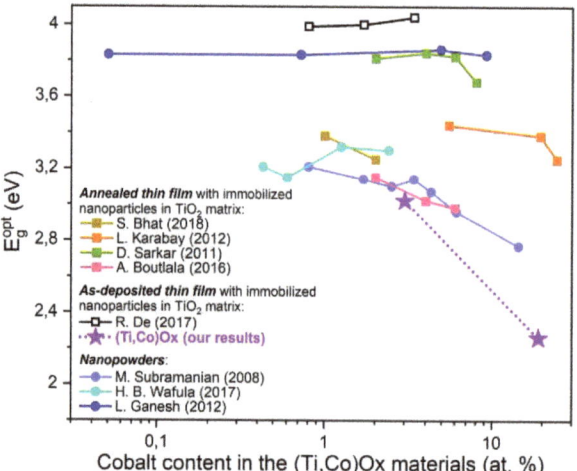

Figure 10. Influence of cobalt content on the light transmission coefficient of oxide materials based on Ti [7,21,25,27,52–54,57]. On the basis of data from the publication, the cobalt content was converted from wt.% to at.%.

3.3. Structural Characterisation of Thin Films of (Ti,Co)Ox

In Figure 11, SEM images of the surface and cross-sectional topography of the prepared oxide films are shown. It can be stated that the microstructure of all coatings from the GIMS processes was very homogeneous.

Cross-sectional images indicate that the TiO_x and $(Ti,Co)Ox$ films had a columnar character. Their microstructure was densely packed and free of cracks or gaps between columns. The increase in Co content resulted in a decrease in the width of the columns from 30 to 15 nm. The lack of titanium in the film resulted in a microstructure of a different nature, i.e., grainy. This means that the presence of titanium has a key influence on the process of nucleation of (Ti,Co)Ox coatings in the GIMS process.

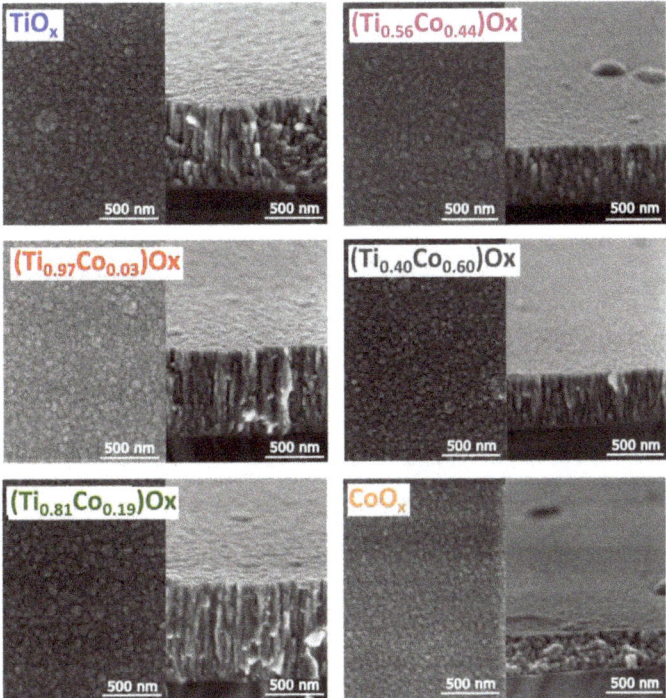

Figure 11. SEM images of the surface and cross-sectional topography of TiO_x, $(Ti,Co)O_x$, and CoO_x thin films.

The structure of the as-deposited $(Ti,Co)O_x$ coatings was examined using X-ray diffraction (XRD) in the grazing incidence mode (GIXRD). In Figure 12, the GIXRD patterns of the films are shown.

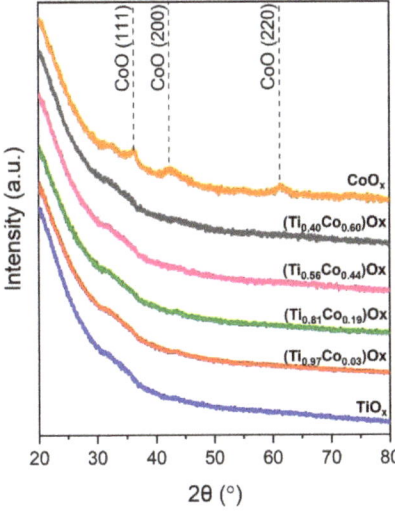

Figure 12. GIXRD patterns of TiO_x, $(Ti,Co)O_x$, and CoO_x thin films prepared by gas impulse magnetron sputtering.

As can be seen, all coatings prepared by the GIMS technique were amorphous except for CoO_x, which was nanocrystalline. In this case, the crystalline form of CoO was identified, but it should be emphasised that the intensity of the peaks in the pattern is very weak.

Furthermore, broad peaks were also exposed in patterns related to the SiO_2 substrate. However, there is a lack of peaks that could testify about the crystal form of titanium, cobalt, or their oxides. Similar results were also obtained for other oxide materials prepared using the GIMS technique [44–46], which is a consequence of this innovative sputtering method [42]. Detailed results of the XRD analysis are collected in Table 2.

Table 2. Structural properties of TiO_x, $(Ti,Co)O_x$, and CoO_x thin films, based on GIXRD measurements.

Thin Film	Phase (-)	(hkl) (-)	D (nm)	d (nm)	d_{PDF} (nm)	a (Å) [1]
CoO_x	CoO	(111)	4.1	0.24597	0.24807	4.26033
	CoO	(200)	4.0	0.21302	0.21687	4.33740
	CoO	(220)	4.1	0.15062	0.15125	4.26018
$(Ti_{0.40}Co_{0.60})O_x$						
$(Ti_{0.56}Co_{0.44})O_x$						
$(Ti_{0.81}Co_{0.19})O_x$			amorphous			
$(Ti_{0.97}Co_{0.03})O_x$						
TiO_x						

D—average crystallite size; d—interplanar distance; d_{PDF}—standard interplanar distance, a—lattice parameters;
[1] Similar results were obtained in publications [58–60].

The influence of the Co content on the hardness of nonstoichiometric Ti-based films with the aid of nanoindentation was also investigated. Figure 13 shows the results of hardness as a function of the relative intender displacement for all prepared TiO_x, $(Ti,Co)O_x$, and CoO_x coatings. There was no significant change in the hardness value (H) with increasing cobalt. The value of H was in the range of 6.3 GPa to 6.9 GPa. Only for the CoO_x film was a lower value (4.8 GPa) observed (Figure 14). These results may testify to the influence of the type of microstructure on the hardness. Thus, the TiO_x and $(Ti,Co)O_x$ films with columnar microstructure had a similar hardness, whereas the grainy microstructure of the CoO_x film resulted in ca. 25% lower hardness. However, there is a lack of data related to the hardness of such mixed Ti-Co oxide materials, especially non-stoichiometric.

Because the GIXRD results did not reveal the forms in which cobalt of titanium occurred in the $(Ti,Co)O_x$ coatings, we decided to perform photoelectron spectroscopy (XPS). In Figure 15, the XPS survey spectra recorded for composite bonding are shown [61–64].

In Figure 16 the XPS spectra recorded for Ti2p peaks recorded for $(Ti,Co)O_x$ and TiO_x thin films are shown. For all films, the presence of a characteristic doublet of Ti2p peaks can be observed. The positions of both $Ti2p_{3/2}$ and $Ti2p_{1/2}$ photoelectron peaks correspond to the +4 oxidation state of titanium [61,62,65,66]. The differences in the positions of the Ti2p doublet peaks are very small, i.e., below 0.1 eV (Figure 17). Therefore, the amount of Co addition does not affect the position of the Ti2p doublet. This suggests that titanium only forms its distinct oxide forms and not compounds with cobalt. It should also be noted that the difference in the position of the peaks in the doublet (ΔBE) itself, which is in the range of 5.6 to 5.7 eV, indicates the presence of the TiO_2 form (Figure 17).

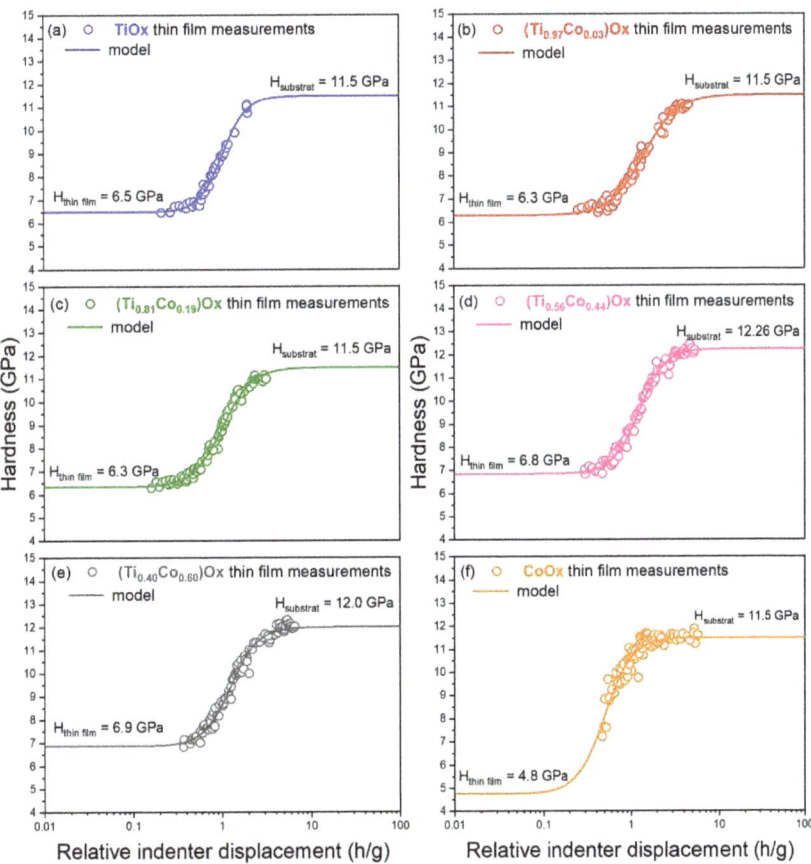

Figure 13. Hardness of (**a**) TiO$_x$, (**b**) (Ti$_{0.93}$Co$_{0.03}$)O$_x$, (**c**) (Ti$_{0.81}$Co$_{0.19}$)O$_x$, (**d**) (Ti$_{0.56}$Co$_{0.44}$)O$_x$, (**e**) (Ti$_{0.40}$Co$_{0.60}$)O$_x$ and (**f**) CoO$_x$ oxide thin films.

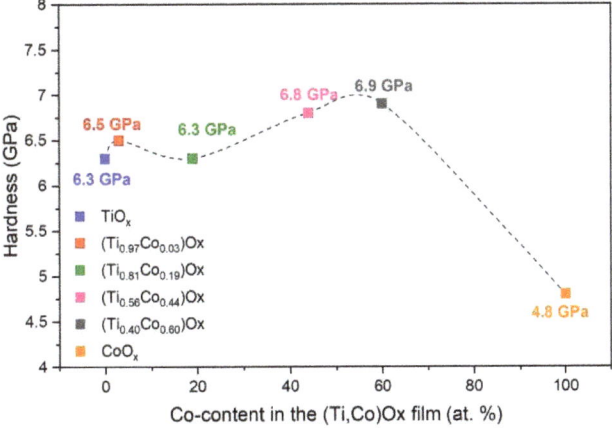

Figure 14. The effect of cobalt on the hardness of TiO$_x$ films.

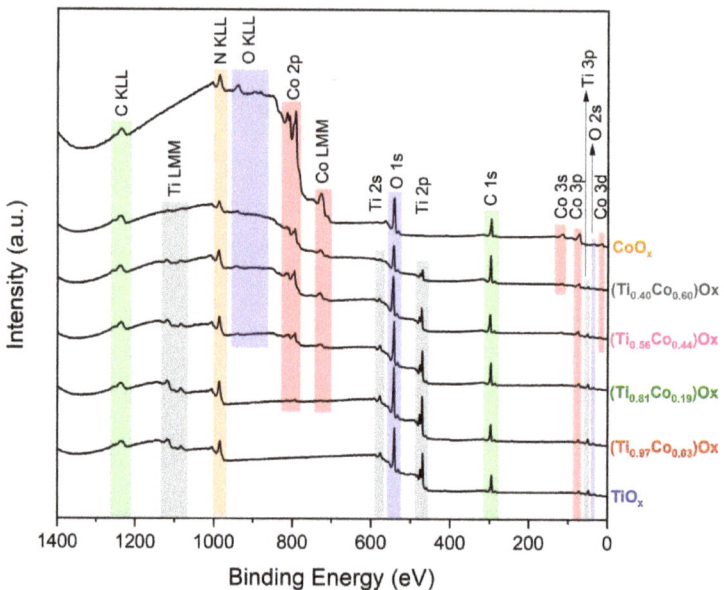

Figure 15. XPS survey spectra of the as-deposited TiO_x, $(Ti,Co)O_x$, and CoO_x thin films.

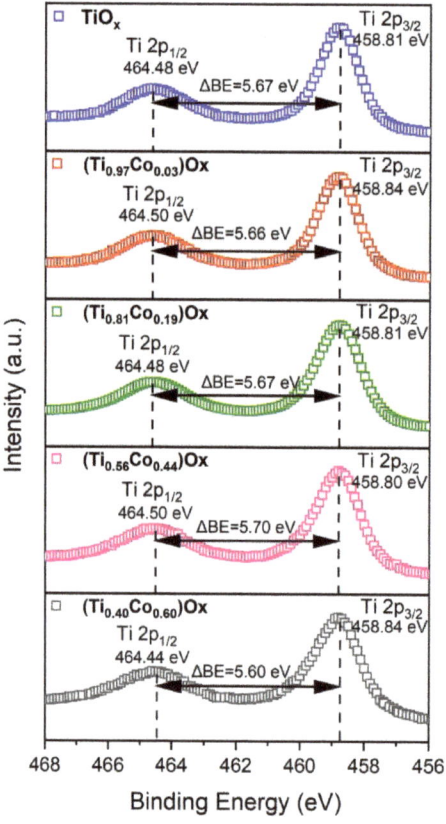

Figure 16. The XPS spectra of the Ti2p state for TiO_x and $(Ti,Co)O_x$ thin films.

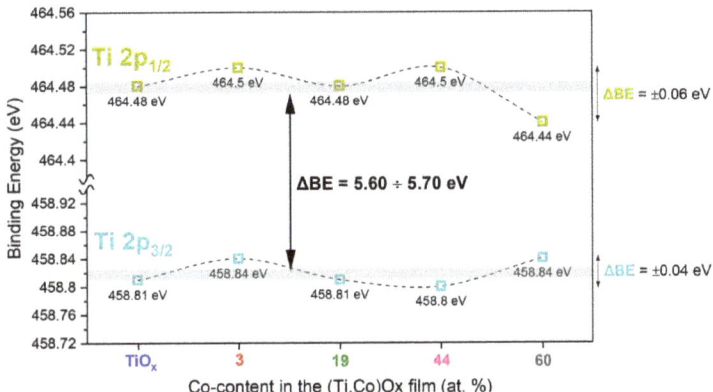

Figure 17. The influence of Co content on the position of Ti2p doublet peak.

Figure 18 presents the XPS spectra for the O1s peaks recorded for the prepared TiO_x, $(Ti,Co)O_x$, and CoO_x thin films. Deconvolution of the multipeak (Figure 19a) showed that it consisted of three peaks, centred on: (i) 530.3 eV, attributed to lattice oxygen [61,62,66,67]; (ii) 532.1 eV, related to the presence of hydroxyl groups (OH^-) on the surface [61,62,66,67]; and (iii) 533.6 eV, related to the water molecules adsorbed on the surface (H_2O_{ads}) [61,62,66]. The differences in the positions of the O1s doublet peaks are very small, i.e., below 0.5 eV. Taking into account the intensity of the multipeak O1s (Figure 19b), it could be concluded that the surfaces of $(Ti,Co)O_x$ with 44 at.% and 60 at.% cobalt, as well as CoO_x thin films, are more likely to absorb OH^- from the surrounding environment than TiO_x and $(Ti,Co)O_x$ with 19 at.% or 3 at.% Co. Furthermore, with increasing Co content, the level of H_2O_{ads} also increased from 2% to 15%.

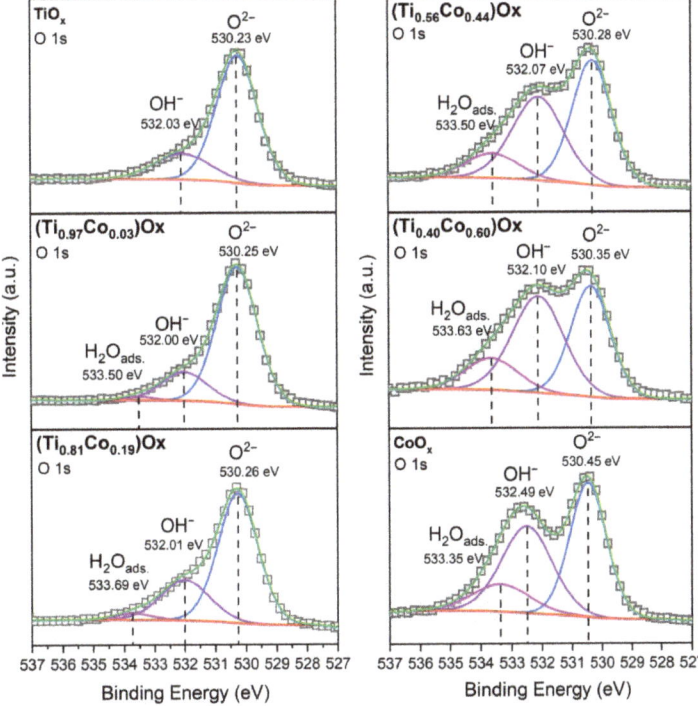

Figure 18. XPS spectra for the O1 state of TiO_x, $(Ti,Co)O_x$, and CoO_x thin films.

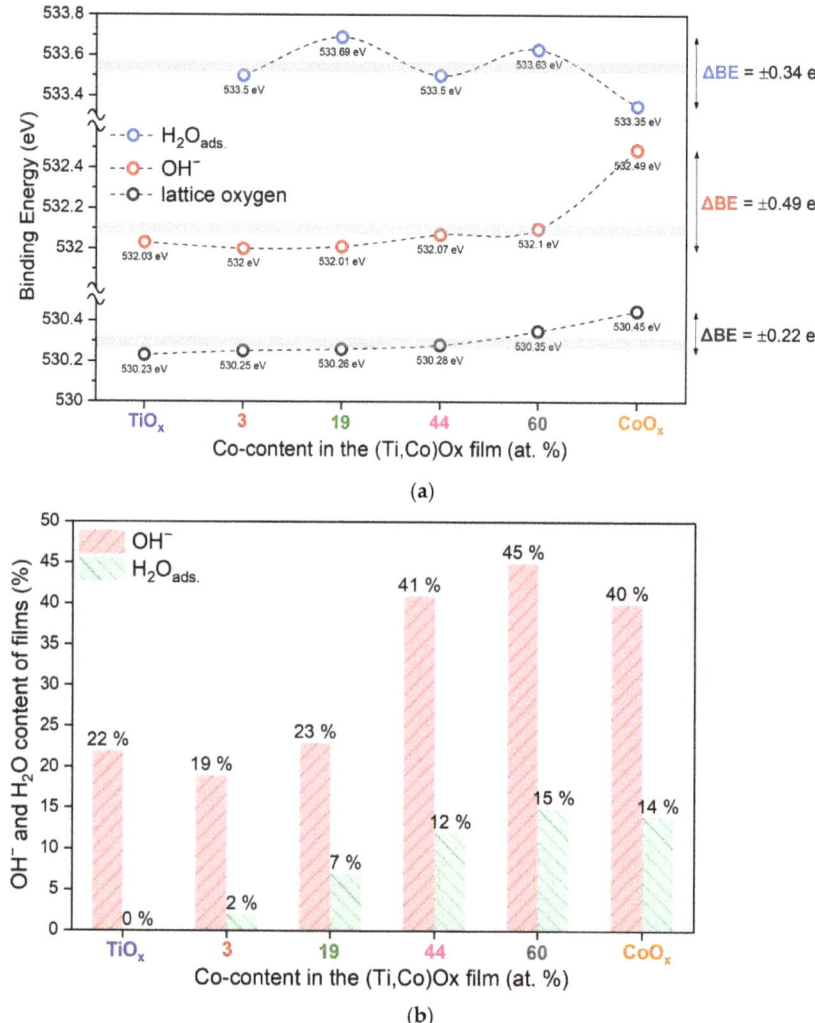

Figure 19. The influence of cobalt content on the position of O1 multipeak (**a**); the relative content of OH groups and H_2O_{ads} molecules adsorbed on the surface of prepared coatings to the O1 signal (**b**).

The most important part of the XPS research was the analysis of the spectra recorded for the Co2p state. We aimed to reveal the degree of oxidation of cobalt ions and thus the forms in which its oxides may occur in the coatings. In Figure 8a, the $Co2p_{3/2}$ spectra of prepared (Ti,Co)Ox and CoO_x thin films are shown. In each spectrum, a multipeak can be observed (Figure 20). It was deconvoluted into a satellite (786.5 eV) and two main peaks.

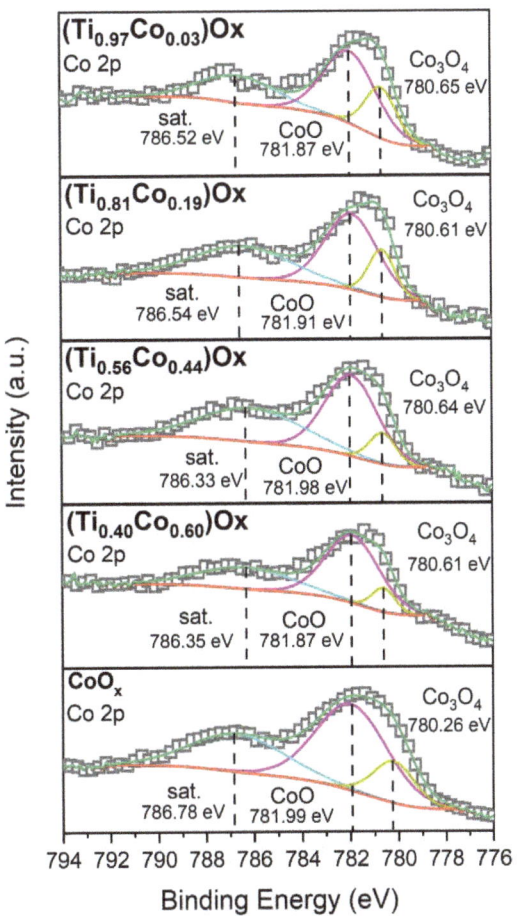

Figure 20. XPS spectra for Co2p$_{3/2}$ state of (Ti,Co)Ox and CoOx thin films.

The first Co2p$_{3/2}$ peak centred at approximately 781.8 eV is related to Co^{2+} ions; thus, the presence of the CoO form can be confirmed [61,62,68–70]. The second Co2p$_{3/2}$ peak of Co2p$_{3/2}$ located at approximately 780.6 eV corresponds to Co^{2+} and Co^{3+} ions and can be related to the appearance of Co$_3$O$_4$ [61,62,68–71]. However, some authors argue that, based on XPS measurements of Co$_2$O$_3$ and Co$_3$O$_4$, which contain both Co^{2+} and Co^{3+}, cannot be distinguished by chemical shift or satellite [21]. This suggests that the surface films of Co$_2$O$_3$ and Co$_3$O$_4$ are similar. The binding energy values estimated for Co2p$_{3/2}$ core levels are in agreement with those reported for Co oxides in the literature [61,62,68–70]. There are no references to mixed oxides in the literature, while quite a few examples involve single oxides. Due to the very low number of literature reports on such materials, it was relatively difficult to compare the peaks to Co2p$_{3/2}$ spectra. Comparison of the areas under the peaks revealed that the percentage of Co ions in the +2 oxidation state is about 3–4 times that of +3. Differences in the positions of the Co2p$_{3/2}$ doublet peaks are very small, i.e., below 0.5 eV (Figure 21a). Furthermore, with an increase in the Co content in the (Ti,Co)Ox films, the percentage content of the CoO form increases at the expense of Co$_3$O$_4$ (Figure 21b).

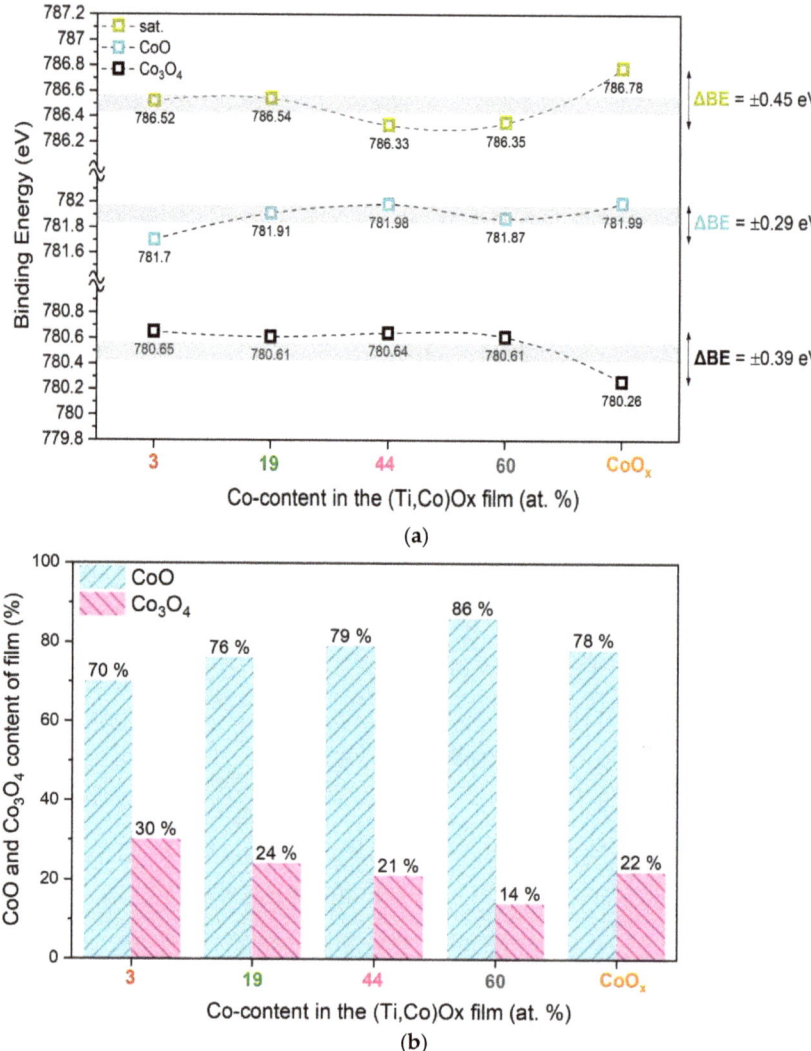

Figure 21. The influence of cobalt on the multipeak position of Co2p$_{3/2}$ (**a**); the content of adsorbed hydroxy groups and water molecules on the surface of the examined coatings (**b**).

4. Conclusions

In this work, the influence of the Co content on the properties of non-stoichiometric (Ti,Co)Ox thin films prepared by a GIMS process has been described. The aim of our research was to obtain a broad change in the material composition for coatings prepared under the same conditions. The application of gas impulse sputtering resulted in the manufacture of amorphous (Ti,Co)Ox films with a homogeneous distribution of cobalt. The change in Co content resulted in modification of the morphology and optical properties. The most significant changes were observed for the optical parameters. A large drop in the transparency level and the optical band gap was observed along with the increase in the cutoff wavelength ('red shift'). Structural studies revealed that, except for the nanocrystalline CoOx film (with crystallites of <5 nm in size) with fine-grained microstructure, the morphology of all amorphous TiOx and (Ti,Co)Ox coatings had a columnar nature. The hardness of TiOx and (Ti,Co)Ox films (6.5 GPa) was higher compared to CoOx (4.8 GPa).

The XPS analysis has revealed that the prepared films consisted of mixed oxides. For all (Ti,Co)Ox films, the occurrence of TiO_2 form was observed, while in the cobalt, the additive was present in its own forms. The occurrence of CoO and Co_3O_4 forms was identified. The quantity of cobalt ions in the +2 state was 3–4 times higher compared to the +3 ions. However, modification of the composition of the material resulted in a decrease in CoO at the expense of Co_3O_4 with an increase in the cobalt content of the film.

Author Contributions: Conceptualization, P.P. and D.W.; methodology, P.P. and D.W.; validation, D.W.; investigation, P.P., M.S., M.K., P.M. and D.W.; resources, D.W.; writing—original draft preparation, P.P. and D.W.; writing—review and editing, P.P. and D.W.; supervision, D.W. All authors have read and agreed to the published version of the manuscript.

Funding: This work was co-financed from the sources given by the Polish National Science Centre (NCN) as a research project number 2018/29/B/ST8/00548.

Institutional Review Board Statement: Not applicable.

Informed Consent Statement: Not applicable.

Data Availability Statement: Not applicable.

Conflicts of Interest: The authors declare no conflict of interest.

Abbreviations

The following table shows all abbreviations used in this work.

a	lattice parameters
d	interplane distance
d_{PDF}	standard interplane distance (according to PDF card)
D	average crystallite size
E	energy
E_g	width of the energy gap
$E_g^{opt.}$	width of the optical energy gap
$H_{substart}$	hardness of the substrate
$H_{thin\ film}$	hardness of the thin film
h	Planck's constant
T_λ	light transmission coefficient
T_{550}	average value of the light transmission coefficient for the wavelength λ = 550 nm
$T_{\lambda a}$	average transmission in the range of 300 to 900 nm
t	thickness of the films
α	absorption coefficient
λ	wavelength of electromagnetic radiation
$\lambda_{cut-off}$	position of the edge of optical absorption
EDS	energy-dispersive spectroscopy
$(Ti_{0.97}Co_{0.03})Ox$	abbreviated notation of the thin film as a mixture of oxides in which the atomic content is: 97% at. Ti to 3% at. Co
PDF	powder diffraction files
SEM	transmission electron microscopy
XPS	X-ray photoelectron spectroscopy
XRD	grazing incidence X-ray diffraction

References

1. Venturini, J.; Bonatto, F.; Guaglianoni, W.C.; Lemes, T.; Arcaro, S.; Alves, A.K.; Bergmann, C.P. Cobalt-doped titanium oxide nanotubes grown via one-step anodization for water splitting applications. *Appl. Surf. Sci.* **2019**, *464*, 351–359. [CrossRef]
2. Liu, C.; Wang, F.; Zhu, S.; Xu, Y.; Liang, Q.; Chen, Z. Controlled charge-dynamics in cobalt-doped TiO_2 nanowire photoanodes for enhanced photoelectrochemical water splitting. *J. Colloid Interface Sci.* **2018**, *530*, 403–411. [CrossRef] [PubMed]
3. Yang, C.; Makabu, C.M.; Du, X.; Li, J.; Sun, D.; Liu, G. Cobalt nanorods decorated titanium oxide arrays as efficient and stable electrocatalyst for oxygen evolution reaction. *Electrochim. Acta* **2021**, *396*, 139213. [CrossRef]
4. Khan, W.; Ahmad, S.; Hassan, M.M.; Naqvi, A.H. Structural phase analysis, band gap tuning and fluorescence properties of Co doped TiO_2 nanoparticles. *Opt. Mater.* **2014**, *38*, 278–285. [CrossRef]

5. Yamada, Y.; Toyosaki, H.; Tsukazaki, A.; Fukumura, T. Epitaxial growth and physical properties of a room temper ferromagnetic semiconductor: Anatase phase $Ti_{1-x}Co_xO_2$. *J. Appl. Phys.* **2004**, *96*, 5097. [CrossRef]
6. Husain, S.; Alkhtaby, L.A.; Giorgetti, E.; Zoppi, A.; Miranda, M.M. Influence of cobalt doping on the structural, optical and luminescence properties of sol-gel derived TiO_2 nanoparticles. *Philos. Mag.* **2017**, *97*, 17–27. [CrossRef]
7. Bhat, S.; Sandeep, K.M.; Kumar, P.; Dharmaprakash, S.M.; Byrappa, K. Characterization of transparent semiconducting cobalt doped titanium dioxide thin films prepared by sol–gel process. *J. Mater. Sci. Mater. Electron.* **2018**, *29*, 1098–1110. [CrossRef]
8. Jiang, P.; Xiang, W.; Kuang, J.; Liu, W.; Cao, W. Effect of cobalt doping on the electronic, optical and photocatalytic properties of TiO_2. *Solid State Sci.* **2015**, *46*, 27–32. [CrossRef]
9. Musa, M.Z.; Ameran, Z.F.; Mamat, M.H.; Malek, M.F.; Rasheid, B.A.; Noor, U.M.; Rusop, M. Effects of cobalt doping concentration on the structural, electrical, and optical properties of titanium dioxide thin films. In Proceedings of the 2011 International Conference on Electronic Devices, Systems and Applications (ICEDSA), Kuala Lumpur, Malaysia, 25–27 April 2011; pp. 339–342. [CrossRef]
10. Wojcieszak, D.; Mazur, M.; Pokora, P.; Wrona, A.; Bilewska, K.; Kijaszek, W.; Kotwica, T.; Posadowski, W.; Domaradzki, J. Properties of metallic and oxide thin films based on Ti and Co prepared by magnetron sputtering from sintered targets with different Co-content. *Materials* **2021**, *14*, 3797. [CrossRef]
11. Wojcieszak, D.; Domaradzki, J.; Pokora, P.; Sikora, M.; Mazur, M.; Chodasewicz, P.; Morgiel, J.; Gibson, D. Optical and structural properties of gradient (Ti,Co)Ox thin-film coatings with a resistive switching effect. *Appl. Opt.* **2022**, *34*, 10283–10289. [CrossRef]
12. Mei, J.; Liao, T.; Ayoko, G.A.; Bell, J.; Sun, Z. Cobalt oxide-based nanoarchitectures for electrochemical energy applications. *Prog. Mater. Sci.* **2019**, *103*, 596–677. [CrossRef]
13. Verma, M.; Mitan, M.; Kim, H.; Vaya, D. Efficient photocatalytic degradation of malachite green dye using facilely synthesized cobalt oxide nanomaterials using citric acid and oleic acid. *J. Phys. Chem. Solids* **2021**, *155*, 110125. [CrossRef]
14. Gajbhiye, N.S.; Sharma, S.; Nigam, A.K.; Ningthoujam, R.S. Tuning of single to multi-domain behavior for monodispersed ferromagnetic cobalt nanoparticles. *Chem. Phys. Lett.* **2008**, *466*, 181–185. [CrossRef]
15. Kaloyeros, A.E.; Pan, Y.; Goff, J.; Arles, B. Cobalt thin films: Trends in processing technologies and emerging applications. *ECS J. Solid State Sci. Technol.* **2019**, *8*, 119–152. [CrossRef]
16. Hirohata, A.; Yamada, K.; Nakatani, Y.; Prejbeanu, I.L.; Dieny, B.; Pirro, P.; Hillebrands, B. Review on spintronics: Principles and device applications. *J. Magn. Magn. Mater.* **2020**, *509*, 166711. [CrossRef]
17. Bewan, S.; Ndolomingo, M.J.; Meijboom, R.; Bingwa, N. Cobalt oxide promoted tin oxide catalysts for highly selective glycerol acetalization reaction. *Inorg. Chem. Commun.* **2021**, *128*, 108578. [CrossRef]
18. Svegl, F.; Orel, B.; Hutchins, M.G.; Kalcher, K. Structural and Spectroelectrochemical Investigations of sol-gel derived electrochromic spinel Co_3O_4 films. *J. Electrochem. Soc.* **1996**, *143*, 1532. [CrossRef]
19. Bahlawane, N.; Rivera, E.F.; Hoinghaus, K.K.; Brechling, A.; Kleineberg, U. Characterization and tests of planar Co_3O_4 model catalysts prepared by chemical vapor deposition. *Appl. Catal. B Environ.* **2004**, *53*, 245–255. [CrossRef]
20. Yao, W.; Fang, H.; Ou, E.; Wang, J.; Yan, Z. Highly efficient catalytic oxidation of cyclohexane over cobalt-doped mesoporous titania with anatase crystalline structure. *Catal. Commun.* **2006**, *7*, 387–390. [CrossRef]
21. Ganesh, I.; Gupta, A.K.; Kumar, P.P.; Sekhar, P.S.C.; Radha, K.; Padmanabham, G.; Sundararajan, G. Preparation and characterization of Co-doped TiO_2 materials for solar light induced current and photocatalytic applications. *Mater. Chem. Phys.* **2012**, *135*, 220–234. [CrossRef]
22. Wang, H.; Wang, J.; Wu, Z.; Liu, Y. NO catalytic oxidation behaviors over CoO_x/TiO_2 catalysts synthesized by sol–gel method. *Catal. Lett.* **2010**, *134*, 295–302. [CrossRef]
23. Yang, W.H.; Kim, M.H.; Ham, S.W. Effect of calcination temperature on the low-temperature oxidation of CO over CoO_x/TiO_2 catalysts. *Catal. Today* **2007**, *123*, 94–103. [CrossRef]
24. Sadanandam, G.; Lalitha, K.; Kumari, V.D.; Shankar, M.V.; Subrahmanyam, M. Cobalt doped TiO_2: A stable and efficient photocatalyst for continuous hydrogen production from glycerol: Water mixtures under solar light irradiation. *Int. J. Hydrogen Energy* **2013**, *38*, 9655–9664. [CrossRef]
25. Wafula, H.B.; Musembi, R.J.; Juma, A.; Tonui, P.; Simiyu, J.; Sakwa, T.; Prakash, D.; Verma, K.D. Compositional analysis and optical properties of Co doped TiO_2 thin films fabricated by spray pyrolysis method for dielectric and photocatalytic applications. *Optik* **2017**, *128*, 212–217. [CrossRef]
26. Li, J.G.; Buchel, R.; Isobe, M.; Mori, T.; Ishigaki, T. Cobalt-doped TiO_2 nanocrystallites: Radio-frequency thermal plasma processing, phase structure, and magnetic properties. *J. Phys. Chem. C* **2009**, *113*, 8009–8015. [CrossRef]
27. Subramanian, M.; Vijayalakshmi, S.; Venkataraj, S.; Jayavel, R. Effect of cobalt doping on the structural and optical properties of TiO_2 films prepared by sol-gel process. *Thin Solid Films* **2008**, *516*, 3776–3782. [CrossRef]
28. Seong, N.J.; Yoon, S.G. Effects of Co-doping level on the microstructural and ferromagnetic properties of liquid-delivery metalorganic-chemical-vapor-deposited $Ti_{1-x}Co_xO_2$ thin films. *Appl. Phys. Lett.* **2002**, *81*, 4209. [CrossRef]
29. Song, H.Q.; Mei, L.M.; Zhang, Y.P.; Yan, S.S.; Ma, X.; Wang, Y.; Zhang, Z.; Chen, L.Y. Magneto-optical Kerr rotation in amorphous TiO_2/Co magnetic semiconductor thin films. *Phys. B Condens. Matter* **2007**, *388*, 130–133. [CrossRef]
30. Logacheva, V.A.; Lukin, A.N.; Afonin, N.N.; Serbin, O.V. Synthesis and optical properties of cobalt-modified titanium oxide films. *Opt. Spectrosc.* **2019**, *126*, 674–680. [CrossRef]

31. Ko, Y.; Park, D.S.; Seo, B.S.; Yang, H.J.; Shin, H.J.; Kim, J.Y.; Lee, J.H.; Lee, W.H.; Reucroft, P.J.; Lee, J.G. Studies of cobalt thin films deposited by sputtering and MOCVD. *Mater. Chem. Phys.* **2003**, *80*, 560–564. [CrossRef]
32. Quiroz, H.P.; Calderon, J.A.; Dussan, A. Magnetic switching control in Co/TiO$_2$ bilayer and TiO$_2$:Co thin films for magnetic-resistive random access memories (M-RRAM). *J. Alloys Compd.* **2020**, *840*, 155674. [CrossRef]
33. Quiroz, H.; Galindez, E.F.; Dussan, A. Ferromagnetic-like behavior of Co doped TiO$_2$ flexible thin films fabricated via co-sputtering for spintronic applications. *Heliyon* **2020**, *6*, e03338. [CrossRef]
34. Afonin, N.N.; Logacheva, V.A. Cobalt modification of thin rutile films magnetron-sputtered in vacuum. *Tech. Phys.* **2018**, *63*, 605–611. [CrossRef]
35. Griffin, K.A.; Pakhomov, A.B. Cobalt-doped anatase TiO$_2$: A room temperature dilute magnetic dielectric material. *J. Appl. Phys.* **2005**, *97*, 10D320. [CrossRef]
36. Ahmad, M.K.; Rasheid, N.A.; Ahmed, A.Z.; Abdullah, S.; Rusop, M. Study of cobalt doping on the electrical and optical properties of titanium dioxide thin film prepared by sol-gel method. In Proceedings of the 2008 IEEE International Conference on Semiconductor Electronics, Johor Bahru, Malaysia, 25–27 November 2008; pp. 561–565. [CrossRef]
37. Park, W.; Ortega-Hertogs, R.J.; Moodera, J.S. Semiconducting and ferromagnetic behavior of sputtered Co-doped TiO$_2$ thin films above room temperature. *J. Appl. Phys.* **2002**, *91*, 8093. [CrossRef]
38. Tian, J.; Deng, H.; Sun, L.; Kong, H.; Yang, P.; Chu, J. Effects of Co doping on structure and optical properties of TiO$_2$ thin films prepared by sol–gel method. *Thin Solid Films* **2012**, *520*, 5179–5183. [CrossRef]
39. Quiroz, H.P.; Dussan, A. Synthesis temperature dependence on magnetic properties of cobalt doped TiO$_2$ thin films for spintronic applications. *Appl. Surf. Sci.* **2019**, *484*, 688–691. [CrossRef]
40. Park, Y.R.; Kim, K.J. Structural and optical properties of rutile and anatase TiO$_2$ thin films: Effects of Co doping. *Thin Solid Films* **2005**, *484*, 34–38. [CrossRef]
41. Chodun, R.; Dypa, M.; Wicher, B.; Langier, K.N.; Okrasa, S.; Minikayev, R.; Zdunek, K. The sputtering of titanium magnetron target with increased temperature in reactive atmosphere by gas injection magnetron sputtering technique. *Appl. Surf. Sci.* **2022**, *574*, 151597. [CrossRef]
42. Zdunek, K.; Nowakowska-Langier, K.; Dora, J.; Chodun, R. Gas injection as a tool for plasma process control during coating deposition. *Surf. Coat. Technol.* **2013**, *228*, S367–S373. [CrossRef]
43. Skowronski, L.; Zdunek, K.; Nowakowska-Langier, K.; Chodun, R.; Trzcinski, M.; Kobierski, M.; Kustra, M.; Wachowiak, A.; Wachowiak, W.; Hiller, T.; et al. Characterization of microstructural, mechanical and optical properties of TiO2 layers deposited by GIMS and PMS methods. *Surf. Coat. Technol.* **2015**, *282*, 16–23. [CrossRef]
44. Wiatrowski, A.; Mazur, M.; Obstarczyk, A.; Wojcieszak, D.; Kaczmarek, D.; Morgiel, J.; Gibson, D. Comparison of the physico-chemical properties of TiO$_2$ thin films obtained by magnetron sputtering with continuous and pulsed gas flow. *Coatings* **2018**, *8*, 412. [CrossRef]
45. Mazur, M. Analysis of the properties of functional titanium dioxide thin films deposited by pulsed DC magnetron sputtering with various O$_2$: Ar ratios. *Opt. Mater.* **2017**, *69*, 96–104. [CrossRef]
46. Mazur, M.; Wojcieszak, D.; Wiatrowski, A.; Kaczmarek, D.; Lubanska, A.; Domaradzki, J.; Mazur, P.; Kalisz, M. Analysis of amorphous tungsten oxide thin films deposited by magnetron sputtering for application in transparent electronics. *Appl. Surf. Sci.* **2021**, *570*, 151151. [CrossRef]
47. Lei, C.; Du, Y.; Zhu, M.; Huo, W.; Wu, H.; Zhang, Y. Microstructure and mechanical properties of in situ TiC/Ti composites with a laminated structure synthesized by spark plasma sintering. *Mater. Sci. Eng. A* **2021**, *812*, 141136. [CrossRef]
48. Wimler, D.; Lindemann, J.; Gammer, C.; Spoerk-Erdely, P.; Stark, A.; Clemens, H.; Mayer, S. Novel intermetallic-reinforced near-α Ti alloys manufactured by spark plasma sintering. *Mater. Sci. Eng. A* **2020**, *792*, 139798. [CrossRef]
49. Lis, M.; Wrona, A.; Mazur, J.; Dupont, C.; Kamińska, M.; Kopyto, D.; Kwarciński, M. Fabrication And Properties Of Silver Based Multiwall Carbon Nanotube Composite Prepared By Spark Plasma Sintering Method. *Arch. Metall. Mater.* **2015**, *60*, 1351–1355. [CrossRef]
50. Oliver, W.; Pharr, G. An improved technique for determining hardness and elastic modulus using load and displacement sensing indentation experiments. *J. Mater. Res.* **1992**, *7*, 1564–1583. [CrossRef]
51. Jung, Y.-G.; Lawn, B.; Martyniuk, M.; Huang, H.; Hu, X. Evaluation of elastic modulus and hardness of thin films by nanoindentation. *J. Mater. Res.* **2004**, *19*, 3076–3080. [CrossRef]
52. Karabay, I.; Yuksel, S.A.; Ongul, F.; Ozturk, S.; Asli, M. Structural and optical characterization of TiO$_2$ thin films prepared by sol-gel process. *Acta Phys. Pol. A* **2012**, *121*, 265–267. [CrossRef]
53. Sarkar, D.; Ghosh, C.K.; Maiti, U.N.; Chattopadhyay, K.K. Effect of spin polarization on the optical properties of Co-doped TiO$_2$ thin films. *Phys. B* **2011**, *406*, 1429–1435. [CrossRef]
54. De, R.; Tripathi, S.; Naidu, S.C.; Prathap, C.; Tripathu, J.; Singh, J.; Haque, S.M.; Rao, K.D.; Sahoo, N.K. Investigation on optical properties of spin coated TiO$_2$/Co composite thin films. *AIP Conf. Proc.* **2017**, *1832*, 080038. [CrossRef]
55. Adewinbi, S.A.; Buremoh, W.; Owoeye, V.A.; Ajayeoba, Y.A.; Salau, A.O.; Busari, H.K.; Tijani, M.A.; Taleatu, B.A. Preparation and characterization of TiO$_2$ thin film electrode for optoelectronic and energy storage potentials: Effects of Co incorporation. *Chem. Phys. Lett.* **2021**, *779*, 138854. [CrossRef]
56. Islam, M.N.; Podder, J. The role of Al and Co co-doping on the band gap tuning of TiO$_2$ thin films for applications in photovoltaic and optoelectronic devices. *Mater. Sci. Semicond. Process.* **2021**, *121*, 105419. [CrossRef]

57. Boutlala, A.; Bourfaa, F.; Mahtili, M.; Bouaballou, A. Deposition of Co-doped TiO$_2$ thin films by sol-gel method. *IOP Conf. Ser. Mater. Sci. Eng.* **2016**, *108*, 012048. [CrossRef]
58. Waseda, Y.; Matsubara, E.; Shinoda, K. *X-ray Diffraction Crystallography: Introduction, Examples and Solved Problems*; Springer: Berlin/Heidelberg, Germany, 2011; ISBN 978-3-642-16634-1. [CrossRef]
59. Jauch, W.; Reehuls, M.; Bleif, H.J.; Kubanek, F.; Pattison, P. Crystallographic symmetry and magnetic structure of CoO. *Phys. Rev. B* **2001**, *64*, 052102. [CrossRef]
60. Wdowik, U.D.; Parlinski, K. Lattice dynamics of CoO from first principles. *Phys. Rev. B* **2007**, *75*, 104306. [CrossRef]
61. Moulder, J.F.; Stickle, W.F.; Sobol, P.E.; Bomben, K.D. *Handbook of X-ray Photoelectron Spectroscopy*; Physical Electronics, Inc.: Eden Prairie, MN, USA, 1995.
62. Crist, B.V. *Handbooks of Monochromatic XPS Spectra: Volume 1—The Elements and Native Oxides*; XPS International, Inc.: Ames, IA, USA, 1999.
63. Cole, K.M.; Kirk, D.; Thorpe, S. Co$_3$O$_4$ nanoparticles characterized by XPS and UPS. *Surf. Sci. Spectrs.* **2021**, *28*, 014001. [CrossRef]
64. Velhal, N.B.; Yun, T.H.; Ahn, J.; Kim, T.; Kim, J.; Yim, C. Tailoring cobalt oxide nanostructures for stable and high-performance energy storage applications. *Ceram. Int.* **2023**, *49*, 4889–4897. [CrossRef]
65. Gong, B.; Luo, X.; Bao, N.; Ding, J.; Li, S.; Yi, J. XPS study of cobalt TiO$_2$ films prepared by pulsed laser deposition. *Surf. Interface Anal.* **2014**, *46*, 1043–1046. [CrossRef]
66. Naseem, S.; Pinchuk, I.V.; Luo, Y.K.; Kawakami, R.K.; Khan, S.; Husain, S.; Khan, W. Epitaxial growth of cobalt doped TiO$_2$ thin films on LaAlO3(100) substrate by molecular beam epitaxy and their opto-magnetic based applications. *Appl. Surf. Sci.* **2019**, *493*, 691–702. [CrossRef]
67. Kim, J.; Livonen, T.; Hamalainen, J.; Kemell, M.; Meinander, K.; Mizohata, K.; Wang, L.; Raisanen, J.; Beranek, R.; Leskela, M.; et al. Low-temperature atomic layer deposition of cobalt oxide as effective catalyst for photoelectrochemical water-splitting devices. *Chem. Mater.* **2017**, *29*, 5796–5805. [CrossRef]
68. Biesinger, M.C.; Payne, B.P.; Grosvenor, A.P.; Lau, L.W.M.; Gerson, A.R.; Smart, R.S.C. Resolving surface chemical states in XPS analysis of first row transition metals, oxides and hydroxides: Cr, Mn, Fe, Co and Ni. *Appl. Surf. Sci.* **2011**, *257*, 2717–2730. [CrossRef]
69. Liu, T.; Guo, Y.F.; Yan, Y.M.; Wang, F.; Deng, C.; Rooney, D.; Sun, K.N. CoO nanoparticles embedded in the tree-dimensional nitrogen/sulfur co-doped carbon nanofiber networks as a bifunctional catalyst for oxygen reduction/evolution reactions. *Carbon* **2016**, *106*, 84–92. [CrossRef]
70. Li, X.C.; She, F.S.; Shen, D.; Liu, C.P.; Chen, L.H.; Li, Y.; Deng, Z.; Chen, Z.H.; Wang, H.E. Coherent nanoscale cobalt/cobalt oxide heterostructures embedded in porous carbon for the oxygen reduction reaction. *R. Soc. Chem.* **2018**, *8*, 28625. [CrossRef] [PubMed]
71. Chuang, T.J.; Brundle, C.R.; Rice, D.W. Interpretation of the X-ray photoemission spectra of cobalt oxides and cobalt oxide surfaces. *Surf. Sci.* **1976**, *59*, 413–429. [CrossRef]

Disclaimer/Publisher's Note: The statements, opinions and data contained in all publications are solely those of the individual author(s) and contributor(s) and not of MDPI and/or the editor(s). MDPI and/or the editor(s) disclaim responsibility for any injury to people or property resulting from any ideas, methods, instructions or products referred to in the content.

Article

Annealing Studies of Copper Indium Oxide (Cu₂In₂O₅) Thin Films Prepared by RF Magnetron Sputtering

Giji Skaria [1,*], Ashwin Kumar Saikumar [1], Akshaya D. Shivprasad [2] and Kalpathy B. Sundaram [1]

[1] Department of Electrical and Computer Engineering, University of Central Florida, Orlando, FL 32816, USA; saikumarashwin1991@knights.ucf.edu (A.K.S.); Kalpathy.Sundaram@ucf.edu (K.B.S.)
[2] School of Electrical Engineering and Computer Science, The Pennsylvania State University, University Park, PA 16802, USA; ads6067@psu.edu
* Correspondence: giji.skaria@ucf.edu

Abstract: Copper indium oxide ($Cu_2In_2O_5$) thin films were deposited by the RF magnetron sputtering technique using a $Cu_2O:In_2O_3$ target. The films were deposited on glass and quartz substrates at room temperature. The films were subsequently annealed at temperatures ranging from 100 to 900 °C in an O_2 atmosphere. The X-ray diffraction (XRD) analysis performed on the samples identified the presence of $Cu_2In_2O_5$ phases along with $CuInO_2$ or In_2O_3 for the films annealed above 500 °C. An increase in grain size was identified with the increase in annealing temperatures from the XRD analysis. The grain sizes were calculated to vary between 10 and 27 nm in films annealed between 500 and 900 °C. A morphological study performed using SEM further confirmed the crystallization and the grain growth with increasing annealing temperatures. All films displayed high optical transmission of more than 70% in the wavelength region of 500–800 nm. Optical studies carried out on the films indicated a small bandgap change in the range of 3.4–3.6 eV during annealing.

Keywords: $Cu_2In_2O_5$; RF sputtering; annealing studies; optical characteristics; XRD; morphology studies; optical bandgap

Citation: Skaria, G.; Saikumar, A.K.; Shivprasad, A.D.; Sundaram, K.B. Annealing Studies of Copper Indium Oxide (Cu₂In₂O₅) Thin Films Prepared by RF Magnetron Sputtering. *Coatings* **2021**, *11*, 1290. https://doi.org/10.3390/coatings11111290

Academic Editors: Rafal Chodun and Roman A. Surmenev

Received: 21 September 2021
Accepted: 20 October 2021
Published: 24 October 2021

Publisher's Note: MDPI stays neutral with regard to jurisdictional claims in published maps and institutional affiliations.

Copyright: © 2021 by the authors. Licensee MDPI, Basel, Switzerland. This article is an open access article distributed under the terms and conditions of the Creative Commons Attribution (CC BY) license (https://creativecommons.org/licenses/by/4.0/).

1. Introduction

Transparent conducting oxides (TCOs) have a unique ability to allow visible light to pass through, and to conduct electricity. TCOs find many applications in solar cells, optical displays, reflective coatings, light emission devices, low-emissivity windows, electrochromic mirrors, UV sensors and windows, defrosting windows, electromagnetic shielding, and transparent electronics [1–6]. The bulk of research conducted on TCOs involves n-type TCOs for creating devices [7–12]. Many transparent electronic applications require the necessity of p-type TCOs. An early attempt to synthesize a p-type TCO from $CuAlO_2$ by the laser ablation method was reported by Kawazoe et al. [13]. Recently, researchers have been looking at the copper indium oxide- and copper gallium oxide-based thin film materials for possible p-type TCO applications [14–27]. Nair et al. fabricated a transparent thin-film p–n junction consisting of Ca and tin-doped $CuInO_2$ [28]. Further, Nair et al. reported the potential use of $CuInO_2$ for thermoelectric applications [29]. $Cu_2In_2O_5$ is another phase of Cu-In-based oxide. There has not been a lot of work conducted on $Cu_2In_2O_5$. Synthesis of $Cu_2In_2O_5$ has been reported only using either smeltering or chemical processes [20,21]. The chemical processes include synthesizing $Cu_2In_2O_5$ from aqueous solutions of nitrates, chlorides, and sulfates of Cu, In, and Ga [20]. At this moment, there have not been many attempts to investigate $Cu_2In_2O_5$ deposited by RF magnetron sputtering. RF magnetron sputtering allows films to be deposited with high uniformity and homogeneity as well as providing the capability to control the film thickness and deposition rate [30]. It has an additional advantage of having a low cost as well as the ability to achieve large-area deposition. In this work, the focus was on the deposition of $Cu_2In_2O_5$ by RF magnetron sputtering using a single target of $Cu_2O:In_2O_3$ in the ratio of 1:1. The structural and optical properties of $Cu_2In_2O_5$ thin films were investigated.

2. Experimental Details

2.1. Deposition of Copper Indium Oxide Thin Films

Copper indium oxide films were deposited by a radio frequency magnetron sputtering system using a CTI 100 cryogenic high-vacuum pump. A 600 W, 13.56 MHz RF power supply (Dressler Cesar 136 FST RF Generator, Denver, CO, USA) was used to power the MAK 2 sputter gun (San Jose, CA, USA). The Advance Energy VarioMatch-1000 matching network (Fort Collins, CO, USA) was used to match the source and the load impedance. All the depositions were performed at a power of 50 W. The power was ramped up at the rate of 1 W/s. Glass substrates were used to deposit the films for annealing studies up to 400 °C, whereas quartz substrates were used for annealing studies above 500 °C. The substrates were cleaned with acetone and methanol in an ultrasonic bath followed by rinsing using DI water. The samples were dried with nitrogen gas before loading them into the vacuum system. A 2″ powder pressed target of Cu_2O/In_2O_3 (1/1 mol%, 99.9% purity), ACI Alloy Inc. (San Jose, CA, USA), was used to deposit the films. A gap of 5 cm between the target and the substrate was maintained to achieve a uniform film thickness. A base pressure of 5×10^{-6} Torr was achieved before initiating the deposition. During the deposition, the pressure was maintained at 10 mTorr with an argon flow of 10 sccm. The deposition rate was found to be approximately 270 Å per minute. This was measured using a Veeco Dektak-150 profilometer (Plainview, NY, USA). All the depositions were conducted for 7.5 min to achieve approximately 2000 Å of film thickness. Post-deposition annealing was conducted from 100 to 900 °C for 90 min in O_2 gas flow.

2.2. Characterization of Copper Indium Oxide Thin Films

The XRD measurements were performed using a PANalytical Empyrean XRD system (Malvern Panalytical, Westborough, MA, USA), using radiation from a Cu source at 45 kV and 40 mA. The diffraction patterns were recorded between 2θ angles of 15° and 60°, and the phase information was analyzed using HighScore Plus software (Malvern Panalytical, Westborough, MA, USA). The surface morphology of the film was assessed using a field-emission scanning electron microscope, Zeiss ULTRA-55 FEG SEM (Zeiss Microscopy, White Plains, NY, USA). The optical transmission studies were performed using a Cary 100 UV–Vis spectrophotometer (Varian Analytical Instruments, Walnut Creek, CA, USA).

3. Results and Discussions

3.1. XRD Analysis

As-deposited films and films annealed up to 400 °C did not reveal any diffraction peaks indicating an amorphous nature. Although the XRD measurement was conducted between 15° and 60°, due to the presence of a broad amorphous peak related to the quartz substrates, the 2θ angle reported in Figure 1 is limited between 25° and 60°. Figure 1 shows the X-ray diffractograms of the films annealed at 500–900 °C in an O_2 atmosphere. The film annealed at 500 °C started showing low-intensity peaks related to the ($\bar{2}10$), ($\bar{5}03$), and ($\bar{3}13$) planes that have been attributed to the $Cu_2In_2O_5$ phase (JCPDSPDF# 30-0479). The films annealed at 600 °C and above showed more peaks related to $Cu_2In_2O_5$. It was observed that the peak intensity and peak sharpness increased with an increase in annealing temperature, denoting an increase in crystallinity. These identified planes match very well with the $Cu_2In_2O_5$-synthesized nanoparticles reported by Su et al. [20]. In addition to the $Cu_2In_2O_5$ phase, a peak at 30.7° (006) was identified. This can be attributed to either the In_2O_3 or $CuInO_2$ phase [31]. As-deposited films (not shown in Figure 1) did not display any diffraction peaks.

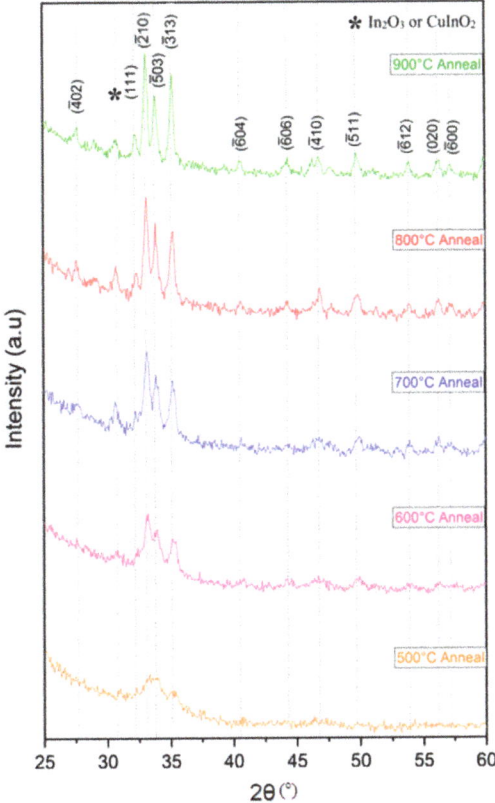

Figure 1. X-ray diffraction of the films annealed at 500–900 °C in O$_2$ for 90 min.

The grain sizes of the films annealed at 500–900 °C were calculated using the Debye–Scherrer equation [32].

$$D = \frac{0.9\lambda}{\beta \cos \theta} \quad (1)$$

where θ is the Bragg angle, β is the full width at half maximum of the peak, λ is the wavelength of the X-ray, and D is the average grain size. The ($\bar{3}$13) peak was used for the calculation of grain size. The average grain sizes of the Cu$_2$In$_2$O$_5$ films annealed at 500, 600, 700, 800, and 900 °C were calculated to be 10, 13, 17, 21, and 27 nm, respectively.

3.2. Morphology Studies

Figure 2 shows the SEM images of as-deposited Cu$_2$In$_2$O$_5$ thin films as well as those annealed at temperatures varying from 500 to 900 °C. Changes in the morphology were identified for the film annealed at 500–900 °C. As-deposited films and the film annealed at 500 °C displayed the presence of very small grains, as shown in Figure 2a,b. However, the films annealed at 600 °C and above showed an increase in grain size. This coincided with the results from the XRD analysis where the diffraction peaks started to appear for films annealed at 500 °C and above. Both the SEM and the XRD analysis studies indicated that a minimum of 500 °C is required to initiate nanocrystalline growth. Continuous growth in grain size was subsequently observed for the films annealed at 600–900 °C. It is worth mentioning that a pinhole-like appearance was detected in films annealed at 800 and 900 °C. Elemental analysis of all the samples was performed using EDAX incorporated in the FESEM. Nearly equal ratios of Cu:In were identified in all films.

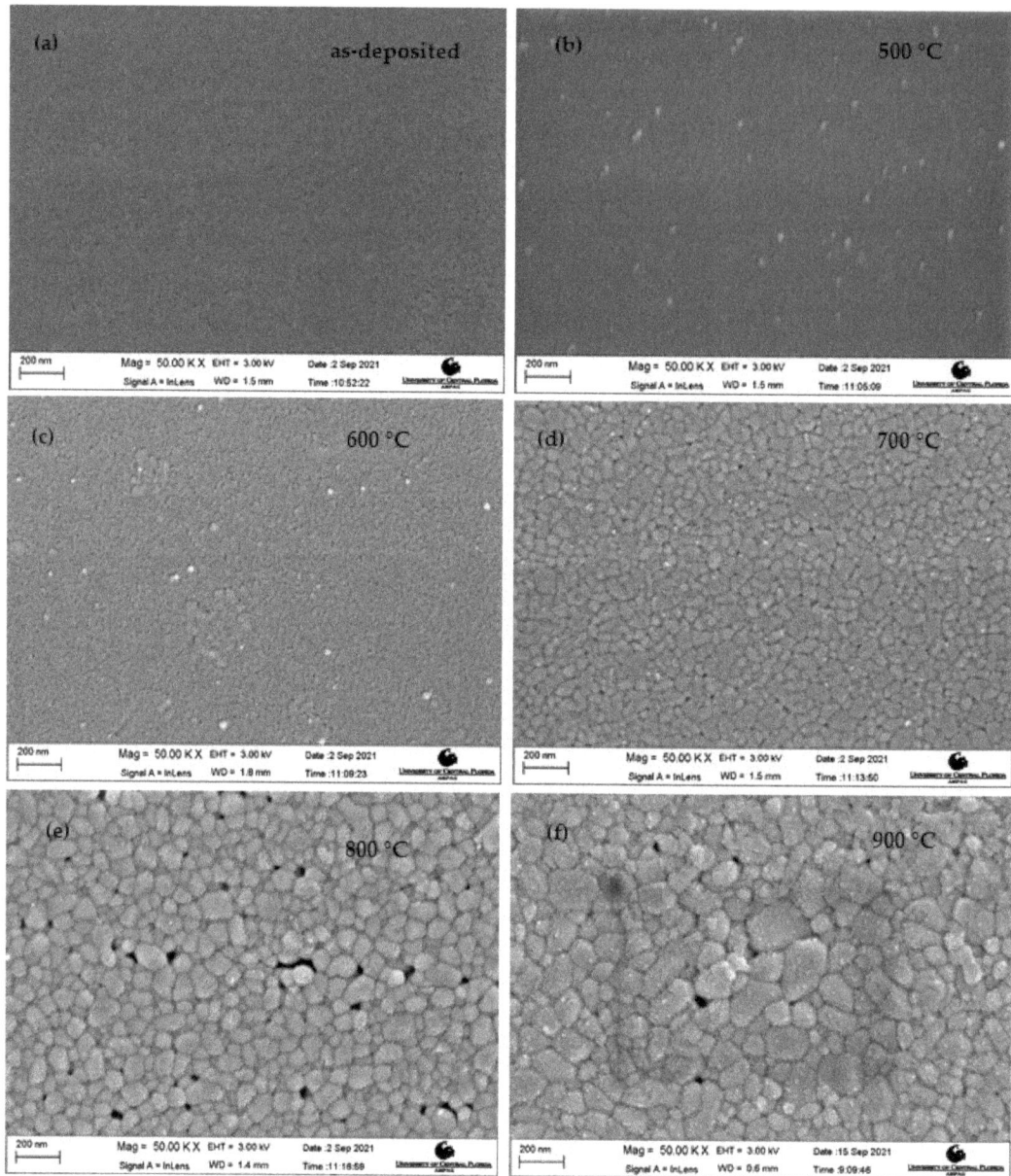

Figure 2. SEM images of $Cu_2In_2O_5$ films (**a**) as deposited and annealed at (**b**) 500 °C, (**c**) 600 °C, (**d**) 700 °C, (**e**) 800 °C, and (**f**) 900 °C.

3.3. Optical Studies

A Cary 100 UV–Vis spectrophotometer (Agilent Technologies, Santa Clara, CA, USA) was used to perform the optical characterization of the annealed $Cu_2In_2O_5$ thin films. Figure 3 shows the percent of transmission for the films deposited on glass and quartz substrates and subsequently annealed from 100 to 900 °C. Overall, the optical transmission increased for the films annealed at different temperatures. The transmission values of the annealed films were observed to vary between 70% and 90%. At a 450 nm wavelength, the films annealed at 900 °C displayed the highest transmission of 80%. However, the films annealed at 900 °C subsequently showed a decreasing trend in transmission beyond a 500 nm wavelength. This could possibly be attributed to an increase in grain size with annealing, as reported in [33,34].

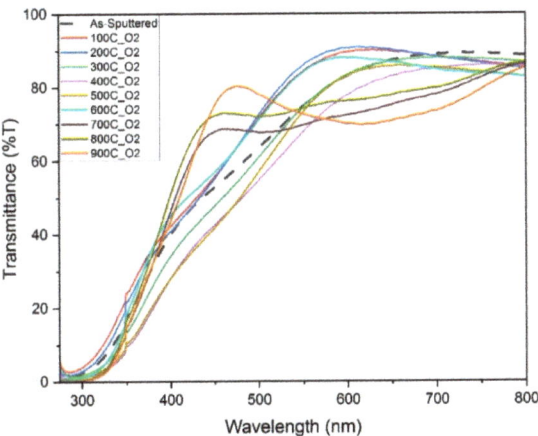

Figure 3. Optical transmission spectra of the $Cu_2In_2O_5$ thin films annealed at various temperatures.

3.4. Optical Bandgap

The optical transmission data were used to calculate the optical band gap of $Cu_2In_2O_5$ thin films using the Tauc plot method [35–37]. Since the reflectance was identified to be less than 5%, the absorption coefficient α was calculated directly from the transmission data [36]. The absorption coefficient α was calculated using Equation (2), where d is the thickness of the film, and T is the percent of transmission. The optical bandgap (Eg) was estimated from Equation (3).

$$\alpha = \frac{1}{d} \ln\left(\frac{1}{T}\right) \qquad (2)$$

$$(\alpha h\nu)^{1/n} = B(h\nu - Eg) \qquad (3)$$

where $h\nu$ is the photon energy, B is a constant, Eg is the optical bandgap, and n = 1/2 for the direct bandgap transition. Figure 4a–j show the Tauc plot generated using the above equations. The linear region of the curve was extrapolated to the x-axis to identify the Eg value. The extrapolated values of the bandgap are listed in Table 1. The bandgap for $Cu_2In_2O_5$ thin films is reported for the first time in this work. The bandgap was in the range of 3.4–3.6 eV. It is worth mentioning that an increase in the annealing temperature did not have any major effect on the bandgap.

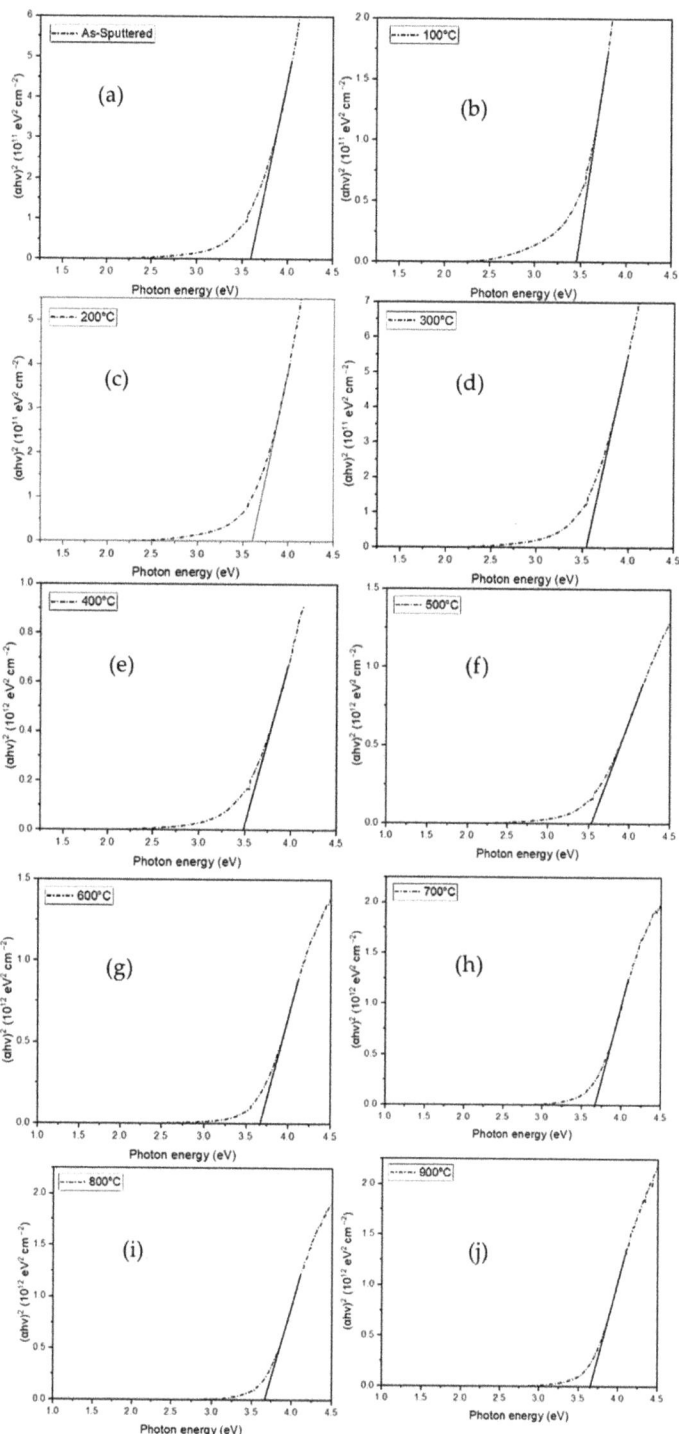

Figure 4. (**a–j**): Tauc plots of the $Cu_2In_2O_5$ thin films annealed at different temperatures.

Table 1. Optical bandgap values obtained for $Cu_2In_2O_5$ thin films.

Annealing Temperature (°C)	Bandgap (eV)
As deposited	3.59
100	3.45
200	3.6
300	3.54
400	3.48
500	3.53
600	3.66
700	3.66
800	3.66
900	3.64

4. Conclusions

$Cu_2In_2O_5$ thin films were deposited by the RF magnetron sputtering technique. The effects of structural, morphological, and optical properties due to post-deposition annealing at 100–900 °C with a constant O_2 flow were studied. Both the XRD analysis and the FESEM images concluded that a minimum annealing temperature of 500 °C was required to initiate the crystallization and grain growth. A further increase in the annealing temperature resulted in an increase in crystallization and grain size. The largest average grain size was observed in films annealed at 900 °C. In addition to the $Cu_2In_2O_5$ phases, the XRD results reveal the presence of an additional phase corresponding to either In_2O_3 or $CuInO_2$. Optical studies showed a bandgap of 3.4–3.6 eV for the films.

Author Contributions: Conceptualization, G.S. and K.B.S.; methodology, G.S. and A.K.S.; validation, G.S. and A.D.S.; formal analysis, G.S., A.K.S. and K.B.S.; data curation, G.S., A.K.S. and K.B.S.; writing—original draft preparation, G.S. and K.B.S.; writing—review and editing, G.S., A.K.S. and K.B.S.; visualization, G.S. and A.K.S.; supervision, K.B.S.; project administration, K.B.S. All authors have read and agreed to the published version of the manuscript.

Funding: Article processing charges were provided in part by the UCF College of Graduate Studies Open Access Publishing Fund.

Institutional Review Board Statement: Not applicable.

Informed Consent Statement: Not applicable.

Data Availability Statement: Not applicable.

Conflicts of Interest: The authors declare no conflict of interest.

References

1. Chopra, K.; Major, S.; Pandya, D. Transparent conductors—A status review. *Thin Solid Film.* **1983**, *102*, 1–46. [CrossRef]
2. Dawar, A.; Joshi, J. Semiconducting transparent thin films: Their properties and applications. *J. Mater. Sci.* **1984**, *19*, 1–23. [CrossRef]
3. Gordon, R.G. Criteria for choosing transparent conductors. *MRS Bull.* **2000**, *25*, 52–57. [CrossRef]
4. Saikumar, A.K.; Skaria, G.; Sundaram, K.B. ZnO gate based MOSFETs for sensor applications. *ECS Trans.* **2014**, *61*, 65. [CrossRef]
5. Szyszka, B.; Loebmann, P.; Georg, A.; May, C.; Elsaesser, C. Development of new transparent conductors and device applications utilizing a multidisciplinary approach. *Thin Solid Film.* **2010**, *518*, 3109–3114. [CrossRef]
6. Zhang, B.; Yu, B.; Jin, J.; Ge, B.; Yin, R. Performance of InSnZrO as transparent conductive oxides. *Phys. Status Solidi* **2010**, *207*, 955–962. [CrossRef]
7. Khan, A.; Rahman, F. Study of microstructural and optical properties of nanocrystalline indium oxide: A transparent conducting oxide (TCO). *AIP Conf. Proc.* **2019**, *2115*, 030091.
8. Nehate, S.D.; Prakash, A.; Mani, P.D.; Sundaram, K.B. Work function extraction of indium tin oxide films from MOSFET devices. *ECS J. Solid State Sci. Technol.* **2018**, *7*, P87. [CrossRef]
9. Sundaram, K.; Khan, A. Work function determination of zinc oxide films. *J. Vac. Sci. Technol. A Vac. Surf. Film.* **1997**, *15*, 428–430. [CrossRef]
10. Sundaram, K.; Khan, A.J.T.S.F. Characterization and optimization of zinc oxide films by RF magnetron sputtering. *Thin Solid Film.* **1997**, *295*, 87–91. [CrossRef]

11. Sundaresh, S.; Nehate, S.D.; Sundaram, K.B. Electrical and optical studies of reactively sputtered indium oxide thin films. *ECS J. Solid State Sci. Technol.* **2021**, *10*, 065016. [CrossRef]
12. Noh, J.H.; Ryu, S.Y.; Jo, S.J.; Kim, C.S.; Sohn, S.-W.; Rack, P.D.; Kim, D.-J.; Baik, H.K. Indium oxide thin-film transistors fabricated by RF sputtering at room temperature. *IEEE Electron. Device Lett.* **2010**, *31*, 567–569. [CrossRef]
13. Kawazoe, H.; Yasukawa, M.; Hyodo, H.; Kurita, M.; Yanagi, H.; Hosono, H.J.N. P-type electrical conduction in transparent thin films of $CuAlO_2$. *Nat. Cell Biol.* **1997**, *389*, 939–942. [CrossRef]
14. Biswas, S.K.; Sarkar, A.; Pathak, A.; Pramanik, P.J.T. Studies on the sensing behaviour of nanocrystalline $CuGa_2O_4$ towards hydrogen, liquefied petroleum gas and ammonia. *Talanta* **2010**, *81*, 1607–1612. [CrossRef]
15. Fenner, L.; Wills, A.; Bramwell, S.; Dahlberg, M.; Schiffer, P. Zero-point entropy of the spinel spin glasses $CuGa_2O_4$ and $CuAl_2O_4$. *J. Phys. Conf. Ser.* **2008**, *145*, 012029. [CrossRef]
16. Makhova, L.; Wett, D.; Lorenz, M.; Konovalov, I. X-ray spectroscopic investigation of forbidden direct transitions in $CuGaO_2$ and $CuInO_2$. *Phys. Status* **2006**, *203*, 2861–2866. [CrossRef]
17. Pilliadugula, R.; Nithya, C.; Krishnan, N.G.; Nammalwar, G. Influence of Ga_2O_3, $CuGa_2O_4$ and Cu_4O_3 phases on the sodium-ion storage behaviour of CuO and its gallium composites. *Nanoscale Adv.* **2020**, *2*, 1269–1281. [CrossRef]
18. Saikumar, A.K.; Sundaresh, S.; Nehate, S.D.; Sundaram, K.B.J.C. Properties of RF magnetron-sputtered copper gallium oxide ($CuGa_2O_4$). *Thin Film.* **2021**, *11*, 921.
19. Shi, J.; Liang, H.; Xia, X.; Li, Z.; Long, Z.; Zhang, H.; Liu, Y. Preparation of high-quality $CuGa_2O_4$ film via annealing process of Cu/β-Ga_2O_3. *J. Mater. Sci.* **2019**, *54*, 11111–11116. [CrossRef]
20. Su, C.-Y.; Chiu, C.-Y.; Chang, C.-H.; Ting, J.-M. Synthesis of $Cu_2In_2O_5$ and $CuInGaO_4$ nanoparticles. *Thin Solid Film.* **2013**, *531*, 42–48. [CrossRef]
21. Su, C.-Y.; Mishra, D.K.; Chiu, C.-Y.; Ting, J.-M. Effects of Cu_2S sintering aid on the formation of $CuInS_2$ coatings from single crystal $Cu_2In_2O_5$ nanoparticles. *Surf. Coat. Technol.* **2013**, *231*, 517–520. [CrossRef]
22. Wang, J.; Ibarra, V.; Barrera, D.; Xu, L.; Lee, Y.-J.; Hsu, J.W.P. Solution synthesized p-type copper gallium oxide nanoplates as hole transport layer for organic photovoltaic devices. *J. Phys. Chem. Lett.* **2015**, *6*, 1071–1075. [CrossRef] [PubMed]
23. Wei, H.; Chen, Z.; Wu, Z.; Cui, W.; Huang, Y.; Tang, W. Epitaxial growth and characterization of $CuGa_2O_4$ films by laser molecular beam epitaxy. *AIP Adv.* **2017**, *7*, 115216. [CrossRef]
24. Yin, H.; Shi, Y.; Dong, Y.; Chu, X. Synthesis of spinel-type $CuGa_2O_4$ nanoparticles as a sensitive non-enzymatic electrochemical sensor for hydrogen peroxide and glucose detection. *J. Electroanal. Chem.* **2021**, *885*, 115100. [CrossRef]
25. Zardkhoshoui, A.M.; Davarani, S.S. Designing a flexible all-solid-state supercapacitor based on $CuGa_2O_4$ and FeP-rGO electrodes. *J. Alloys Compd.* **2019**, *773*, 527–536. [CrossRef]
26. Zhang, K.H.L.; Xi, K.; Blamire, M.; Egdell, R.G. P-type transparent conducting oxides. *J. Phys. Condens. Mater.* **2016**, *28*, 383002. [CrossRef]
27. Singh, M.; Singh, V.N.; Mehta, B.R. Synthesis and properties of nanocrystalline copper indium oxide thin films deposited by RF magnetron sputtering. *J. Nanosci. Nanotechnol.* **2008**, *8*, 3889–3894. [CrossRef] [PubMed]
28. Nair, B.G.; Rahman, H.; Aijo, J.K.; Keerthi, K.; Shaji, G.S.O.; Sharma, V.; Philip, R.R. Calcium incorporated copper indium oxide thin films—A promising candidate for transparent electronic applications. *Thin Solid Film.* **2019**, *693*, 137673. [CrossRef]
29. Nair, B.G.; Okram, G.S.; Naduvath, J.; Shripathi, T.; Fatima, A.; Patel, T.; Jacob, R.; Keerthi, K.; Remillard, S.K.; Philip, R.R. Low temperature thermopower and electrical conductivity in highly conductive $CuInO_2$ thin films. *J. Mater. Chem. C* **2014**, *2*, 6765–6772. [CrossRef]
30. Saikumar, A.K.; Nehate, S.D.; Sundaram, K.B. Review—RF sputtered films of Ga_2O_3. *ECS J. Solid State Sci. Technol.* **2019**, *8*, Q3064–Q3078. [CrossRef]
31. Yang, B.; He, Y.; Polity, A.; Meyer, B.K. Structural, optical and electrical properties of transparent conducting $CuInO_2$ thin films prepared by RF sputtering. *MRS Proc.* **2005**, *865*, 865. [CrossRef]
32. Holzwarth, U.; Gibson, N. The Scherrer equation versus the 'Debye-Scherrer equation'. *Nat. Nanotechnol.* **2011**, *6*, 534. [CrossRef] [PubMed]
33. Ramana, C.V.; Smith, R.J.; Hussain, O.M. Grain size effects on the optical characteristics of pulsed-laser deposited vanadium oxide thin films. *Phys. Status* **2003**, *199*, R4–R6. [CrossRef]
34. Shakti, N.; Gupta, P. Structural and optical properties of sol-gel prepared ZnO thin film. *Appl. Phys. Res.* **2010**, *2*, 19. [CrossRef]
35. Duta, M.; Anastasescu, M.; Calderon-Moreno, J.M.; Predoana, L.; Preda, S.; Nicolescu, M.; Stroescu, H.; Bratan, V.; Dascalu, I.; Aperathitis, E.; et al. Sol–gel versus sputtering indium tin oxide films as transparent conducting oxide materials. *J. Mater. Sci. Mater. Electron.* **2016**, *27*, 4913–4922. [CrossRef]
36. Spence, W. The UV absorption edge of tin oxide thin films. *J. Appl. Phys.* **1967**, *38*, 3767–3770. [CrossRef]
37. Lv, J.; Huang, K.; Chen, X.; Zhu, J.; Cao, C.; Song, X.; Sun, Z. Optical constants of Na-doped ZnO thin films by sol-gel method. *Opt. Commun.* **2011**, *284*, 2905–2908. [CrossRef]

Article

P-type (CuTi)Ox Thin Films Deposited by Magnetron Co-Sputtering and Their Electronic and Hydrogen Sensing Properties

Ewa Mańkowska, Michał Mazur *, Jarosław Domaradzki and Damian Wojcieszak

Faculty of Electronic, Photonics and Microsystems, Wrocław University of Science and Technology, Janiszewskiego 11/17, 50-372 Wrocław, Poland
* Correspondence: michal.mazur@pwr.edu.pl

Abstract: Thin films of copper oxide (Cu_xO), titanium oxide (TiO_x), and several mixtures of copper and titanium oxides ((CuTi)Ox) were deposited using magnetron sputtering. X-ray diffraction analysis of the as-deposited TiO_x thin film revealed the presence of TiO crystallites, while in the case of (CuTi)Ox with the lowest amount of copper, metallic Cu crystallites were found. In the case of $(Cu_{0.77}Ti_{0.23})Ox$ and Cu_xO thin films, characteristic peaks for metallic copper and copper oxides were observed in their diffractograms. It was found that post-process annealing at 473 K considerably affects the microstructure of (CuTi)Ox thin films. After annealing, anatase phase was observed in $(Cu_{0.23}Ti_{0.77})Ox$ and $(Cu_{0.41}Ti_{0.59})Ox$ thin films. In turn, the $(Cu_{0.77}Ti_{0.23})Ox$ and Cu_xO films were formed only in the copper oxide phase. The $(Cu_{0.77}Ti_{0.23})Ox$ film annealed at 473 K showed the best opto-electronic performance, as it had the highest transmission and the lowest resistivity. However, the greatest advantage of this thin film was the p-type semiconducting behavior, which was the strongest of all of the thin films in this work, as indicated by the measurement of the Seebeck coefficient. All deposited thin films were sensitive to hydrogen exposure, while the best sensor response of 10.9 was observed for the $(Cu_{0.77}Ti_{0.23})Ox$ thin film annealed at 473 K.

Keywords: thin film; transparent oxide semiconductor; copper-titanium mixed oxides; magnetron sputtering; p-type semiconductors; optical and electrical properties; gas sensor

1. Introduction

Over the decades, there has been a steadily growing interest in metal oxides as materials for gas sensors, transparent electronics, lithium-ion batteries, solar cells, and self-cleaning or electrochromic coatings [1–6]. The reasons for such a high interest include the relatively low cost of metal oxides, the ability to create an extensive surface area with respect to the volume of the material, and the catalytic efficiency. In addition, the deposition technologies of such oxides in the form of thin films are compatible with production technology such as MOS [7], which makes it possible to easily integrate various oxide-based active devices, such as sensors, with electronics circuits in a single device. Copper and titanium oxides are nontoxic compounds commonly found in the Earth's crust. They exhibit many interesting optical, electrical, and chemical properties that can be customized due to their different application areas, e.g., high optical transmittance, semiconducting properties, and photocatalytic properties [8,9].

There are two stable forms of copper oxide: cuprous oxide (Cu_2O) and cupric oxide (CuO) [10]. The former has a cubic crystal structure and the latter crystallizes in the monoclinic phase. The band gap of copper oxides (Cu_xO)—depending on the oxide form—is in the range of 1.2–2.1 eV and 2.1–2.6 eV for CuO and Cu_2O, respectively [10]. Such narrow band gaps make them promising materials for solar energy conversion [11]. Moreover, Cu_xO can be an interesting material for gas sensing applications [8,12] or for transparent electronic devices [3,13] since both copper oxides are well-known p-type oxide semiconductors. Their hole-type conduction is due to the presence of negatively charged Cu vacancies. The great advantage of Cu_2O is one of the highest reported hole mobility (exceeding 100 cm^2/Vs), which is caused by its unique band structure. Unlike other metal oxides, the valence band of the cupric oxide is formed of hybridized Cu 3d and O 2p orbitals that leads to the creation of a less localized hole transport pathway [14–16].

One of the simplest methods to obtain Cu_xO thin films is thermal oxidation of metallic copper foil. Annealing at 473 K initiates partial oxidation of Cu_2O, a further increase in temperature to 573 K results in the formation of a mixed phase of Cu_2O and CuO, and finally above 623 K, a single cupric oxide is formed [8].

On the other hand, titanium dioxide (TiO_2) is a well-known photocatalyst [4,17]. It may crystallize in two stable tetragonal forms, anatase and rutile. A moderately wide band gap of 3.0–3.4 eV contributes to excellent transparency in the visible wavelength range. However, in the case of optoelectronic applications, an improvement of electrical parameters is necessary [3] as stoichiometric titanium dioxide at room temperature is an insulator [18]. The deposition of non-stoichiometric oxide with oxygen vacancies that cause intrinsic n-type semiconducting behavior could be one of the methods for modification of electronic properties. Therefore, the combination of good optical properties of titanium oxide with p-type conduction of copper oxides could result in formation of an attractive material for optoelectronic application.

For the production of technologically advanced transparent electronic devices or for the practical realization of such a sensor device as an electronic nose, it is necessary to combine p-type and n-type oxides. Recently, n-type metal oxides such as SnO_2 or In_2O_3 have been successfully applied in commercial electronic devices, but their p-type counterparts still lack performance. Therefore, the development of p-type semiconducting metal oxides is still a strategic issue. The (CuTi)Ox thin films appear to be one of the most attractive systems. For example, Mor et al. [15] successfully manufactured the (CuTi)Ox(p)–TiO_2(n) junction and showed that (CuTi)Ox, depending on the copper concentration, it can possess n-type properties for the lowest copper concentration [15] and p-type properties when the amount of copper increases [3,19]. Likewise, the optical properties of mixed copper and titanium oxide depend on the concentration of copper oxides [20]. However, it is possible to produce transparent [3] or semitransparent [13,20] semiconducting (CuTi)Ox thin films called transparent oxide semiconductors (TOS). Furthermore, the aforementioned p-type composite system exhibits gas sensing properties to NO_2, O_3, Cl_2, H_2, CH_3COH, and Cl_2 at moderate working temperatures [21–23] and the interactions between the sensor surface and the gas can be advantageously altered by doping with Li [22] or Au [23]. Barreca et al. [23] reported that the addition of titanium oxide can improve the performance of copper oxide as an electrode for lithium-ion batteries. Incorporation of Cu_xO into the TiO_2 matrix improved photocatalytic activity in the case of pollution decomposition [24–27] including CO_2 reduction [28] and hydrogen generation in the photoinduced water splitting process [15,29–32]. Antimicrobacterial studies show that copper addition to TiO_2 has a positive effect in reducing bacterial (e.g., E. coli) viability [2,33]. In turn, the fabrication of coatings with gradient elemental distribution resulted in the formation of copper-reach regions of (CuTi)Ox, which acted as a semiconducting material, and copper-poor regions, which acted as an insulating material. This structure exhibited memristive properties without the necessity of depositing multilayer structures [19,34,35].

Mixed copper and titanium oxides have recently been prepared by microwave-assisted synthesis [24], ion beam sputtering [21,36], anodization of Cu-Ti films deposited by magnetron sputtering [15], low pressure CVD [23,37], or sol–gel [30,38]. In our previous works [3,19,34,35], preparation of (CuTi)Ox thin films using magnetron sputtering has also been described. In this paper, we propose a two-step preparation route: (1) deposition of not fully oxidized thin films using magnetron sputtering method and (2) additional post-process annealing at 473 K for tailoring of the thin film properties. To our knowledge, the composite material of mixed copper and titanium oxides with such a variety of copper concentrations (23–77 at.%) has never been thoroughly analyzed. The aim of the work was to obtain material with good optoelectronic properties, which is not an obvious task, since enhancing the electrical properties usually results in deterioration of the material's transparency. The idea was to obtain a composite material in which one component possesses excellent transparency and the other is a desirable p-type oxide semiconductor, in order to enhance the dominance of the hole conduction type in the material, lowering its resistivity, while obtaining moderate transparency. Furthermore, results of investigations of the prepared sensor response to hydrogen have been demonstrated at a low operating temperature of 473 K.

2. Materials and Methods

The copper and titanium (CuTi)Ox mixed oxides thin films were deposited by the magnetron co-sputtering method using two circular titanium targets (99.999%) and one copper target (99.999%), each with a diameter of 28.5 mm and a thickness of 3 mm. A constant flow of argon and oxygen was provided during the deposition process, acting as the working and reactive gases, respectively. The amount of oxygen in the $Ar - O_2$ gas mixture was equal to 15%. The base pressure was on the order of 10^{-5} mbar, while the pressure in the vacuum chamber during the sputtering was kept at approximately 1.25×10^{-2} mbar. The magnetrons were powered using an MSS2 2 kW pulsed DC power supply unit (DORA Power System, Wilczyce, Poland). The power applied to each magnetron was a decisive factor in obtaining thin films with various material compositions, i.e., the copper–titanium ratio. The most crucial deposition parameters are summarized in Table 1.

Table 1. Summary of the main process parameters for magnetron co-sputtering deposition of copper oxide (Cu_xO), titanium oxide (TiO_x), and mixed copper and titanium oxides ($(CuTi)Ox$).

Thin Film	Deposition Rate (Å/s)	Power Applied to Targets (W)			Distance between Magnetrons and Substrate Holder (cm)	Pressure of $Ar + O_2$ in the Vacuum Chamber (10^{-2} mbar)	Thickness of the Thin Film (nm)
		Ti	Cu	Ti			
TiO_x	4.5	590	-	465	10	1.25	540
$(Cu_{0.23}Ti_{0.77})Ox$	5.7	560	70	520			620
$(Cu_{0.41}Ti_{0.59})Ox$	6.3	525	140	580			570
$(Cu_{0.56}Ti_{0.44})Ox$	8.1	505	140	550			580
$(Cu_{0.77}Ti_{0.23})Ox$	4.8	250	250	260			430
Cu_xO	2.9	-	255	-			430

No additional heating or electrical biasing of the substrate was used during the deposition. However, after the deposition process, (CuTi)Ox samples were additionally annealed in an air ambient at 473 K for 4 h. Thin films deposited on fused silica (SiO_2) were used for structural and optical investigations, while scanning electron microscopy and energy-dispersive X-ray spectroscopy were performed for thin films deposited on n-type silicon substrates. For electrical measurements, ceramic substrates with interdigitated platinum-gold electrodes were used. The thickness of the thin films was measured using the Taylor Hobson optical profilometer (Talysurf CCI Lite, Leicester, UK) and was in the range of 430 to 620 nm.

The structure of the thin films was investigated by X-ray diffraction in the grazing incidence mode (GIXRD) using an Empyrean X-ray diffractometer (PANalytical, Malvern, UK) with a PIXel3D detector. The diffraction pattern was collected in the 2θ range of 20 to 80° using Cu Kα radiation (λ = 1.5406 Å). By comparing the obtained pattern with PDF cards (Cu #04-0836 [39], Cu_2O #65-3288 [40], CuO #65-2309 [41], TiO #65-2900 [42], anatase #21-1272 [43]) a phase structure was determined. With the aid of MDI JADE 5.0 software, the average crystallite size was calculated using Scherrer's equation [44].

The morphology and elemental composition were investigated using the SEM/Ga-FIB FEI Helios NanoLab 600i scanning electron microscope (FEI, Hillsboro, OR, USA) equipped with the energy dispersive X-ray spectrometer (EDS). Surface SEM imaging was expanded to cross-sectional profile studies.

The optical measurement workstation consisted of Ocean Optics QE 65000 and NIR 256-2.1 spectrophotometers (Ocean Optics, Largo, FL, USA) and a coupled deuterium–halogen light source. The transmission coefficient was determined from transmittance spectra measured in the wavelength range of 250 ÷ 2000 nm. The average transmission was evaluated by calculating the integral in visible wavelength range of 380 to 760 nm and in near infrared wavelength range of 760 to 2000 nm, while the position of the fundamental absorption edge was designated by fitting the descending to zero linear part of the transmission characteristic.

For electrical characterization, a M100 Cascade Microtech probe station (Cascade Microtech, Beaverton, OR, USA) and a Keithley SCS4200 system were used. Current–voltage characteristics were measured in a shielded Faraday cage at room temperature. Measurements were performed for temperature changing from 303 K to 353 K that allowed us to determine the activation energy. Thermoelectric characteristics were obtained using the FLUKE 8846A voltmeter (Fluke, Everett, WA, USA) and the Instek mK1000 temperature controller (GW Instek, Taipei, Taiwan). For the determination of the Seebeck coefficient [45] a temperature gradient (ΔT) between 'hot' and 'cold' electrical contacts was applied in the range from 0 to 50 K, as the 'cold' contact was kept at room temperature. The Seebeck coefficient (S_c) was calculated according to Equation (1) and the conduction type was evaluated based on the sign of the obtained values:

$$S_c = \lim_{\Delta T \to \infty} \frac{\Delta U}{\Delta T} \qquad (1)$$

where S_c is the Seebeck coefficient, ΔU is the difference in thermoelectric voltage, and ΔT is the temperature gradient.

Furthermore, the gas sensing properties of annealed (CuTi)Ox thin films were investigated for 3.5% of H_2 diluted in Ar. Measurements were performed in a self-assembled system equipped with an Agilent 34,901A data acquisition system; Instec mK1000 temperature controller; and a mass flow controller. Before injection of H_2, the thin films were heated on a heating table to an operating temperature equal to 473 K and stabilized in an air environment for an hour.

3. Results

3.1. Microstructure and Morphology

Figure 1 presents results of X-ray microanalysis investigations for as-deposited thin oxide films of copper (Cu_xO), titanium (TiO_x) and of four oxide mixtures with various concentrations of copper and titanium ((CuTi)Ox). As one can see, in the energy range of 4 to 10 keV only TiK_α, TiK_β peaks were recognized in the spectrum attributed to the TiO_x thin film and only CuK_α and CuK_β peaks were found in the Cu_xO spectrum. Furthermore, as the copper concentration increased, the ratio of CuK_α to TiK_α increased simultaneously. The atomic content of copper in the thin films of oxide mixtures was estimated to be 23 at.%, 41 at.%, 56 at.%, and 77 at.%. The distribution maps of Cu, Ti, and O presented in Figure 1b show a homogeneous dispersion of each element in the thin films. A decisive factor in obtaining various elemental compositions in the deposited thin films was a proper regulation of the sputtering power supplying of each magnetron. Figure 2 presents the copper concentration in the thin film as a function of the ratio of the sputtering power at the Cu target (P_{Cu}) to the total power at the three targets ($P_{Cu} + P_{Ti}$). As expected, the increase in the $P_{Cu}/(P_{Cu} + P_{Ti})$ ratio resulted in a higher amount of copper in the thin film.

The morphology of the deposited thin films was investigated using scanning electron microscopy. The elemental composition strongly affects the morphology of the thin films, as can be deduced from the SEM images presented in Figure 3. The surface of the as-deposited TiO_x thin film was relatively smooth with some visible voids. The cross-sectional image revealed that it was composed of densely packed columnar grains. In contrast, the morphology of oxide mixtures with the 23, 41, and 56 at.% of copper differs significantly from TiO_x, as the columnar character changed to elongated and coarse grains. The surfaces were relatively flat and crack-free. Furthermore, single nanowires were formed at the top of $(Cu_{0.23}Ti_{0.77})Ox$. It is worth mentioning that when the copper content increased to 77 at.%, the morphology of oxides changed considerably. The grains were bulb-shaped, smaller at the bottom, and widen upward. Each bulb-shaped grain had a granular surface that resembles cauliflower. Such a grain shape resulted in the largest active surface of all of the as-deposited thin films, which is highly desirable in the case of photocatalytic activity or gas-sensing properties [23]. Compared to other thin films, $(Cu_{0.77}Ti_{0.23})Ox$ had the largest grains of ca. 250–300 nm and intergrain spaces (voids). The surface of copper oxide thin film was more homogeneous, the grains were still clearly visible, but the voids between them were smaller. Furthermore, the shape of the grains was also different—that is, they were elongated, granular, and more densely packed as compared to the oxide mixture with 77 at.% Cu.

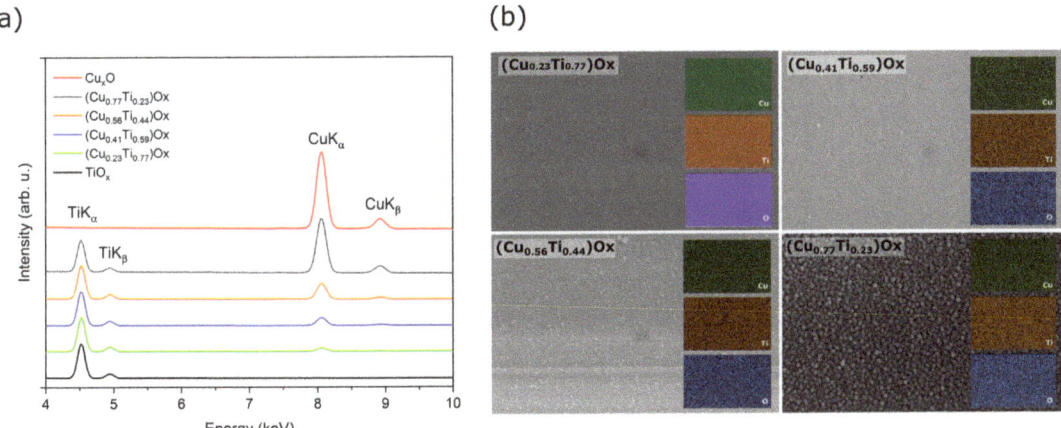

Figure 1. Results of X-ray microanalysis investigations for as-deposited copper and titanium oxide thin films: (**a**) EDS spectra and (**b**) maps of elemental distribution of mixed oxides.

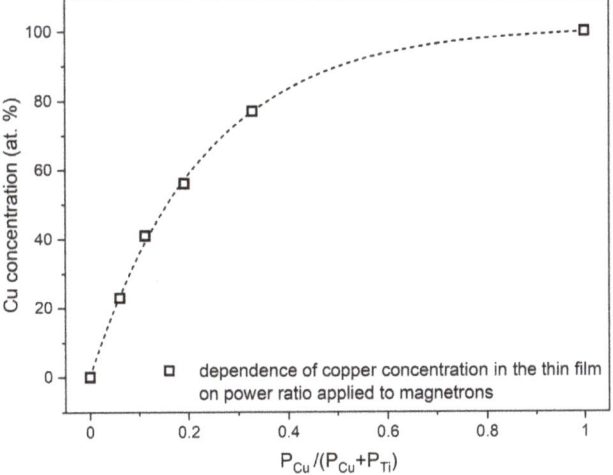

Figure 2. Dependence of copper content in as-deposited thin films on the power ratio $P_{Cu}/(P_{Cu} + P_{Ti})$ applied to magnetrons with Cu and Ti targets.

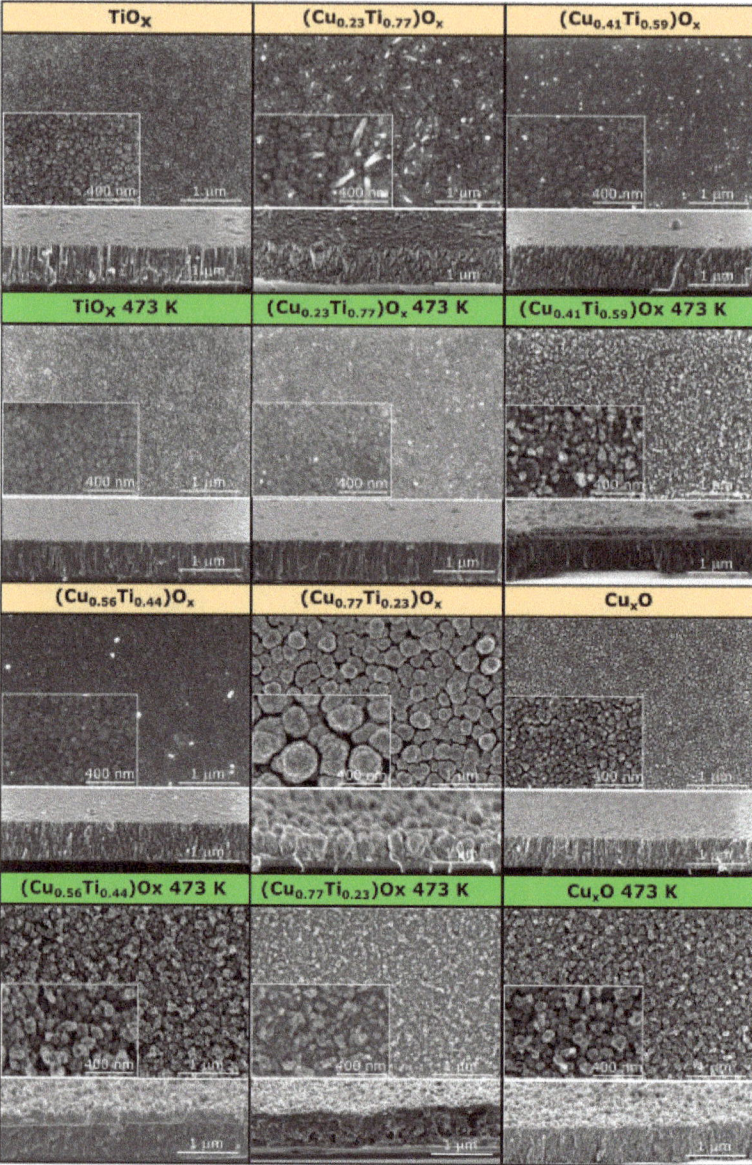

Figure 3. Scanning electron microscopy images of the surface and cross-section of as-deposited and annealed at 473 K TiO$_x$, (CuTi)Ox, and Cu$_x$O thin films.

After annealing at 473 K, no significant changes occurred on the surface of the TiO$_x$ thin film, but thermal oxidation considerably influenced the morphology of the oxide mixtures. It can be clearly seen from the cross-sectional images of the annealed thin films containing 23, 41, and 56 at.%, that at the bottom the elongated grains grow perpendicularly to the substrate surface, while close to the surface, they became grainy and rough. In addition, the grainy, sponge-shaped structure was thicker as the amount of copper in the thin film increased. The thin films were made up of grains smaller than 100 nm. The (Cu$_{0.77}$Ti$_{0.23}$)Ox was spongy in nature throughout its thickness. There were apparent interlayer voids of various shapes and sizes. For annealed Cu$_x$O thin films, the structure remains porous.

However, the vertical orientation of the grains was favorable. Furthermore, the voids were smaller and uniformly distributed.

XRD analysis was performed to determine the structural properties of the prepared thin films (Figure 4a). In the diffraction pattern of the TiO_x thin film, three characteristic peaks attributed to the (111), (200), and (220) planes of the TiO phase were observed, while no characteristic peaks of TiO_2 anatase or rutile were found. The size of the crystallites of the TiO phase was in the range from 4.9 to 11.8 nm, as calculated according to Scherrer's formula (Equation (2)):

$$D = \frac{k\lambda}{B \cos\theta} \quad (2)$$

where $k = 0.89$ is the shape factor, λ is the CuK_α X-ray radiation wavelength (1.5406 Å), B is the full width at half maximum intensity of the peak (FWHM), and θ is a diffraction angle.

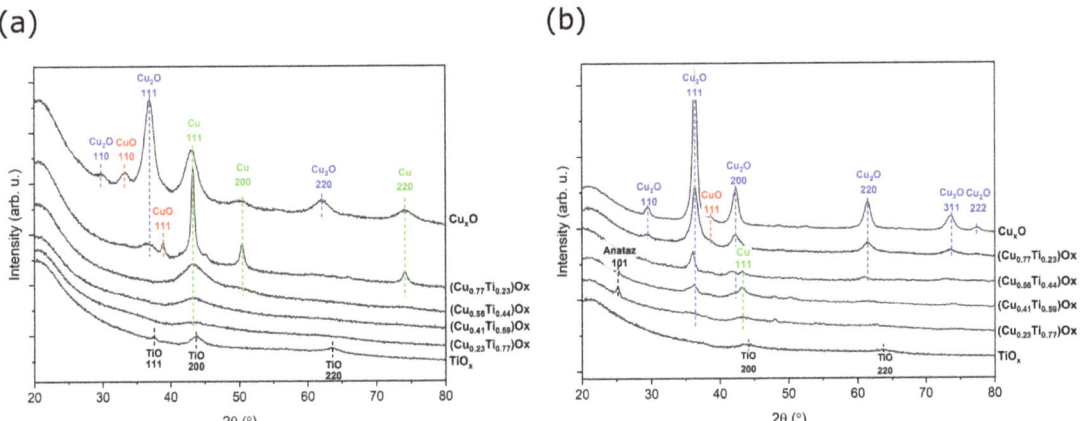

Figure 4. X-ray diffraction patterns of (**a**) as-deposited and (**b**) annealed at 473 K (CuTi)Ox thin films.

In each diffractogram of the copper and titanium oxide mixtures, a peak occurring at $2\theta = 42.7°$ is associated with the metallic copper (111) lattice plane. As the amount of copper in the thin film increased, the character of this peak changed significantly from very wide and poorly defined to sharp and intense. The crystallite size of the Cu (111) phase varied from ca. 3.0 to 13.4 nm. For $(Cu_{0.77}Ti_{0.23})Ox$ thin film, in addition to the most dominant Cu (111) peak, reflections from Cu (200), (220), Cu_2O (111), and CuO (111) were detected. The Cu_2O-cubic phase was characterized by the smallest crystallite size, which was equal to 5.1 nm, since the peak was broad, while metallic copper crystallites were equal to 13.4 nm. The copper oxide thin film consisted of the same metallic Cu peaks, but they are less intense, and the crystallite size was considerably smaller. In addition to the most intense peak of Cu_2O (111), there are also weaker reflections of Cu_2O (110) and Cu_2O (220) observed at $2\theta = 29.88°$ and $61.86°$, respectively. Furthermore, a peak centered at $33.1°$ was attributed to the crystalline phase of CuO (110). However, as the titanium concentration in the thin films of oxide mixtures was decreasing, the copper was more oxidized.

Post-process annealing at 473 K did not significantly affect the TiO_x thin film microstructure. The sample was still mostly amorphous in nature, but a TiO crystalline phase was observed. The most intense peak was attributed to the (200) crystal plane. Moreover, the crystallite size remained similar (ca. 5 nm) to that before annealing. Thermal oxidation of the thin films of oxide mixtures with a Ti content greater than 50% resulted in crystallization of the anatase. A strong peak of the anatase crystal plane (101) was found at $2\theta = 25.2°$, but when the amount of copper exceeded that of titanium, no anatase phase was observed. Similarly to the as-deposited thin films, the metallic copper reflections from the (111) plane was found in the diffractogram of the annealed thin films containing 23 at.%, 41 at.%, and 56 at.% of Cu; however, the intensity of the peaks decreased. Furthermore, characteristic Cu_2O peaks were found in addition to metallic copper, suggesting that annealing at 473 K causes oxidation of the metallic Cu to Cu_2O. It should be emphasised that in the $(Cu_{0.77}Ti_{0.23})Ox$ thin film, the phase transition was completed, as no Cu peaks were observed. The (111) crystal plane was the most preferable orientation for Cu_2O and was observed at $2\theta = 29.63°$ and its intensity increased with increasing amount of copper in the thin film. The diffraction pattern of the thin films with the highest copper concentration additionally consisted of Cu_2O peaks corresponding to the (110), (200), (220), (311), and (222) planes. All cuprous oxide crystallites were smaller than 13 nm. Moreover, a minor phase of CuO was also detected. The results of the X-ray investigations are summarized in Table 2.

Table 2. Summary x-ray diffraction structure analysis of as-deposited and annealed (CuTi)Ox thin films.

Thin Film	As-Deposited			Annealed at 473 K		
	Phase	hkl	D (nm)	Phase	hkl	D (nm)
TiO_x	TiO	111	11.8	TiO	200	5.4
	TiO	200	5.3	TiO	220	4.5
	TiO	220	4.9			
$(Cu_{0.23}Ti_{0.77})Ox$	Cu	111	3.1	TiO_2—anatase	101	19.1
				Cu_2O	111	5.4
				Cu	111	4.8
$(Cu_{0.41}Ti_{0.59})Ox$	Cu	111	2.9	TiO_2—anatase	101	26.5
				Cu_2O	111	8.4
				Cu	111	5.8
$(Cu_{0.56}Ti_{0.44})Ox$	Cu	111	3.1	Cu_2O	111	10.7
				Cu_2O	200	9.5
				Cu	111	21.4
$(Cu_{0.77}Ti_{0.23})Ox$	Cu_2O	111	5.1	Cu_2O	110	12.2
	CuO	111	20.0	Cu_2O	111	8.2
	Cu	111	13.4	CuO	111	17.3
	Cu	200	12.8	Cu_2O	200	7.2
	Cu	220	13.6	Cu_2O	220	7.7
				Cu_2O	311	6.3
Cu_xO	Cu_2O	110	8.5	Cu_2O	110	12.1
	CuO	110	9.0	Cu_2O	111	11.1
	Cu_2O	111	6.1	CuO	111	13.4
	Cu	111	4.6	Cu_2O	200	8.8
	Cu	200	5.3	Cu_2O	220	8.8
	Cu_2O	220	4.7	Cu_2O	311	8.5
	Cu	220	5.0	Cu_2O	222	10.8

Both copper and titanium strongly affect the crystal structure and the oxidation state in the oxide mixtures, as in as-deposited and annealed thin films. With decreasing amount of titanium, the phase transformation process from Cu to Cu_2O and finally to CuO is more advanced. This can be explained by the fact that Ti exhibits high affinity to oxygen and, consequently, a high Ti concentration limits the oxidation of copper [46–48]. In turn, after post-process treatment, the peaks from the anatase phase were observed only in the oxide mixtures with the lowest copper concentration (23 at.% and 41 at.%), confirming that the high concentration of copper hinders the crystallization of anatase or rutile [13,20,33,49].

3.2. Electrical and Optical Properties

The type of electrical conduction was determined by observation of the sign of the Seebeck coefficient (S_c) calculated from the thermoelectrical power measurements (Figure 5). A positive sign of the S_c indicates the hole type of conduction, while the negative value testifies about electron type of electrical conduction. The TiO_x thin film exhibited the strongest n-type properties of all as-deposited thin films and the Seebeck coefficient was equal to −25 μV/K. The S_c value gradually increased as the amount of copper in the thin film increased and the change in the type of conduction was observed when the copper concentration exceeded 56 at.%. The thin films with the highest copper content were characterized by weak p-type conduction. For the Cu_xO thin film, the Seebeck coefficient was equal to +6.1 μV/K. Post-process annealing significantly influenced the conduction type of thin films consisting of crystalline copper and copper oxides, as samples were characterized by a hole type of electrical conduction (Figure 5c). Furthermore, it should be noted that the studies of copper and titanium oxide mixtures annealed at 473 K have shown that the Seebeck coefficient of the $(Cu_{0.56}Ti_{0.44})Ox$ and $(Cu_{0.77}Ti_{0.23})Ox$ thin films was significantly higher compared to Cu_xO thin film for which the S_c coefficient was +245.1 μV/K. Copper oxides are well-known semiconductors with strong p-type properties [10], while TiO_2 is reported to be an n-type semiconductor [18]. When it comes to mixed copper-titanium oxides phase, Mor [15] et al. showed for Cu-Ti-O nanotubes fabricated by anodization of Cu-Ti films that the conduction type depends on the copper concentration. The sample with a Cu:Ti ratio equal to 24:76 was characterized by n-type conduction, while for ratios of 60:40 and 74:26, the samples were p-type semiconductors.

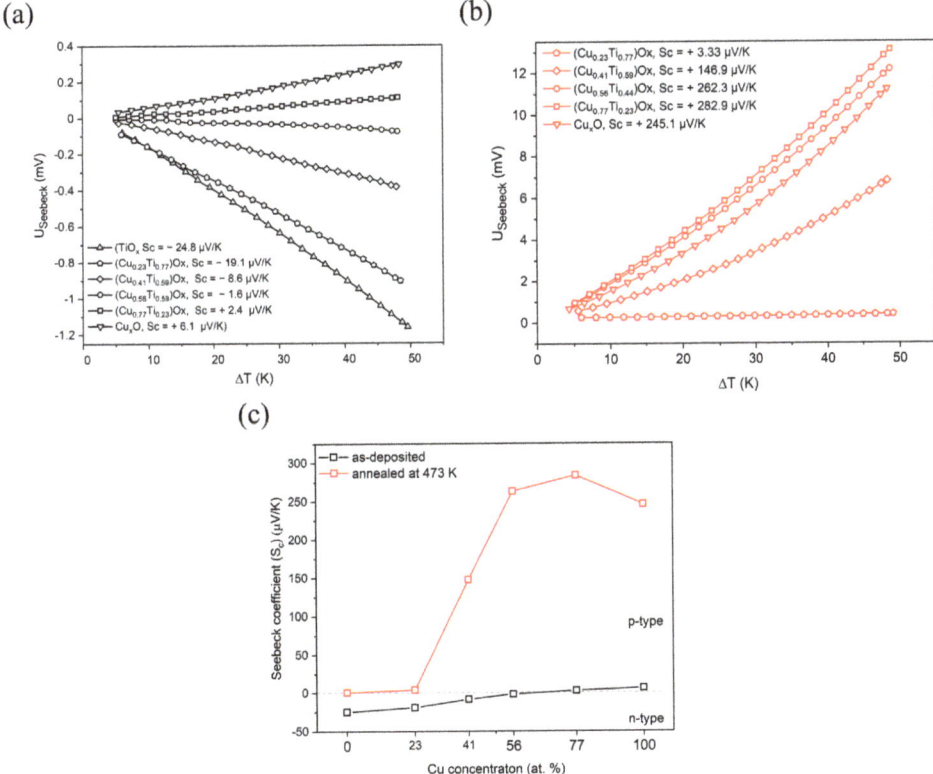

Figure 5. Thermoelectric characteristics of: (**a**) as-deposited, (**b**) annealed at 473 K (CuTi)Ox thin films, and (**c**) Seebeck coefficient changes depending on the elemental composition and annealing of the thin films.

Investigations of electrical properties were extended with measurements of the current–voltage characteristics, performed at room temperature and at elevated temperature (from 303 to 353 K). For all samples, the characteristics were linear and high repeatability was observed. Based on the Arrhenius formula (Equation (3)) and the slope of the fit of log (ρ) = f(1000/T) plot, the activation energy was calculated from

$$\rho = \rho_0 \exp\left(\frac{E_a}{kT}\right) \qquad (3)$$

where k is the Boltzmann constant and T is the temperature.

The obtained electrical parameters are presented in Table 3. With the increase in the copper concentration in the thin film, the resistivity of the (CuTi)Ox gradually decreased and for thin films with 56 and 77 at.% of Cu was equal to $2.5 \cdot 10^{-2}$ and $3.0 \cdot 10^{-3}$ Ω cm, respectively (Figure 6). Annealing at 473 K caused an increase in the resistivity by ca. three orders of magnitude for all thin films except (Cu$_{0.77}$Ti$_{0.23}$)Ox for which the resistivity increased almost four orders of magnitude. The resistivity of the TiO$_x$ sample after thermal treatment was 3.8×10^3 Ω cm. Titanium monoxide is generally perceived as an excellent conductor with a resistivity value in the range of 10^{-4}–10^{-3} Ω cm [50–53]. However, there are also reports showing that the resistivity value of nanocrystalline TiO can be as high as 10^5 Ω cm [54]. Therefore, the resistivity value obtained in this work falls within the wide range reported in the literature, while its quite high value may be explained by a large amount of the amorphous phase present in the thin film, as evidenced by the low intensity of the peaks visible in the XRD (Figure 4). The Cu$_x$O and (Cu$_{0.23}$Ti$_{0.77}$)Ox thin films after annealing had a resistivity equal to ca. 7×10^2 Ω cm. The addition of 41 at.% of Cu caused a decrease in ρ by an order of magnitude compared to the TiO$_x$ sample. A further increase in the Cu content resulted in a gradual decrease in resistivity, which was equal to 41.5 Ω cm and 28.1 Ω cm for (Cu$_{0.56}$Ti$_{0.44}$)Ox and (Cu$_{0.77}$Ti$_{0.23}$)Ox.

Table 3. Electrical parameters of as-deposited and annealed at 473 K (CuTi)Ox thin films.

Thin Films	Seebeck Coefficient (µV/K)		Resistivity (Ωcm)		E_a (eV/K)	
	As-Deposited	Annealed at 473 K	As-Deposited	Annealed at 473 K	As-Deposited	Annealed at 473 K
TiO$_x$	−24.8	-	4.3	$3.8 \cdot 10^3$	0.12	0.26
(Cu$_{0.23}$Ti$_{0.77}$)Ox	−19.1	+3.3	$3.1 \cdot 10^{-1}$	$7.4 \cdot 10^2$	0.10	0.32
(Cu$_{0.41}$Ti$_{0.59}$)Ox	−8.6	+146.9	$1.1 \cdot 10^{-1}$	$3.5 \cdot 10^2$	0.04	0.26
(Cu$_{0.56}$Ti$_{0.44}$)Ox	−1.6	+262.3	$2.5 \cdot 10^{-2}$	$4.2 \cdot 10^1$	0.00	0.26
(Cu$_{0.77}$Ti$_{0.23}$)Ox	+2.4	+282.9	$3.0 \cdot 10^{-3}$	$2.8 \cdot 10^1$	0.00	0.29
Cu$_x$O	+6.1	+249.1	$6.0 \cdot 10^{-1}$	$7.1 \cdot 10^2$	0.01	0.35

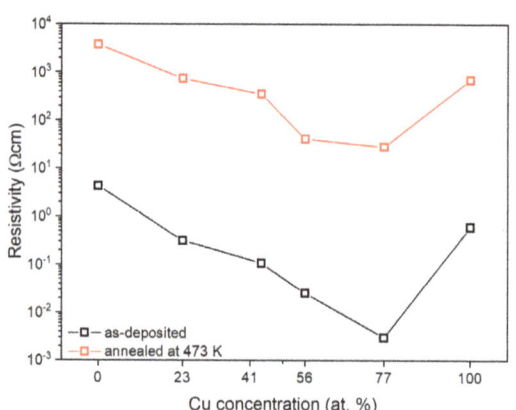

Figure 6. Resistivity changes depending on elemental composition and annealing temperature.

Figure 7 presents the dependence of the resistivity vs. the temperature in the range from 303 K to 353 K. For the thin films with the lowest copper concentration, the resistivity gradually decreased with increasing temperature, while remaining constant when the Cu concentration in the film exceeded 50%. Almost no change in resistivity as a function of temperature indicated the very low activation energy of electrical conduction processes that might be connected with the presence of metallic Cu species in those thin films.

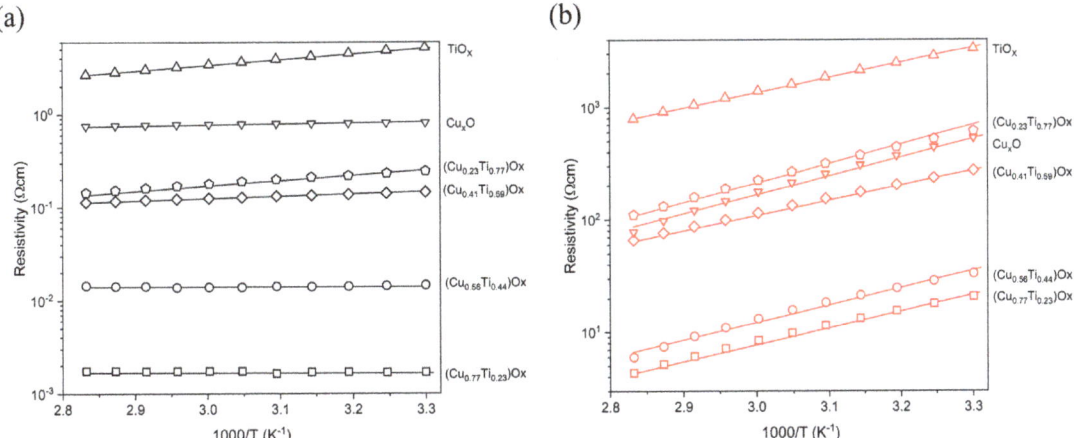

Figure 7. Temperature dependent resistivity of (**a**) as-deposited and (**b**) annealed at 473 K thin films.

In each case of annealed thin films, the resistivity was found to decrease with increasing temperature. The activation energy was in the range from 0.26 to 0.35 eV/K indicating the semiconducting nature of the (CuTi)Ox thin films after thermal oxidation.

The optical properties of the (CuTi)Ox thin films were determined on the basis of the transmission measurements of light in the wavelength range from 250 nm to 2000 nm. All as-deposited thin films were opaque, which could be the consequence of the low oxygen concentration during the sputtering process [55], resulting in the presence of metallic copper crystallites in the Cu_xO and in (CuTi)Ox thin films. In the visible wavelength range, the transmission of the as-deposited thin films was less than 2%. As can be seen in Figure 8, the post-process thermal treatment strongly influenced the optical properties of the TiO_x, $(Cu_{0.77}Ti_{0.23})Ox$ and Cu_xO thin films as they oxidized and became transparent, mainly in the infrared region. Moreover, interference fringes began to appear. However, the samples with 23, 41, and 56 at.% of Cu remained opaque. The absorption edge of the TiO_x thin film was positioned at the cut-off wavelength ($\lambda_{cut-off}$) equal to 339 nm, while a redshift to ca. 490 nm was observed for the $(Cu_{0.77}Ti_{0.23})Ox$ and pure copper oxide films. This observation is consistent with the data reported for $\lambda_{cut-off}$ for non-stoichiometric TiO_x and Cu_xO thin films [56,57]. The average transmittance of TiO_x in the visible wavelength range was equal to 11%, while for the $(Cu_{0.77}Ti_{0.23})Ox$ and Cu_xO thin films it was equal to 22% and 27%, respectively. In turn, in the near infrared wavelength range the average transmittance increased slightly for the TiO_x and was equal to 21%, but for the $(Cu_{0.77}Ti_{0.23})Ox$ and Cu_xO thin films it was even ca. 70%.

3.3. Hydrogen Gas Sensing

Figure 9 presents the dynamic sensor response of the annealed titanium oxide, oxide mixtures, and copper oxide thin films exposed to 3.5% hydrogen diluted in argon at an operating temperature of 473 K. Titanium oxide showed the electron conduction type as its resistance decreased in the presence of the reducing gas (H_2). The opposite effect was observed for the (CuTi)Ox and Cu_xO, indicating that holes were the dominant electrical charge carriers. Therefore, the conduction type designated on the basis of the gas sensing characteristics was consistent with the thermoelectric measurements. The sensing mechanism in metal oxide semiconductors is a complex issue as it depends not only on the material composition and conduction type, but also on their morphology, grain size, gas concentration, operating temperature, and humidity condition [58]. In general, in the case of metal oxide structures, the dominant role is in adsorption and desorption processes on the surface of the thin film. In air, oxygen molecules are adsorbed at the surface of sensing material in the form of O_2^-, O^-, or O^{2-} depending on working temperature. Temperatures above 423 K dominate the adsorption of O^- or O^{2-} species [59]. The oxygen adsorption might follow as in Equations (4) and (5):

$$O_2 \text{ (gas)} + e^- \leftrightarrow O_2^- \quad (4)$$

$$O_2 \text{ (gas)} + 2e^- \leftrightarrow 2O^- \quad (5)$$

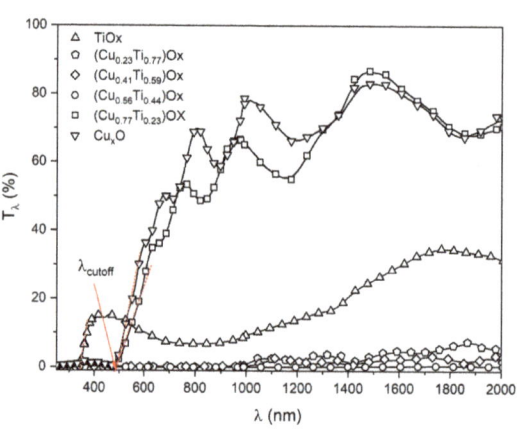

Figure 8. Transmission characteristic of thin films annealed at 473 K.

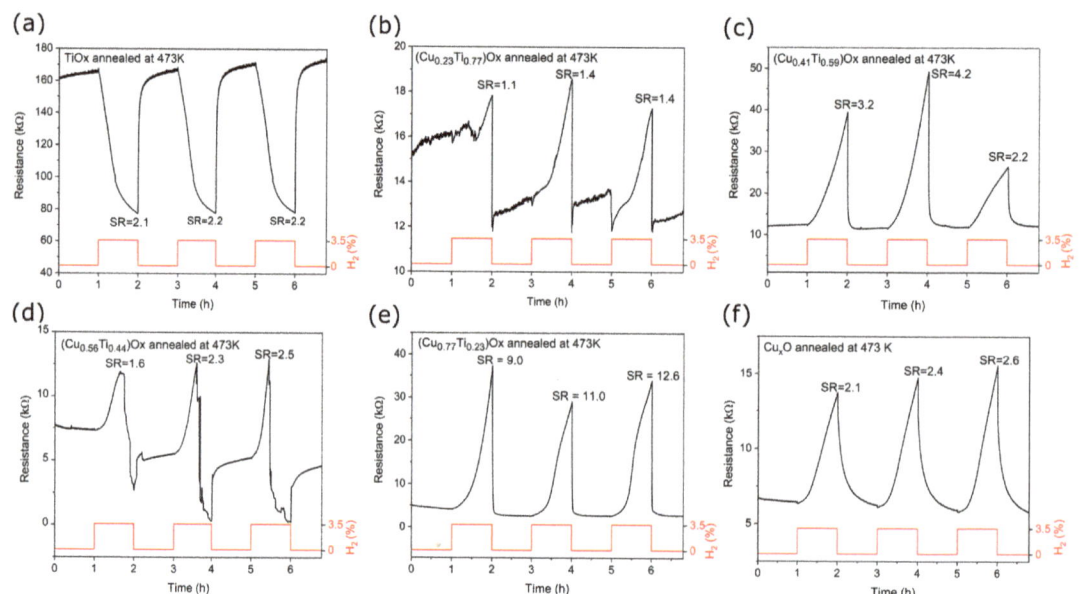

Figure 9. Resistance changes of: (**a**) TiO$_x$, (**b**–**e**) (CuTi)O$_x$, and (**f**) Cu$_x$O thin films annealed at 473 K upon exposure to 3.5% H$_2$.

Chemisorbed oxygen extracts electrons from the surface, forming an electron depletion layer. For n-type semiconductors, the resistance of the depleted layer is higher than that of the bulk material, causing the current to follow through the bulk grain. However, for p-type materials, the opposite phenomenon is observed: the hole accumulation area is less resistive [6]. The introduction of hydrogen (a reducing gas) causes a reaction with chemisorbed oxygen according to the equation [60]

$$H_2 \text{ (gas)} + O^- \leftrightarrow H_2O(\text{gas}) + e^- \quad (6)$$

As a result of the interaction of hydrogen with metal oxide, electrons are donated back to the conduction band, which occurs as an increase in the resistance for p-type materials and a decrease in the resistance for n-type materials.

The sensor response (SR) was calculated as the ratio of the resistance of the thin film in the air (R_{air}) to the resistance in the presence of hydrogen (R_{H2}) for the TiO_x thin film [61] and as the ratio of R_{H2} to R_{air} for the other thin films. Furthermore, the response and recovery times defined as the time required to reach 90% of the difference between R_{air} and R_{H2} were determined and summarised in Table 4.

Table 4. Summary of sensing parameters of TiO_x, Cu_xO, and $(CuTi)Ox$ thin films annealed at 473 K.

Thin Film	Operating Temperature	Concentration of Target Gas	SR	$t_{response}$ (min)	$t_{recovery}$ (s)
TiO_x			2.2 ± 0.1	36 ± 1	322 ± 31
$(Cu_{0.23}Ti_{0.77})Ox$			1.3 ± 0.1	56 ± 1	-
$(Cu_{0.41}Ti_{0.59})Ox$	473 K	3.5% H_2	3.8 ± 1.1	56 ± 2	187 ± 146
$(Cu_{0.56}Ti_{0.44})Ox$			2.1 ± 0.4	-	-
$(Cu_{0.77}Ti_{0.23})Ox$			10.9 ± 1.5	56 ± 1	67 ± 14
Cu_xO			2.4 ± 0.2	55 ± 1	1444 ± 79

The SR value for TiO_x and Cu_xO was similar and equal to 2.2 and 2.4, respectively. As shown in Figure 9a, the titanium oxide response began to saturate after ca. 30 min of hydrogen flow and after ca. 5 min of air flow. The consequence of this was the short response and recovery time compared to those of Cu_xO, whose signal was far from saturation. Metal oxides that exhibit the p-type of electrical conduction typically respond worse than their n-type counterparts [6,62], however, the slightly better Cu_xO response may be caused by a considerable difference in the morphology and structure of the two thin films. As shown in the SEM images, titanium oxide annealed at 473 K had a dense and flat surface morphology, while copper oxide after the same thermal treatment was porous, resulting in a larger active surface area. In addition, the copper oxide crystallites were well-defined, whereas the titanium oxide thin film was mostly amorphous.

As can be concluded, the preparation of a composite of metal oxides can noticeably improve sensing capabilities—e.g., sensitivity and selectivity [58]—compared to a sensing material consisting of one of the components. In the case of heterostructures, the sensing response is determined by changes in the width of the charge depletion layer that is formed when the Fermi levels of the two metal oxides equilibrate and charge transfer occurs [63]. The undoubted advantage of oxide mixture thin films over double-layer structures is the better interaction of the reducing/oxidizing gas with p–n areas, since both the p-type and the n-type oxides are present at the surface simultaneously. This leads to a stronger manipulation of the p–n junction performance. It should be noted that in mixed composite structures not only the elemental composition of the thin film but also the dispersion state of the component are important [58].

In this work, a significant improvement in the gas detection response was observed for the $(Cu_{0.77}Ti_{0.23})Ox$ and $(Cu_{0.41}Ti_{0.59})Ox$ thin films compared to TiO_x and Cu_xO. Sensor response of the thin film containing 77 at.% of Cu was more than 4.5 times higher than that of Cu_xO. The thin film with 23 at.% of Cu showed the worst H_2 detection abilities, while in the case of $(Cu_{0.56}Ti_{0.44})Ox$, the unique behavior was observed. As shown in Figure 9d under exposure to 3.5% H_2, the resistance of the $(Cu_{0.56}Ti_{0.44})Ox$ thin film initially increased, but after a specific time decreased. Kosc et al. [64] reported a similar inversion of the response type in TiO_2 modified with NiO structures in the presence of hydrogen. Inversion was observed when the hydrogen concentration exceeded a specific critical point. This unique behavior was explained by the fact that, when the amount of electrons from broken bonds of adsorbed oxygen exceeds the amount of holes in the Debaye length, the conduction type of heterostructure changes. In the ambient air atmosphere, the oxygen is re-adsorbed on the film surface causing the resistance to increase and reach the initial base value [58,64].

The resistance of the (CuTi)Ox thin films was constantly increasing throughout the hydrogen flow, and no signal plateau was observed, resulting in the response time exceeding 55 min. However, recovery times were relatively short. The best sensing performance to 3.5% H_2 showed a thin film with 77 at.% of Cu as the sensor response was equal to 10.9 and the recovery time was the shortest of all measured thin films.

Comparing the sensing parameters of the $(Cu_{0.77}Ti_{0.23})Ox$ thin films to hydrogen with previously published works presented in Table 5, it can be seen that the sensor response is the highest among all the works reported so far. Similarly to this work, Park et al. [48] reported that the response time was shortened when TiO_2 was added to the CuO nanorods. Furthermore, the recovery time increased after titanium dioxide modification and was in the range from 300 to 400 s. In turn, Alev et al. [63] reported that the response of TiO_2 nanotubes covered with a CuO layer to 1000 ppm of H_2 saturates in 15 min and the recovery time was equal to 10 min. The operating temperature applied in all the measurements of the response of copper and titanium oxide composites to hydrogen was in the range of 473 K to 673 K. Park et al. [48] and Alev [63] et al. presented findings that the optimal working temperature for their structures was 473 K, while for other structures higher temperatures were more desirable. In all articles summarized in Table 5, cupric oxide was the dominant phase of copper oxides. Our aim was to show that oxide mixtures of Cu_2O and TiO_2 can exhibit a good response to H_2, while higher working temperature could cause oxidation to CuO, thereby hindering the SR value.

Table 5. Gas sensing performance of copper oxide and titanium oxide composites toward hydrogen.

Structure	Preparation Method	H_2 Concentration (ppm)	Working Temperature (K)	SR	Ref.
CuO-TiO_2 nanocomposites	Chemical vapor deposition	1000	473–673	2–0.4	[23]
TiO_2 decorated CuO nanorods	Thermal oxidation and solvothermal method	0.1–5	573	1.94–8.57	[48]
CuO/TiO_{2-y} heterostructure	MF magnetron sputtering	10–320	573–673	1.0–1.5	[61]
CuO thin film/TiO_2 nanotubes	Electrochemical anodization, thermal evaporation, oxidation	100–1000	373–473	2.0	[63]
TiO_2/Cu_xO	Spray pyrolysis, magnetron sputtering, annealing	100	523–623	0.2–0.6	[65]
Li-doped CuO-TiO_2 heterostructure	Ion beam sputtering, annealing	1000	573	1.16	[22]
TiO_2-CuO (50%–50% and (25%–75%)	Ion beam sputtering, annealing	100	573	0.98	[21]
mixed (CuTi)Ox	Magnetron sputtering	35,000	473	1.3–10.9	This work

4. Summary

The purpose of this study was to investigate the selected properties of thin films of copper and titanium oxide mixtures. In this regard, the (CuTi)Ox thin films with various Cu concentrations were prepared using magnetron co-sputtering. In addition, post-process annealing at 473 K was applied to additionally oxidize deposited thin films. The as-deposited thin films had dense and well-packed structures. The copper concentration was equal to 23, 41, 56, and 77 at.%. XRD studies showed that the as-deposited oxide mixtures consisted of metallic copper and copper oxides crystallites. Post-process annealing resulted in a change in morphology to partially porous structures and caused crystallization of anatase and partial oxidation of copper to Cu_2O or CuO. All annealed (CuTi)Ox thin films exhibited strong p-type semiconducting properties. Moreover, stronger hole-type conductivity was obtained in the (CuTi)Ox composite material than in the Cu_xO thin film. The $(Cu_{0.77}Ti_{0.23})Ox$ annealed at 473 K exhibited the best optoelectronic properties, as it was semitransparent in the visible radiation range, exhibited the strongest p-type conductivity and had the lowest resistivity of all deposited samples. Moreover, all the thin films responded to 3.5% hydrogen. These responses were characterized by long response times and very short recovery times. Again, the $(Cu_{0.77}Ti_{0.23})Ox$ sample exhibited the highest response (10.9 times).

As has been shown in this work, the hydrogen sensing properties of (CuTi)Ox thin films are promising, nevertheless, there is much to be done in the future works, e.g., further investigation of gas sensing properties to hydrogen of concentrations smaller than 1000 ppm and determination of chemical state of the surface are planned. The authors' aim is to establish the mechanism responsible for the change in resistance under the influence of hydrogen. For that purpose, in-situ XPS measurements without and under the influence of hydrogen are scheduled to determine changes in oxidation states on the surface.

Author Contributions: Conceptualization, E.M. and M.M.; Methodology, E.M. and M.M.; Validation, E.M., M.M. and J.D.; Formal analysis, E.M.; Investigation, E.M., M.M., J.D. and D.W.; Resources, M.M.; Data curation, E.M.; Writing—original draft preparation, E.M., M.M. and J.D.; Writing—review and editing, E.M. and M.M.; Visualization, E.M.; Supervision, M.M.; Project administration, M.M.; Funding acquisition, M.M. All authors have read and agreed to the published version of the manuscript.

Funding: This work was co-financed from the sources given by the Polish National Science Center (NCN) as a research project number UMO-2018/29/B/ST8/00548 and UMO-2020/39/D/ST5/00424.

Institutional Review Board Statement: Not applicable.

Informed Consent Statement: Not applicable.

Data Availability Statement: Not applicable.

Acknowledgments: Authors would like to thank Danuta Kaczmarek for discussion on the investigation results.

Conflicts of Interest: The authors declare no conflict of interest.

References

1. Wojcieszak, D.; Domaradzki, J.; Mazur, M.; Kotwica, T.; Kaczmarek, D. Investigation of a Memory Effect in a Au/(Ti–Cu)Ox-Gradient Thin Film/TiAlV Structure. *Beilstein J. Nanotechnol.* **2022**, *13*, 265–273. [CrossRef] [PubMed]
2. Wojcieszak, D.; Mazur, M.; Kaczmarek, D.; Poniedziałek, A.; Osekowska, M. An Impact of the Copper Additive on Photocatalytic and Bactericidal Properties of TiO_2 Thin Films. *Mater. Sci. Pol.* **2017**, *35*, 421–426. [CrossRef]
3. Domaradzki, J. Perspectives of Development of TCO and TOS Thin Films Based on (Ti-Cu)Oxide Composites. *Surf. Coatings Technol.* **2016**, *290*, 28–33. [CrossRef]
4. Nakata, K.; Fujishima, A. TiO_2 Photocatalysis: Design and Applications. *J. Photochem. Photobiol. C Photochem. Rev.* **2012**, *13*, 169–189. [CrossRef]
5. Livage, J.; Ganguli, D. Sol-Gel Electrochromic Coatings and Devices: A Review. *Sol. Energy Mater. Sol. Cells* **2001**, *68*, 365–381. [CrossRef]
6. Maziarz, W. TiO_2/SnO_2 and TiO_2/CuO Thin Film Nano-Heterostructures as Gas Sensors. *Appl. Surf. Sci.* **2019**, *480*, 361–370. [CrossRef]
7. Nunes, D.; Pimentel, A.; Goncalves, A.; Pereira, S.; Branquinho, R.; Barquinha, P.; Fortunato, E.; Martins, R. Metal Oxide Nanostructures for Sensor Applications. *Semicond. Sci. Technol.* **2019**, *34*, 043001. [CrossRef]
8. Wojcieszak, D.; Obstarczyk, A.; Mańkowska, E.; Mazur, M.; Kaczmarek, D.; Zakrzewska, K.; Mazur, P.; Domaradzki, J. Thermal Oxidation Impact on the Optoelectronic and Hydrogen Sensing Properties of P-Type Copper Oxide Thin Films. *Mater. Res. Bull.* **2022**, *147*, 111646. [CrossRef]

9. Wiatrowski, A.; Wojcieszak, D.; Mazur, M.; Kaczmarek, D.; Domaradzki, J.; Kalisz, M.; Kijaszek, W.; Pokora, P.; Mańkowska, E.; Lubanska, A.; et al. Photocatalytic Coatings Based on TiO$_x$ for Application on Flexible Glass for Photovoltaic Panels. *J. Mater. Eng. Perform.* **2022**, *31*, 6998–7008. [CrossRef]
10. Zoolfakar, A.S.; Rani, R.A.; Morfa, A.J.; O'Mullane, A.P.; Kalantar-Zadeh, K. Nanostructured Copper Oxide Semiconductors: A Perspective on Materials, Synthesis Methods and Applications. *J. Mater. Chem. C* **2014**, *2*, 5247–5270. [CrossRef]
11. Dahl, M.; Liu, Y.; Yin, Y. Composite Titanium Dioxide Nanomaterials. *Chem. Rev.* **2014**, *114*, 9853–9889. [CrossRef]
12. Rydosz, A. The Use of Copper Oxide Thin Films in Gas-Sensing Applications. *Coatings* **2018**, *8*, 425. [CrossRef]
13. Barreca, D.; Battiston, G.A.; Casellato, U.; Gerbasi, R.; Tondello, E. Low Pressure CVD of Transparent Cu-Al-O and Cu-Ti-O Thin Films. *J. Phys. IV* **2001**, *11*, Pr11-253. [CrossRef]
14. Wang, Z.; Nayak, P.K.; Caraveo-Frescas, J.A.; Alshareef, H.N. Recent Developments in P-Type Oxide Semiconductor Materials and Devices. *Adv. Mater.* **2016**, *28*, 3831–3892. [CrossRef]
15. Mor, G.K.; Varghese, O.K.; Wilke, R.H.T.; Sharma, S.; Shankar, K.; Latempa, T.J.; Choi, K.S.; Grimes, C.A. P-Type Cu-Ti-O Nanotube Arrays and Their Use in Self-Biased Heterojunction Photoelectrochemical Diodes for Hydrogen Generation. *Nano Lett.* **2008**, *8*, 1906–1911. [CrossRef]
16. Fortunato, E.; Figueiredo, V.; Barquinha, P.; Elamurugu, E.; Barros, R.; Gonçalves, G.; Park, S.H.K.; Hwang, C.S.; Martins, R. Thin-Film Transistors Based on p-Type Cu$_2$O Thin Films Produced at Room Temperature. *Appl. Phys. Lett.* **2010**, *96*, 2–5. [CrossRef]
17. Mazur, M.; Wojcieszak, D.; Domaradzki, J.; Kaczmarek, D.; Song, S.; Placido, F. TiO$_2$/SiO$_2$ Multilayer as an Antireflective and Protective Coating Deposited by Microwave Assisted Magnetron Sputtering. *Opto-Electron. Rev.* **2013**, *21*, 233–238. [CrossRef]
18. Nair, P.B.; Justinvictor, V.B.; Daniel, G.P.; Joy, K.; Ramakrishnan, V.; Thomas, P.V. Effect of RF Power and Sputtering Pressure on the Structural and Optical Properties of TiO$_2$ Thin Films Prepared by RF Magnetron Sputtering. *Appl. Surf. Sci.* **2011**, *257*, 10869–10875. [CrossRef]
19. Domaradzki, J.; Wiatrowski, A.; Kotwica, T.; Mazur, M. Analysis of Electrical Properties of Forward-to-Open (Ti,Cu)O$_x$ Memristor Rectifier with Elemental Gradient Distribution Prepared Using (Multi)Magnetron Co-Sputtering Process. *Mater. Sci. Semicond. Process.* **2019**, *94*, 9–14. [CrossRef]
20. Zhang, W.; Li, Y.; Zhu, S.; Wang, F. Copper Doping in Titanium Oxide Catalyst Film Prepared by Dc Reactive Magnetron Sputtering. *Catal. Today* **2004**, *93–95*, 589–594. [CrossRef]
21. Torrisi, A.; Horák, P.; Vacík, J.; Cannavò, A.; Ceccio, G.; Yatskiv, R.; Kupcik, J.; Fara, J.; Fitl, P.; Vlcek, J.; et al. Preparation of Heterogenous Copper-Titanium Oxides for Chemiresistor Applications. *Mater. Today Proc.* **2019**, *33*, 2512–2516. [CrossRef]
22. Torrisi, A.; Ceccio, G.; Cannav, A.; Lavrentiev, V.; Hor, P.; Yatskiv, R.; Vaniš, J.; Grym, J.; Fišer, L.; Hruška, M. Chemiresistors Based on Li-Doped CuO-TiO$_2$ Films. *Chemosensors* **2021**, *9*, 246. [CrossRef]
23. Barreca, D.; Carraro, G.; Comini, E.; Gasparotto, A.; MacCato, C.; Sada, C.; Sberveglieri, G.; Tondello, E. Novel Synthesis and Gas Sensing Performances of CuO-TiO$_2$ Nanocomposites Functionalized with Au Nanoparticles. *J. Phys. Chem. C* **2011**, *115*, 10510–10517. [CrossRef]
24. Kubiak, A.; Bielan, Z.; Kubacka, M.; Gabała, E.; Zgoła-Grześkowiak, A.; Janczarek, M.; Zalas, M.; Zielińska-Jurek, A.; Siwińska-Ciesielczyk, K.; Jesionowski, T. Microwave-Assisted Synthesis of a TiO$_2$-CuO Heterojunction with Enhanced Photocatalytic Activity against Tetracycline. *Appl. Surf. Sci.* **2020**, *520*, 146344. [CrossRef]
25. Ma, Q.; Liu, S.J.; Weng, L.Q.; Liu, Y.; Liu, B. Growth, Structure and Photocatalytic Properties of Hierarchical Cu-Ti-O Nanotube Arrays by Anodization. *J. Alloys Compd.* **2010**, *501*, 333–338. [CrossRef]
26. Zhao, Z.; Sun, J.; Zhang, G.; Bai, L. The Study of Microstructure, Optical and Photocatalytic Properties of Nanoparticles(NPs)-Cu/TiO$_2$ Films Deposited by Magnetron Sputtering. *J. Alloys Compd.* **2015**, *652*, 307–312. [CrossRef]
27. Fouad, S.S.; Baradács, E.; Nabil, M.; Parditka, B.; Negm, S.; Erdélyi, Z. Microstructural and Optical Duality of TiO$_2$/Cu/TiO$_2$ Trilayer Films Grown by Atomic Layer Deposition and DC Magnetron Sputtering. *Inorg. Chem. Commun.* **2022**, *145*, 110017. [CrossRef]
28. Mekasuwandumrong, O.; Jantarasorn, N.; Panpranot, J.; Ratova, M.; Kelly, P.; Praserthdam, P. Synthesis of Cu/TiO$_2$ Catalysts by Reactive Magnetron Sputtering Deposition and Its Application for Photocatalytic Reduction of CO$_2$ and H$_2$O to CH$_4$. *Ceram. Int.* **2019**, *45*, 22961–22971. [CrossRef]
29. Zong, M.; Bai, L.; Liu, Y.; Wang, X.; Zhang, X.; Huang, X.; Hang, R.; Tang, B. Antibacterial Ability and Angiogenic Activity of Cu-Ti-O Nanotube Arrays. *Mater. Sci. Eng. C* **2017**, *71*, 93–99. [CrossRef]
30. Pai, M.R.; Banerjee, A.M.; Rawool, S.A.; Singhal, A.; Nayak, C.; Ehrman, S.H.; Tripathi, A.K.; Bharadwaj, S.R. A Comprehensive Study on Sunlight Driven Photocatalytic Hydrogen Generation Using Low Cost Nanocrystalline Cu-Ti Oxides. *Sol. Energy Mater. Sol. Cells* **2016**, *154*, 104–120. [CrossRef]
31. Manjunath, K.; Souza, V.S.; Ramakrishnappa, T.; Nagaraju, G.; Scholten, J.D.; Dupont, J. Heterojunction CuO-TiO$_2$ Nanocomposite Synthesis for Significant Photocatalytic Hydrogen Production. *Mater. Res. Express* **2016**, *3*, 115704. [CrossRef]
32. Wang, Z.; Liu, Y.; Martin, D.J.; Wang, W.; Tang, J.; Huang, W. CuO$_x$-TiO$_2$ Junction: What Is the Active Component for Photocatalytic H$_2$ Production? *Phys. Chem. Chem. Phys.* **2013**, *15*, 14956–14960. [CrossRef] [PubMed]
33. Sreedhar, M.; Reddy, I.N.; Bera, P.; Ramachandran, D.; Gobi Saravanan, K.; Rabel, A.M.; Anandan, C.; Kuppusami, P.; Brijitta, J. Cu/TiO$_2$ Thin Films Prepared by Reactive RF Magnetron Sputtering. *Appl. Phys. A Mater. Sci. Process.* **2015**, *120*, 765–773. [CrossRef]

34. Mazur, M.; Domaradzki, J.; Wojcieszak, D.; Kaczmarek, D. Investigations of Elemental Composition and Structure Evolution in (Ti,Cu)-Oxide Gradient Thin Films Prepared Using (Multi)Magnetron Co-Sputtering. *Surf. Coatings Technol.* **2018**, *334*, 150–157. [CrossRef]
35. Kotwica, T.; Domaradzki, J.; Wojcieszak, D.; Sikora, A.; Kot, M.; Schmeisser, D. Analysis of Surface Properties of Ti-Cu-Ox Gradient Thin Films Using AFM and XPS Investigations. *Mater. Sci. Pol.* **2018**, *36*, 761–768. [CrossRef]
36. Torrisi, A.; Horák, P.; Vacík, J.; Cannavò, A.; Ceccio, G.; Vaniš, J.; Yatskiv, R.; Grym, J. Multilayered Cu–Ti Deposition on Silicon Substrates for Chemiresistor Applications. *Sulfur Silicon Relat. Elem.* **2020**, *195*, 932–935. [CrossRef]
37. Barreca, D.; Carraro, G.; Gasparotto, A.; MacCato, C.; Cruz-Yusta, M.; Gómez-Camer, J.L.; Morales, J.; Sada, C.; Sánchez, L. On the Performances of CuxO-TiO$_2$ (x = 1, 2) Nanomaterials as Innovative Anodes for Thin Film Lithium Batteries. *ACS Appl. Mater. Interfaces* **2012**, *4*, 3610–3619. [CrossRef]
38. Horzum, S.; Gürakar, S.; Serin, T. Investigation of the Structural and Optical Properties of Copper-Titanium Oxide Thin Films Produced by Changing the Amount of Copper. *Thin Solid Films* **2019**, *685*, 293–298. [CrossRef]
39. PDF Card 04-0836; Powder Diffraction File, Joint Commiittee on Powder Diffraction Standards. ASTM: Philadelphia, PA, USA.
40. PDF Card 65-3288; Powder Diffraction File, Joint Commiittee on Powder Diffraction Standards. ASTM: Philadelphia, PA, USA.
41. PDF Card 65-2309; Powder Diffraction File, Joint Commiittee on Powder Diffraction Standards. ASTM: Philadelphia, PA, USA.
42. PDF Card 65-2900; Powder Diffraction File, Joint Commiittee on Powder Diffraction Standards. ASTM: Philadelphia, PA, USA.
43. PDF Card 21-1272; Powder Diffraction File, Joint Commiittee on Powder Diffraction Standards. ASTM: Philadelphia, PA, USA.
44. Holzwarth, U.; Gibson, N. The Scherrer Equation versus the "Debye-Scherrer Equation". *Nat. Nanotechnol.* **2011**, *6*, 534. [CrossRef]
45. Goto, T.; Li, J.H.; Hirai, T.; Maeda, Y.; Kato, R.; Maesono, A. Measurements of the Seebeck Coefficient of Thermoelectric Materials by an AC Method. *Int. J. Thermophys.* **1997**, *18*, 569–577. [CrossRef]
46. Kelkar, G.P.; Carim, A.H. Synthesis, Properties, and Ternary Phase Stability of M6X Compounds in the Ti—Cu—O System. *J. Am. Ceram. Soc.* **1993**, *76*, 1815–1820. [CrossRef]
47. Hickman, J.; Gulbransen, E. Oxide Films Formed on Titanium, Zirconia, and Their Alloys with Nickel, Copper, and Cobalt. *Anal. Chem.* **1948**, *20*, 158–165. [CrossRef]
48. Park, S.; Kim, S.; Kheel, H.; Park, S.E.; Lee, C. Synthesis and Hydrogen Gas Sensing Properties of TiO$_2$-Decorated CuO Nanorods. *Bull. Korean Chem. Soc.* **2015**, *36*, 2458–2463. [CrossRef]
49. Wang, H.; Li, Y.; Ba, X.; Huang, L.; Yu, Y. TiO$_2$ Thin Films with Rutile Phase Prepared by DC Magnetron Co-Sputtering at Room Temperature: Effect of Cu Incorporation. *Appl. Surf. Sci.* **2015**, *345*, 49–56. [CrossRef]
50. Banakh, O.; Schmid, P.E.; Sanjines, R.; Levy, F. Electrical and Optical Properties of TiOx Thin Films Deposited by Reactive Magnetron Sputtering. *Surf. Coatings Technol.* **2002**, *151*, 272–275. [CrossRef]
51. Assim, E.M. Optical Constants of Titanium Monoxide TiO Thin Films. *J. Alloys Compd.* **2008**, *465*, 1–7. [CrossRef]
52. Grigorov, K.G.; Grigorov, G.I.; Drajeva, L.; Bouchier, D.; Sporken, R.; Caudano, R. Synthesis and Characterization of Conductive Titanium Monoxide Films. Diffusion of Silicon in Titanium Monoxide Films. *Vacuum* **1998**, *51*, 153–155. [CrossRef]
53. Wei, Y.; Shi, Y.; Zhang, X.; Li, D.; Zhang, L.; Gong, C.; Zhang, J. Preparation of Black Titanium Monoxide Nanoparticles and Their Potential in Electromagnetic Wave Absorption. *Adv. Powder Technol.* **2020**, *31*, 3458–3464. [CrossRef]
54. Nguyen, T.T.N.; Chen, Y.H.; He, J.L. Preparation of Inkjet-Printed Titanium Monoxide as p-Type Absorber Layer for Photovoltaic Purposes. *Thin Solid Films* **2014**, *572*, 8–14. [CrossRef]
55. Domaradzki, J.; Kaczmarek, D.; Prociow, E.L.; Radzimski, Z.J. Study of Structure Densification in TiO2 Coatings Prepared by Magnetron Sputtering under Low Pressure of Oxygen Plasma Discharge. *Acta Phys. Pol. A* **2011**, *120*, 49–52. [CrossRef]
56. Ju, Y.; Li, L.; Wu, Z.; Jiang, Y. Effect of Oxygen Partial Pressure on the Optical Property of Amorphous Titanium Oxide Thin Films. *Energy Procedia* **2011**, *12*, 450–455. [CrossRef]
57. Figueiredo, V.; Elangovan, E.; Gonçalves, G.; Barquinha, P.; Pereira, L.; Franco, N.; Alves, E.; Martins, R.; Fortunato, E. Effect of Post-Annealing on the Properties of Copper Oxide Thin Films Obtained from the Oxidation of Evaporated Metallic Copper. *Appl. Surf. Sci.* **2008**, *254*, 3949–3954. [CrossRef]
58. Miller, D.R.; Akbar, S.A.; Morris, P.A. Nanoscale Metal Oxide-Based Heterojunctions for Gas Sensing: A Review. *Sens. Actuators B Chem.* **2014**, *204*, 250–272. [CrossRef]
59. Barsan, N.; Weimar, U. Conduction Model of Metal Oxide Gas Sensors. *J. Electroceramics* **2001**, *7*, 143–167. [CrossRef]
60. Lyson-Sypien, B.; Radecka, M.; Rekas, M.; Swierczek, K.; Michalow-Mauke, K.; Graule, T.; Zakrzewska, K. Grain-Size-Dependent Gas-Sensing Properties of TiO$_2$ Nanomaterials. *Sens. Actuators B Chem.* **2015**, *211*, 67–76. [CrossRef]
61. Rydosz, A.; Czapla, A.; Maziarz, W.; Zakrzewska, K.; Brudnik, A. CuO and CuO/TiO$_2$-y Thin-Film Gas Sensors of H$_2$ and NO$_2$. In Proceedings of the 2018 XV International Scientific Conference on Optoelectronic and Electronic Sensors (COE), Warsaw, Poland, 17–20 June 2018; pp. 2016–2019. [CrossRef]
62. Hübner, M.; Simion, C.E.; Tomescu-Stănoiu, A.; Pokhrel, S.; Bârsan, N.; Weimar, U. Influence of Humidity on CO Sensing with P-Type CuO Thick Film Gas Sensors. *Sens. Actuators B Chem.* **2011**, *153*, 347–353. [CrossRef]

63. Alev, O.; Şennik, E.; Öztürk, Z.Z. Improved Gas Sensing Performance of P-Copper Oxide Thin Film/n-TiO$_2$ Nanotubes Heterostructure. *J. Alloys Compd.* **2018**, *749*, 221–228. [CrossRef]
64. Kosc, I.; Hotovy, I.; Rehacek, V.; Griesseler, R.; Predanocy, M.; Wilke, M.; Spiess, L. Sputtered TiO$_2$ Thin Films with NiO Additives for Hydrogen Detection. *Appl. Surf. Sci.* **2013**, *269*, 110–115. [CrossRef]
65. Lupan, O.; Santos-Carballal, D.; Ababii, N.; Magariu, N.; Hansen, S.; Vahl, A.; Zimoch, L.; Hoppe, M.; Pauporté, T.; Galstyan, V.; et al. TiO$_2$/Cu$_2$O/CuO Multi-Nanolayers as Sensors for H$_2$ and Volatile Organic Compounds: An Experimental and Theoretical Investigation. *ACS Appl. Mater. Interfaces* **2021**, *13*, 32363–32380. [CrossRef]

Disclaimer/Publisher's Note: The statements, opinions and data contained in all publications are solely those of the individual author(s) and contributor(s) and not of MDPI and/or the editor(s). MDPI and/or the editor(s) disclaim responsibility for any injury to people or property resulting from any ideas, methods, instructions or products referred to in the content.

Article

Physical Properties of Fe₃Si Films Coated through Facing Targets Sputtering after Microwave Plasma Treatment

Nattakorn Borwornpornmetee [1], Peerasil Charoenyuenyao [1], Rawiwan Chaleawpong [1], Boonchoat Paosawatyanyong [2], Rungrueang Phatthanakun [3], Phongsaphak Sittimart [4], Kazuki Aramaki [4], Takeru Hamasaki [4], Tsuyoshi Yoshitake [4] and Nathaporn Promros [1],*

[1] Department of Physics, Faculty of Science, King Mongkut's Institute of Technology Ladkrabang, Bangkok 10520, Thailand; 62605035@kmitl.ac.th (N.B.); 61605016@kmitl.ac.th (P.C.); 61605015@kmitl.ac.th (R.C.)
[2] Department of Physics, Faculty of Science, Chulalongkorn University, Bangkok 10330, Thailand; paosawat@sc.chula.ac.th
[3] Synchrotron Light Research Institute, Nakhon Ratchasima 30000, Thailand; rungrueang@slri.or.th
[4] Department of Applied Science for Electronics and Materials, Kyushu University, Kasuga, Fukuoka 816-8580, Japan; phongsaphak_sittimart@kyudai.jp (P.S.); kazuki_aramaki@kyudai.jp (K.A.); takeru_hamasaki@kyudai.jp (T.H.); tsuyoshi_yoshitake@kyudai.jp (T.Y.)
* Correspondence: nathaporn_promros@kyudai.jp; Tel.: +66-86-379-8648

Citation: Borwornpornmetee, N.; Charoenyuenyao, P.; Chaleawpong, R.; Paosawatyanyong, B.; Phatthanakun, R.; Sittimart, P.; Aramaki, K.; Hamasaki, T.; Yoshitake, T.; Promros, N. Physical Properties of Fe₃Si Films Coated through Facing Targets Sputtering after Microwave Plasma Treatment. *Coatings* **2021**, *11*, 923. https://doi.org/10.3390/coatings11080923

Academic Editor: Rafal Chodun

Received: 22 June 2021
Accepted: 26 July 2021
Published: 1 August 2021

Publisher's Note: MDPI stays neutral with regard to jurisdictional claims in published maps and institutional affiliations.

Copyright: © 2021 by the authors. Licensee MDPI, Basel, Switzerland. This article is an open access article distributed under the terms and conditions of the Creative Commons Attribution (CC BY) license (https://creativecommons.org/licenses/by/4.0/).

Abstract: Fe₃Si films are deposited onto the Si(111) wafer using sputtering with parallel facing targets. Surface modification of the deposited Fe₃Si film is conducted by using a microwave plasma treatment under an Ar atmosphere at different powers of 50, 100 and, 150 W. After the Ar plasma treatment, the crystallinity of the coated Fe₃Si films is enhanced, in which the orientation peaks, including (220), (222), (400), and (422) of the Fe₃Si are sharpened. The extinction rule suggests that the B₂–Fe₃Si crystallites are the film's dominant composition. The stoichiometry of the Fe₃Si surfaces is marginally changed after the treatment. An increase in microwave power damages the surface of the Fe₃Si films, resulting in the generation of small pinholes. The roughness of the Fe₃Si films after being treated at 150 W is insignificantly increased compared to the untreated films. The untreated Fe₃Si films have a hydrophobic surface with an average contact angle of 101.70°. After treatment at 150 W, it turns into a hydrophilic surface with an average contact angle of 67.05° because of the reduction in the hydrophobic carbon group and the increase in the hydrophilic oxide group. The hardness of the untreated Fe₃Si is ~9.39 GPa, which is kept at a similar level throughout each treatment power.

Keywords: Fe₃Si film; facing targets sputtering; wettability; mechanical property; plasma treatment

1. Introduction

There are a variety of materials that could pair with silicon (Si) to form a silicide composite such as nickel, titanium, chromium, and iron (Fe) to name a few [1]. Among those elements, Fe is an excellent element to merge with Si because both are abundant within the earth [2,3]. Iron silicide (FeSi) possesses various phases, ranging from the nonmagnetic metallic FeSi to the ferromagnetic iron silicide (Fe₃Si), all with unique properties and potential applications of their own [2–8]. Fe₃Si is an outstanding specimen among the phases of FeSi because Fe₃Si owns the following striking features: an almost identical lattice constant to those of gallium arsenide (GaAs) [9], a slight lattice parameter misfit of −2.5% with FeSi₂ owning the β phase and 4.2% with Si [4,5,9,10], an impressive set of magnetic properties of a slight coercive field of 7.5 Oe and a comparatively high spin polarization of 45%, and an impressive thermal stability of over 800 K Curie temperature [5,9,11,12]. These features make Fe₃Si attractive for the use in spin transistor application [2–8]. Aside from its magnetic properties, Fe₃Si also has good physical properties such as high hardness and respectable corrosion resistance [13]. Additionally, Fe₃Si films possess a smooth surface that can be epitaxially produced on the (111) orientation Si wafer [5,9,10]. Hence, it also generates attention as a hard coating material.

Previously, our research group epitaxially created Fe_3Si films onto Si wafers owning a (111) orientation at ambient temperature, relying on sputtering with a facing targets system [5,14–16]. This system provides several merits such as a high plasma density, high-energy particle, stable substrate temperature, low stoichiometry discrepancy, and low plasma damage [5,14–16]. The produced Fe_3Si films also appear to be of a dominant B_2 structure for all existing basic structures (DO_3, A_2 and B_2) of Fe_3Si [5]. We also created the Fe_3Si films at different substrate temperatures, which uncovered that 300 °C is the most suitable substrate temperature for an epitaxial deposition of Fe_3Si films [4]. Employing a 300 °C substrate temperature, Fe_3Si structures are enhanced while they remained at the same phase and kept their electromagnetic traits [4]. Our prior studies on Fe_3Si mainly focused on the structural and magnetic properties of the films [4,5,9–12]. Despite those, there has been little research that involves the wetting angle and the mechanical characteristics, including the hardness and reduced elastic modulus for the Fe_3Si film surfaces.

Many researchers have reported that the physical characteristics of films can be altered via the usage of plasma treatment procedures [17–23]. Plasma treatment is a procedure that changes the material's surface, leading to a change in roughness [17–22]. Plasma ions can chemically react with the film's surface and can also physically bombard some particles or contaminant from the surface, both resulting in the modification of the surface and wetting properties, including an improvement of film surface quality [17–23]. Among the conventional plasma, argon (Ar) plasma possesses a set of intriguing characteristics [19–23]. Ar plasma treatment roughens the surface of the material and can also control surface oxidation, due to the fact that the plasma can either break the oxygen bond with the metal surface or generate an active hydroxyl group on the surface [20,21]. It has been reported that Ar plasma can shift the wetting state of materials from hydrophobic to hydrophilic [21–23]. According to literature, variation of the generating power influences the plasma properties and their interaction above the sample's surface [20,23]. C.C. Surdu-Bob et al. [20] discovered that low power plasma can induce oxidation on films, while high power plasma can sputter etch a GaAs film's surface. L. Ru and C. Jie-Rong [23] studied the effect of plasma power on the wettability of poly-vinyl chloride (PVC). The PVC samples were originally hydrophobic, but the wetting state shifted to hydrophilic after Ar plasma treatment. The hydrophilicity of the treated PVC increased correspondingly to the raising of plasma power. Hence, Ar plasma treatment with power variations should have a potential for the surface modification of Fe_3Si films.

For these reasons, this research work focuses on the modifications of the roughness and chemical composition over the Fe_3Si film's surface through Ar plasma treatment under various powers, to change the wettability and the mechanical properties of the Fe_3Si films. The wetting and mechanical properties for Fe_3Si films are provided, including the effect of microwave (MW) plasma treatment on such properties. Fe_3Si films were formed on Si wafer substrates via facing targets sputtering at 300 °C of substrate temperature, then, separated for Ar plasma treatment at 50 to 150 W. The effect of power on the characteristics of all Fe_3Si samples, untreated and treated, was to be examined through several characterization techniques. It was expected that the roughness and chemical composition of the Fe_3Si surface could be changed by a variation of the plasma power under an Ar ambient, which may lead to the modifications of the wetting angle and hardness. The samples in this research were investigated from a single Fe_3Si sample, while there may be a minor deviation from the results of the other Fe_3Si samples under the same experimental condition.

2. Materials and Methods

2.1. Epitaxial Creation of Fe₃Si Films

The substrate of n-type Si wafers (SUMCO Corp., Tokyo, Japan) (orientation: (111), resistivity: 1000–4000 ohm·cm) was used to produce the Fe$_3$Si films. The Si wafer cleaning was performed by the usage of acetone and methanol (FUJIFILM Wako Pure Chemical Corp., Osaka, Japan) inside an ultrasonic cleaner (As One, Osaka, Japan; model US-1) to remove surface contamination. After that, the Si substrates were dipped into diluted hydrofluoric acid (1%) to remove the native oxide layer. The acid was then rinsed from the substrate surface by using deionized water. Later, the cleansed Si wafers were dried by using nitrogen gas (Iwatani Corp., Osaka, Japan; 99.999% purity) and transferred to a substrate holder in the vacuum chamber.

For the sputtering systems, a couple of Fe$_3$Si alloy targets (TOSHIMA Manufacturing Co., Ltd., Saitama, Japan; 99.9% purity) with an atomic ratio of Fe:Si equal to 3:1 was provided as the sputtering source. The sputtering chamber was connected to a rotary pump (Alcatel Japan, Kanagawa, Japan; model CIT-Alcatel 2030C) and a turbo molecular pump (Osaka Vacuum, Osaka, Japan; model TG1003) to vacuumize the sputtering chamber. The base pressure was evacuated to below 3×10^{-5} Pa. Then, the chamber was filled with Ar gas (Iwatani Corp., Osaka, Japan; 99.9999% purity) at a constant flow rate of 15 sccm via a mass flow controller (KOFLOC, Kyoto, Japan; model 3660), where the pressure was maintained at 1.33×10^{-1} Pa. The temperature controller (OMRON, Kyoto, Japan; model E5CN) heated the Si substrate from the backside at the set temperature of 300 °C. The gas discharge was generated at a voltage of 1 kV and a current of 1.2 mA through a direct current power supply (Micro Denshi, Saitama, Japan; model HD1K-30N). The Fe$_3$Si films were produced at the deposition rate of 1.07 nm/min for 24 h.

Figure 1 illustrates the 2D diagram of facing targets sputtering system for the sputtering of Fe$_3$Si films. The circular Fe$_3$Si targets were provided on both facing cathodes, where the positioning of the substrate located perpendicularly to the targets and outside the ion bombardment confined to over the targets [24]. Permanent magnets were mounted beneath each cathode to control and speed up the charged particles [24]. The inert gas was introduced through the gas feed system, where the gas ions collided with the target planar at the cathode with the same negative voltage [24]. The emitted electrons from the collision were sped up by the electric and magnetic fields [24]. The electrons, which are dominated by Lorentz force, moved toward the opposite target [24]. Consequently, the ionization efficiency for facing targets sputtering could be improved, including deposition rate [24]. The sputtered atoms moved onto the heated substrate surface and condensed into the Fe$_3$Si films [24].

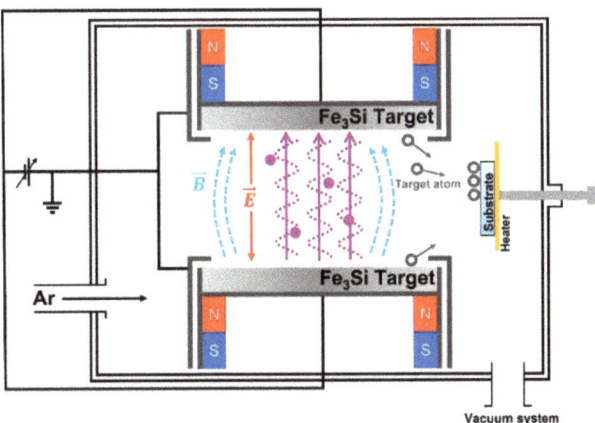

Figure 1. Schematic of facing targets sputtering apparatus used in creating Fe$_3$Si films.

2.2. Surface Modification of Fe₃Si Films

After the production, the Fe$_3$Si films were divided for surface treatment with Ar plasma using a commercial plasma system (Diener Electronic, Ebhausen, Germany; model PICO) equipped with a 2.45 GHz MW generator (Diener Electronic, Ebhausen, Germany; model MWG 1200), as can be seen in Figure 2. Before the treatment process, a rotary pump (ULVAC KIKO Inc., Miyazaki, Japan; model DIS-251) was employed to purge the air inside the chamber of the MW plasma system until the pressure became lower than 2 Pa, which was the base pressure of this arrangement. After that, Ar gas was introduced into the chamber at a 5 sccm flow rate, where the operating pressure for surface treatment was kept at 50 Pa throughout the treatment process. The magnetron head, on the top of the vacuum chamber, generated MW radiation with a consistent power, which was transferred to the chamber through a dielectric quartz window [25]. The channel MW radiation induced Ar gas ionization within the vacuum chamber, causing the generation of Ar plasma that bombarded the film's surface [25]. For the treatment condition, the Fe$_3$Si films were treated for 10 min under the various generating powers of 50, 100, and 150 W. After the process ended, the gas feed was stopped followed by a ventilation process until there was no processing gas remaining and the pressure within vacuum chamber returned to atmospheric pressure. Afterward, the treated samples were safe to be retrieved from the chamber.

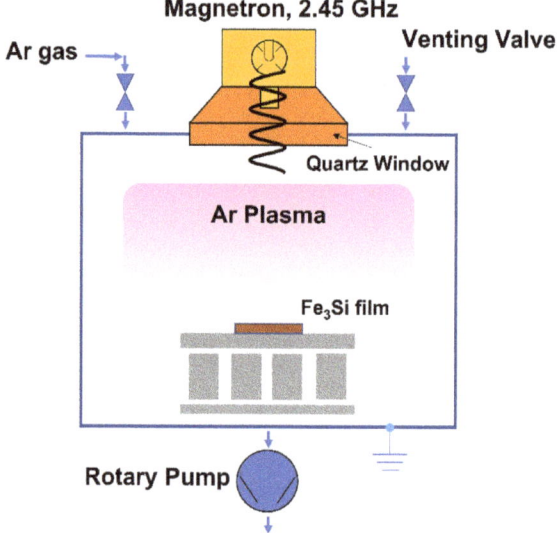

Figure 2. Arrangement of the MW plasma system.

2.3. Investigation of the Properties of Fe₃Si Films

Several properties of the untreated and treated Fe$_3$Si films were inspected with instruments such as X-ray diffractometer (XRD), X-ray photoelectron spectroscopy (XPS), field emission scanning electron microscope (FESEM), atomic force microscope (AFM), contact angle meter, and nanoindenter. A structural investigation of the untreated and treated Fe$_3$Si films, such as crystal orientation and crystallite size, was performed through XRD (Rigaku, Tokyo, Japan; model TTRAX III) under the conventional 2θ-θ scanning mode in the range 20°–90°. The size of the crystallite was simulated by using JADE software (Materials Data, Inc., Livermore, CA, USA; version 9.7.0). The atomic concentration of untreated and Ar-treated Fe$_3$Si films was measured by using an XPS (Kratos Analytical, Manchester, UK; model Axis Ultra DLD) and quantified through CasaXPS software (Casa Software Ltd., Devon, UK; version 2.3.24). The morphology of the Fe$_3$Si film's surface, before and after plasma treatment, was studied through FESEM (Hitachi, Tokyo, Japan; model SU 8230) at 300 kx magnification, under a 10 kV load. The cross-sectional properties for all Fe_3Si films were captured at 30 kx magnification, under a 10 kV load. The surface roughness of the untreated and Ar-treated Fe_3Si was scanned through AFM (Park Systems, Suwon, Korea; model XE-120) in the non-contact scanning mode at 5×5 µm^2 of scanning area, while the root-mean-square roughness (R_{rms}) was evaluated through Gwyddion software (General Public License). The wetting properties for all the Fe_3Si film's surfaces were exposed by a contact angle meter (DataPhysics, San Jose, CA, USA; model OCA20) applying deionized (D.I.) water as the test liquid. The mechanical characteristics of hardness and reduced elastic modulus for the untreated and treated Fe_3Si films were assessed by a nanoindenter (Bruker's Hysitron, Minneapolis, MN, USA; model Ti premier).

3. Results and Discussion

3.1. Structural Properties of Fe_3Si Films

Figure 3 presents the XRD patterns for the Fe_3Si films, created at 300 °C substrate temperature, with before and after Ar plasma treatment at different powers. The sharp diffraction pattern of the untreated Fe_3Si films exhibited the preferential orientations of $Fe_3Si(220)$, $Fe_3Si(222)$, $Fe_3Si(400)$, and $Fe_3Si(422)$ at the positions of 44.42°, 56.26°, 64.70°, and 81.88°, respectively. The plasma-treated Fe_3Si films were also observed for the peaks of $Fe_3Si(220)$, $Fe_3Si(222)$, $Fe_3Si(400)$, and $Fe_3Si(422)$. The untreated Fe_3Si films showed the sharp peaks of $Fe_3Si(220)$ and $Fe_3Si(222)$, including the weak peaks of $Fe_3Si(400)$ and $Fe_3Si(422)$. These obtained peaks are well-known as successful coatings for Fe_3Si films onto Si substrate. The preferential orientations of $Fe_3Si(220)$, $Fe_3Si(222)$, $Fe_3Si(400)$, and $Fe_3Si(422)$ were also reported and confirmed by the literature of S.I. Hirakawa et al. [14] and C.B. Tang et al. [26,27]. For the details, the orientation of $Fe_3Si(222)$ denoted that the films hold a B_2 structure with a superlattice reflection of B_2–Fe_3Si crystallites [5]. The preferred orientations of $Fe_3Si(220)$, $Fe_3Si(400)$, and $Fe_3Si(422)$ showed a fundamental reflection of Fe_3Si films [5]. From the XRD peak profiling, the untreated Fe_3Si exhibited the full width at half maximum (FWHM) of the sharpest diffraction peak with a value of 0.379. After treating with 50, 100, and 150 W, the change in the FWHM was observed with the values of 0.322, 0.342, and 0.361, respectively. Based on the appraisement by Scherrer's equation through the JADE software, the crystallite size for the untreated films was around 30.925 nm. The crystallite size became 35.200, 33.175, and 31.225 nm after treating with Ar plasma at 50, 100, and 150 W, respectively. This behavior shows a rise in the crystallite size of the Fe_3Si films after the 50 W plasma treatment, which may originate from the enhancement in crystallinity of the Fe_3Si films [28–30]. The crystallinity improvement may be due to the defect annihilation progression [28]. The crystallite sizes for the Fe_3Si films treated at 100 and 150 W were both relatively smaller than that of the 50 W treated films due to orientational disarrangement caused by the more energetic ions that bombarded the surface [30,31]. In contrast, the change of grain size scarcely occurred when using Ar plasma treatment due to the nature of the noble gas, which does not cause epitaxial growth [32].

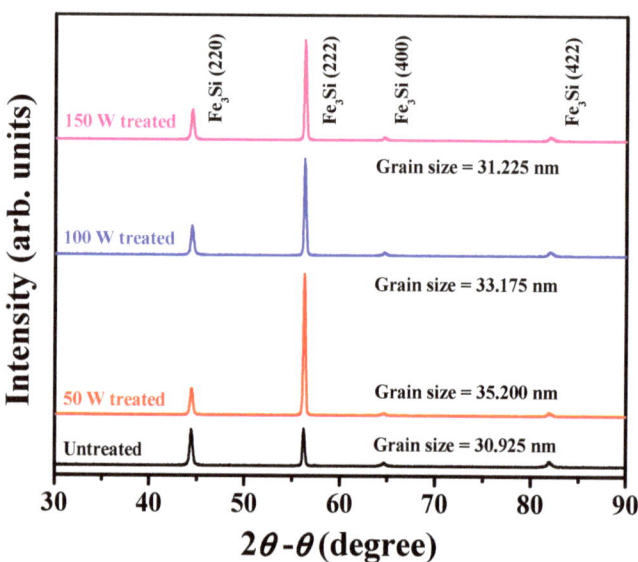

Figure 3. XRD patterns of the untreated Fe$_3$Si film and treated Fe$_3$Si films under the different powers.

3.2. Chemical Properties of Fe$_3$Si Films

Figure 4 demonstrates the XPS results for the surface composition of the untreated Fe$_3$Si films and Fe$_3$Si films treated at different powers. From the XPS results, the peaks of Fe 2p, O 1s, N 1s, C 1s, and Si 2p were observed. These peaks were translated into an atomic concentration of the film's surface, as represented in Table 1. The results showed an opposite behavior between the content of O 1s and C 1s concentrations for all Fe$_3$Si films. In response to the plasma exposure, the C 1s concentration decreased, while O 1s concentration increased as the MW power was raised. The XPS analysis revealed that, aside from Fe and Si, the surface of the Fe$_3$Si films also features abundant carbon and oxygen. Carbon and oxygen are common contaminants found in any material, especially on the metal-contained ones after exposing to the environmental air [33,34]. Fe$_3$Si is an oxidation-prone material that allows the layer of the oxide group to form on its surface easily [33]. The C 1s peak mainly originated from the layer of adventitious carbon, which forms easily on atmospherically exposed metal [34]. For N 1s, the nitrogen content may have come as a part of organic residues on the exposed surface, which is consistent with the copious amounts of C found on the surface [34]. These adventitious carbons are commonly attributed to be the reason behind hydrophobicity [22]. Under the low-pressure MW plasma treatment of Ar, high-energy ions collide with the sample surface, dissociate the organic carbon contaminants, and cause them to volatilize [22]. There is also the contribution from the oxide group, which evidently was the major contributor to hydrophilicity of the treated surface [21,22]. Hydrophilic oxide groups may come from the MW plasma chamber as the Ar gas may contain a tiny amount of oxygen, as well as the highly active groups remaining on the surface after plasma bombardment, which subsequently react with oxygen when exposed to air or plasma impurity [20,21].

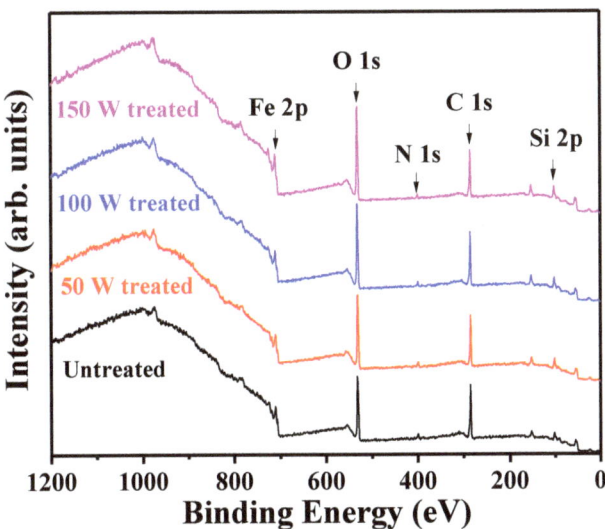

Figure 4. XPS patterns of the untreated Fe_3Si film and treated Fe_3Si films under the different powers.

Table 1. Atomic concentration of Fe_3Si films at different MW plasma treatment conditions.

Sample	Surface Atomic Concentration (at.%)				
	Fe 2p	O 1s	N 1s	C 1s	Si 2p
Untreated	10.30	47.19	3.28	35.68	3.55
Ar treated (50 W; 10 min)	12.78	53.07	2.87	26.47	4.81
Ar treated (100 W; 10 min)	13.61	56.15	2.00	23.48	4.76
Ar treated (150 W; 10 min)	15.49	56.23	2.02	20.35	5.92

3.3. Surface Morphologies of Fe_3Si Films

3.3.1. FESEM Images

Figure 5 shows typical FESEM micrographs from a top view of the untreated and treated Fe_3Si film surfaces by Ar plasma treatment under the various biased powers. These micrographs were captured at the magnification of 300 kx. Figure 5a exposes that the untreated Fe_3Si films presented an abundance of small crystallites over the entire surface area with a uniform surface structure; pinholes with a destructive surface area were not observed. This should be because of the advantages of sputtering with a pair of facing targets. Namely, this coating technique has the benefits from a low increment of substrate temperature, low-different stoichiometry films compared to the sputtering target, and high plasma density [35–37]. Additionally, the plasma's particles were detained within the generated magnetic field from the permanent magnet beneath the sputtering targets [35–37]. For a sputtering technique with facing targets, the surface of the Si wafer substrate was in parallel and situated away from the originated plasma zone during a film's coating, which led to less plasma damage over the film surface [35–37]. The surface structure of the 50 W treated Fe_3Si films was slightly changed from that of the untreated Fe_3Si films. At the higher treating powers of 100 and 150 W, the influence of the plasma treatment on the morphology of the treated Fe_3Si film's surface became more noticeable with the formation of slight bumps, including a non-smooth surface pattern and pinholes. The change of the treated Fe_3Si film's surface increased as the power was increased. This may have originated through the rise in the etching rate because of the Ar ions kinetic energy elevation [20].

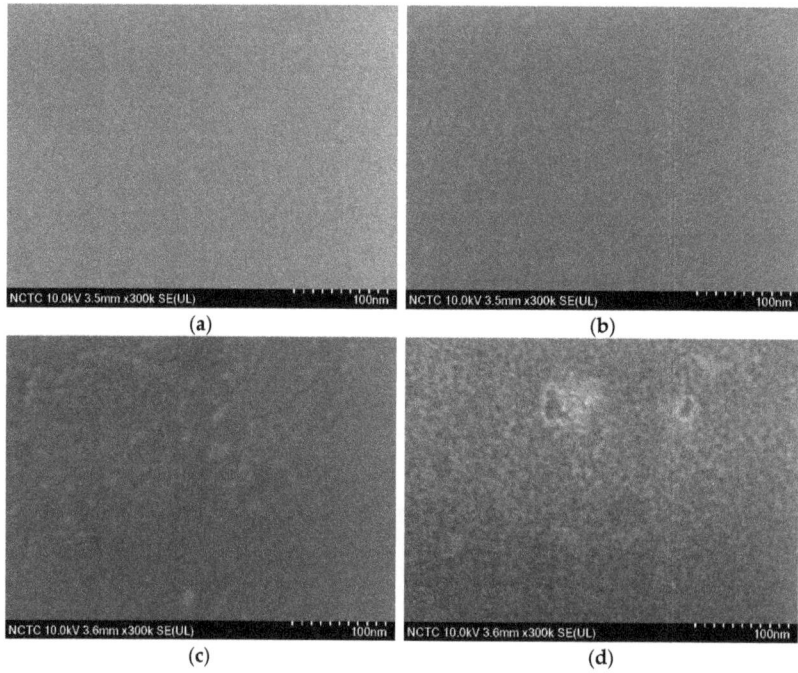

Figure 5. Planar FESEM micrographs of the coated Fe_3Si films (**a**) without and with treatment using Ar plasma at the different applied powers of (**b**) 50 W, (**c**) 100 W, and (**d**) 150 W.

Figure 6 demonstrates the representative cross-sectional FESEM micrographs of the untreated Fe_3Si films and the Fe_3Si films treated by plasma at various powers. All cross-sectional FESEM images were captured with a magnification of 30 kx. Figure 6 uncovers the linear interface between the film layer and the substrate layer. All untreated and treated Fe_3Si films were absent from deformity and discontinuity of interface between the layers. Figure 6a discloses that the constructed Fe_3Si film layer under the untreated condition owned an average thickness of 1.11 µm. The average thickness for the Fe_3Si films treated at the 50 W power was calculated to be around 1.09 µm. By raising the plasma powers to 100 and 150 W, the average thickness values of the treated Fe_3Si films slightly diminished to 1.08 and 1.05 µm, in order. The diminution of the film thickness as the power was increased may be attributable to the rise in high energetic ion bombardment, engendering a higher etching rate [20].

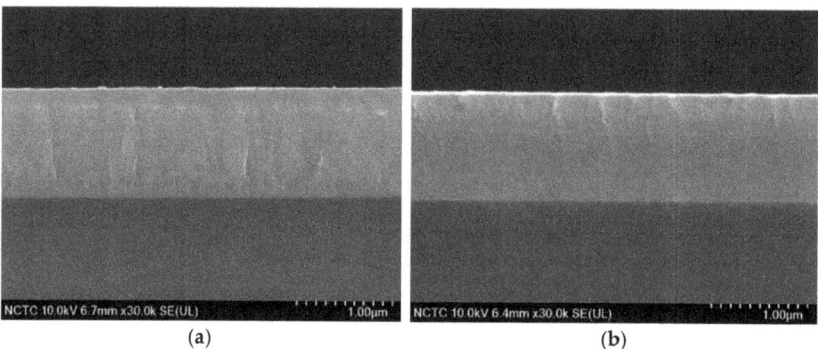

Figure 6. *Cont.*

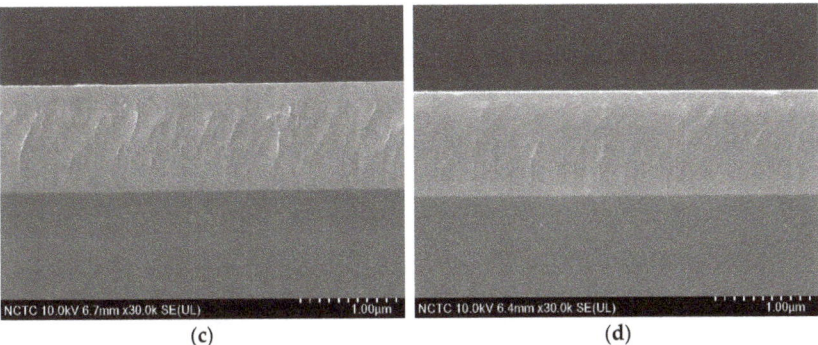

Figure 6. Cross-sectional FESEM micrographs of the Fe$_3$Si films on Si wafers treated at different powers: (**a**) untreated and (**b**) 50 W, (**c**) 100 W, and (**d**) 150 W treated.

3.3.2. AFM Images

Figure 7 represents the AFM scanned images of the untreated and treated Fe$_3$Si films through Ar plasma at the different applied powers. Using AFM analysis, the determination of R_{rms} for the untreated and treated Fe$_3$Si film's surface could be conducted. According to the AFM topography, it was visible that the untreated Fe$_3$Si films exhibited a rather smooth surface with an appraised R_{rms} of 10.63 Å. This outcome was consistent with that of the FESEM result, where the smoothness of the untreated Fe$_3$Si films should be due to the advantages of facing targets sputtering. As a result, the damage from the plasma should have been low and generated less surface roughness over the film's surface plane [35–37]. After the treatment with Ar plasma, the R_{rms} value for the Fe$_3$Si films treated with 50 W power was evaluated to be 10.71 Å, where the R_{rms} value was slightly higher than that of the Fe$_3$Si films without plasma treatment. The treated Fe$_3$Si films at 100 and 150 W manifested the assessed R_{rms} values of 11.07 and 13.06 Å, respectively. Based on the surface topography, the roughness for the Fe$_3$Si films was not drastically changed by the Ar plasma treatment compared to similar materials such as GaAs [20]. In the process of MW plasma treatment, the Fe$_3$Si film surface was bombarded by highly energetic species ejecting some Fe and Si atoms from the Fe$_3$Si surface out, resulting in a physical change in the surface roughness [20]. There was also an occurrence of a dangling bond on the film's surface, which was not permanent as it could either diffuse from the surface or induce oxidization when aging [20,38]. It is acknowledged that elements such as Fe and Si are easily oxidized [33]. As mentioned above, there is a possibility that the Fe$_3$Si surface can be oxidized. Hence, its physical properties, such as the surface roughness, may have changed after the treatment. Oxidation has been proven to hardly affect the roughness of alloy surfaces, where the effect of roughness would not wear off for a considerable amount of time [39,40].

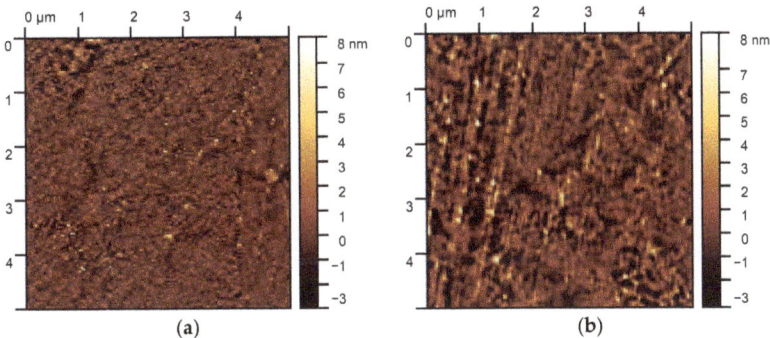

Figure 7. Cont.

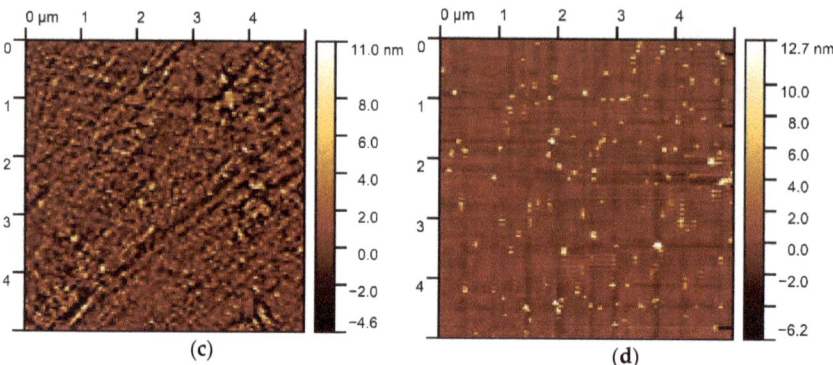

Figure 7. AFM images of the Fe$_3$Si film surfaces (**a**) before and after plasma treatment at (**b**) 50 W, (**c**) 100 W, and (**d**) 150 W.

3.4. Wetting Properties of Fe$_3$Si Films

Figure 8 displays the captured images for the contact angle measurement of the untreated and treated Fe$_3$Si films under the varied powers. For the surfaces of the untreated and treated Fe$_3$Si films, the wetting angles were determined by using D.I. water droplets. For the wetting property, it can be verified, based on the average contact angle (θ_{ca}), into super-hydrophilic ($\theta_{ca} \leq 10°$), hydrophilic ($10° < \theta_{ca} < 90°$), hydrophobic ($90° \leq \theta_{ca} < 150°$), and super-hydrophobic ($150° \leq \theta_{ca} \leq 180°$) [41]. Figure 8a represents the wetting angle between the droplet and the surface of the untreated Fe$_3$Si films, where the evaluated θ_{ca} was determined to be 101.7°. The result shows that the Fe$_3$Si films exhibited a wetting state of hydrophobic. Figure 8b–d presents the captured images between the droplet and the treated Fe$_3$Si films at the increased powers of 50, 100, and 150 W, respectively. After treating with Ar plasma, θ_{ca} was slightly reduced to 87.75° for the Fe$_3$Si films after treated at 50 W, in which the surface was determined as a hydrophilic state. At increasing powers, the θ_{ca} values for the Fe$_3$Si films gradually decreased to 79.20° and 67.05° at 100 W and 150 W, respectively.

It was observed that the surface of the untreated Fe$_3$Si films exhibited their hydrophobicity, while the surfaces of all plasma-treated Fe$_3$Si films turned into the hydrophilic state. The gained results of XPS revealed that more than a third of the untreated Fe$_3$Si film's surface was covered by the carbon functional group following by the oxide group and Fe$_3$Si compositional elements; the hydrophobicity for the surface should be predominantly controlled due to this reason [22]. After the plasma treatment, the XPS results for the treated films revealed a decrease of the hydrophobic carbon chemical composition by selective etching through energetic Ar ions impact [22]. Concurrently, the oxygen concentration also increased due to the highly polarized group left behind after sputter etching at low-pressure MW plasma under an Ar atmosphere [20]. These changes in surface composition resulted in the change in the wetting state from hydrophobic to hydrophilic. The trend continued as the power was increased. Based on the FESEM and AFM results, the surfaces of all the Fe$_3$Si films comprised nano-rough morphology. In physical terms, the surface wetting of the Fe$_3$Si films can be commonly attributed to the wetting models of Wenzel and Cassie-Baxter [42–45]. Wenzel and Cassie-Baxter incorporated the roughness as one of the important parameters in their models, where the former is based on surface alignment and the latter relies on the air groove [42–45]. However, our result shows a significant alteration of the wetting behavior of Fe$_3$Si despite the insignificant change in surface morphology. Hence, the wettability of the Fe$_3$Si films was predominantly dictated by their chemical composition on the surface of the untreated and plasma-treated films.

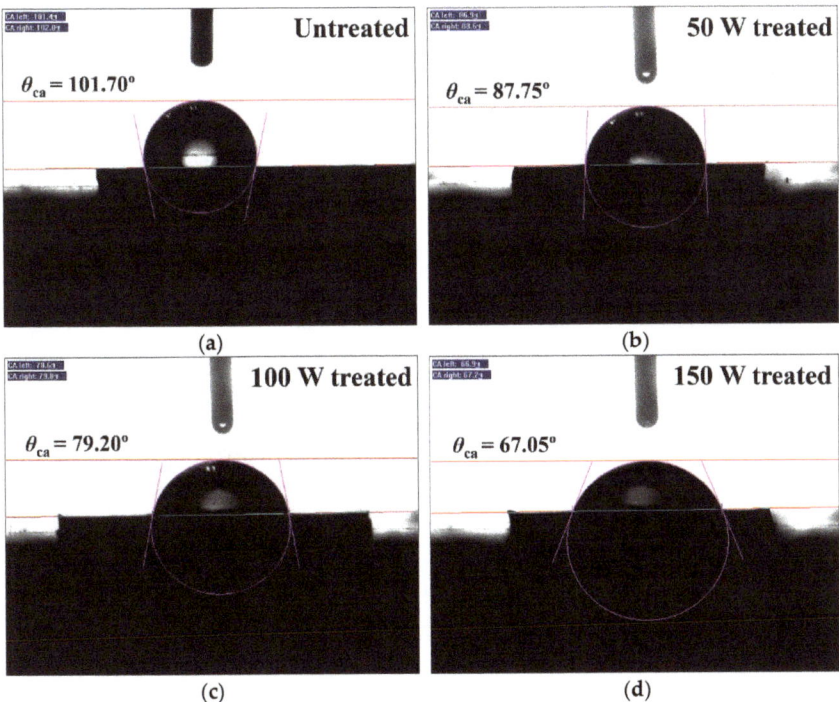

Figure 8. Images for the contact angle measurement on the surface of the Fe$_3$Si films (**a**) before and after treated at (**b**) 50 W, (**c**) 100 W, and (**d**) 150 W.

3.5. Mechanical Properties of Fe$_3$Si Films

For the mechanical properties, the nanoindentation technique using a Berkovich indenter was used to investigate the hardness and reduced elastic characteristics of the untreated and treated Fe$_3$Si film surfaces. The indentation test was carried on by applying an indentation load of 3 mN to all the Fe$_3$Si films, before and after Ar plasma treatment at 50 W–150 W. The maximum depth for the test was controlled at 10% of the thickness of all films (100 nm), where the effect of the substrate could be suppressed [46,47]. The nanoindentation test for the untreated and Ar plasma-treated Fe$_3$Si films was performed repeatedly, five times. Figure 9 presents the plot set of the applied indentation load versus the depth of penetration (load–depth curve) for the Fe$_3$Si films under the conditions of untreated, 50 W treated, 100 W treated, and 150 W treated, respectively.

The average hardness value (H) and reduced elastic modulus value (E_r) of the untreated and Ar plasma-treated Fe$_3$Si films were calculated from the unloading portions of their load–depth curve [48,49]. The H and E_r for all Fe$_3$Si samples are summarized in Table 2. Figure 10 (red line) presents the relative plot between the H for the Fe$_3$Si samples regarding their treatment power, with a standard deviation. The H-power plot shows the decreasing trend of the hardness characteristic of Fe$_3$Si films with increasing Ar plasma treatment power. In the same vein, the E_r decreases, as shown in Figure 10 (blue line), where a plot between the E_r versus power for Fe$_3$Si films is depicted. Based on the nanoindentation result, the H of the untreated Fe$_3$Si films was close to the Fe$_3$Si reported by various sources [25,50]. The H and E_r were almost the same, albeit slightly declined after the Ar plasma treatment. They agreed with the morphology results of the films, where the inconsistency of the film's surface was observed as the standard deviation [51]. As the plasma power increased, the deviation increasingly swayed by the accumulated plasma damage [51]. Considering the fact that the nanoindentation only penetrated 10% of the film's thickness, these changes as a response to surface modification are insignificant when taken together.

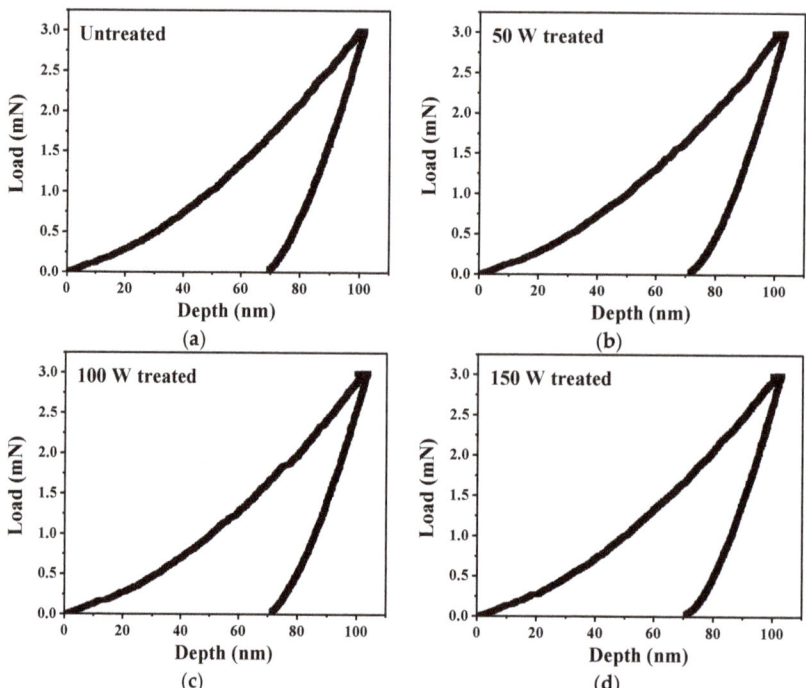

Figure 9. Load versus depth plots of the Fe$_3$Si films (**a**) before and after treated at (**b**) 50 W, (**c**) 100 W, and (**d**) 150 W.

Table 2. H and E$_r$ for untreated Fe$_3$Si and Ar plasma-treated Fe$_3$Si.

Sample	H (GPa)	E$_r$ (GPa)
Untreated	9.392 ± 0.070	204.862 ± 1.226
Ar treated (50 W; 10 min)	8.991 ± 0.069	205.943 ± 4.633
Ar treated (100 W; 10 min)	8.881 ± 0.080	208.085 ± 3.113
Ar treated (150 W; 10 min)	8.857 ± 0.094	212.693 ± 3.211

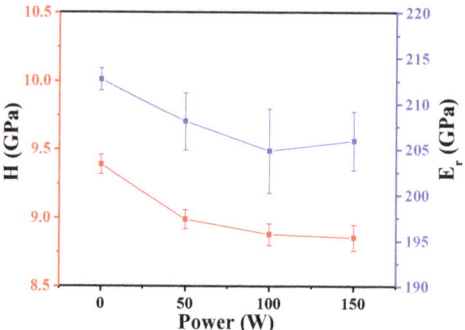

Figure 10. Plots of the (**red**) H and (**blue**) E$_r$ with standard deviation versus plasma-treated power of the surface of the Fe$_3$Si films treated by Ar plasma under different powers.

4. Conclusions

The present study clarified the cause of hydrophobicity of Fe_3Si films epitaxially created through facing targets sputtering at 300 °C heated substrate and the effect of Ar plasma treatment power on the film's properties. The XRD patterns for all Fe_3Si films presented a combination of a B_2 and DO_3 Fe_3Si crystal structure. The orientational peaks of the XRD pattern became much higher at 50 W due to the suppression of surface contamination, resulting in preferable orientations. Meanwhile, the peak intensities went down as the power was increased, due to the sputter etching caused by the higher energy ions. The atomic concentration extracted from the XPS spectrum revealed that the surface of the untreated Fe_3Si was laden with carbon and oxygen, which was the general contamination. The Ar plasma treatment reduced the carbon concentration by the volatilization of the carbon atoms through ion collision. The plasma also left behind a radical reactive site which formed into an oxide layer on the Fe_3Si surface. As the power increased, the atomic concentration of oxygen also increased, while carbon decreased. The morphological outcomes, as gained from both the FESEM and AFM, showed a seamless surface with an R_{rms} of 10.63 Å for the untreated Fe_3Si films. After plasma treatment, the surface was bombarded by high energy ions, causing the appearance of pinholes which roughened the surface to 13.06 Å at 150 W of power. The film's thickness of the constructed Fe_3Si was only slightly decreased by the Ar plasma at different powers. From the contact angle results, the untreated Fe_3Si films possessed θ_{ca} of 101.7°, which changed to 67.05° after Ar plasma treatment at 150 W. The shift in hydrophobicity was likely due to the change in the chemical composition of the surface, namely, the reduction in hydrophobic organic carbon and the augmentation of the hydrophilic oxide group. The H and E_r of the untreated Fe_3Si films were not significantly influenced by the Ar plasma at different powers. For the next settlement, the effect of other plasma treatment parameters on Fe_3Si's properties such as time, pressure, and gas type can be explored. The study will mainly focus on how these parameters can affect the wettability and mechanical properties of Fe_3Si to properly determine the optimized condition for the modification of wettability and mechanical properties.

Author Contributions: Conceptualization, N.P.; methodology, N.B., P.C., R.C., R.P., P.S., K.A., T.H. and N.P.; formal analysis, N.B., P.C., R.C., P.S. and N.P.; investigation, N.B., P.C., R.C., R.P. and N.P.; writing—original draft preparation, N.B., P.C., R.C. and N.P.; writing—review and editing, N.B., P.C., R.C., B.P., P.S., T.Y. and N.P.; Visualization, N.B., P.C. and R.C.; Supervision, B.P. and T.Y. All authors have read and agreed to the published version of the manuscript.

Funding: The allowance for carrying out this research was provided by the Academic Research Fund, Faculty of Science, King Mongkut's Institute of Technology Ladkrabang, with grant numbers 2562-01-05-29, 2562-01-05-44 and 2563-02-05-36.

Institutional Review Board Statement: Not applicable.

Informed Consent Statement: Not applicable.

Data Availability Statement: The data presented in this study are available in this article.

Acknowledgments: The authors also express their gratitude for the contribution and cooperation received from Research Unit of Integrated Science for Electronic, Material and Industry (ISEMI), Faculty of Science, King Mongkut's Institute of Technology Ladkrabang.

Conflicts of Interest: The authors declare no conflict of interest.

References

1. Tam, P.L.; Cao, Y.; Nyborg, L. XRD and XPS characterisation of transition metal silicide thin films. *Surf. Sci.* **2012**, *606*, 329–336. [CrossRef]
2. Promros, N.; Yamashita, K.; Iwasaki, R.; Yoshitake, T. Effects of hydrogen passivation on near-infrared photodetection of n-type β-FeSi$_2$/p-type Si heterojunction photodiodes. *Jpn. J. Appl. Phys.* **2012**, *51*, 108006. [CrossRef]
3. Shaban, M.; Nakashima, K.; Yokoyama, W.; Yoshitake, T. Photovoltaic properties of n-type β-FeSi$_2$/p-type Si heterojunctions. *Jpn. J. Appl. Phys.* **2007**, *46*, L667. [CrossRef]

4. Nakagauchi, D.; Yoshitake, T.; Nagayama, K. Fabrication of ferromagnetic Fe$_3$Si thin films by pulsed laser deposition using an Fe$_3$Si target. *Vacuum* **2004**, *74*, 653–657. [CrossRef]
5. Yoshitake, T.; Nakagauchi, D.; Ogawa, T.; Itakura, M.; Kuwano, N.; Tomokiyo, Y.; Kajiwara, T.; Nagayama, K. Room-Temperature epitaxial growth of ferromagnetic Fe$_3$Si films on Si(111) by facing target direct-current sputtering. *Appl. Phys. Lett.* **2005**, *86*, 262505. [CrossRef]
6. Sittimart, P.; Duangrawa, A.; Onsee, P.; Teakchaicum, S.; Nopparuchikun, A.; Promros, N. Interface state density and series resistance of n-type nanocrystalline FeSi$_2$/p-Type Si heterojunctions formed by utilizing facing-target direct-current sputtering. *J. Nanosci. Nanotechnol.* **2018**, *18*, 1841–1846. [CrossRef]
7. Charoenyuenyao, P.; Promros, N.; Chaleawpong, R.; Saekow, B.; Porntheeraphat, S.; Yoshitake, T. Effect of annealing on surface morphology and wettability of NC-FeSi$_2$ films produced via facing-target direct-current sputtering. *J. Nanosci. Nanotechnol.* **2019**, *19*, 6834–6840. [CrossRef]
8. Mazurenko, V.V.; Shorikov, A.O.; Lukoyanov, A.V.; Kharlov, K.; Gorelov, E.; Lichtenstein, A.I.; Anisimov, V.I. Metal-insulator transitions and magnetism in correlated band insulators: FeSi and Fe$_{1-x}$Co$_x$Si. *Phys. Rev. B* **2010**, *81*, 125131. [CrossRef]
9. Sadoh, T.; Takeuchi, H.; Ueda, K.; Kenjo, A.; Miyao, M. Epitaxial growth of ferromagnetic silicide Fe$_3$Si on Si(111) substrate. *Jpn. J. Appl. Phys.* **2006**, *45*, 3598. [CrossRef]
10. Hamaya, K.; Ueda, K.; Kishi, Y.; Ando, Y.; Sadoh, T.; Miyao, M. Epitaxial ferromagnetic Fe$_3$Si/Si(111) structures with high-quality heterointerfaces. *Appl. Phys. Lett.* **2008**, *93*, 132117. [CrossRef]
11. Herfort, J.; Schönherr, H.P.; Ploog, K.H. Epitaxial growth of Fe$_3$Si/GaAs(001) hybrid structures. *Appl. Phys. Lett.* **2003**, *83*, 3912–3914. [CrossRef]
12. Lin, F.; Jiang, D.; Ma, X.; Shi, W. Structural order and magnetic properties of Fe$_3$Si/Si(100) heterostructures grown by pulsed-laser deposition. *Thin Solid Films* **2007**, *515*, 5353–5356. [CrossRef]
13. Schneeweiss, O.; Pizurova, N.; Jiraskova, Y.; Žák, T.; Cornut, B. Fe$_3$Si Surface coating on SiFe Steel. *J. Magn. Magn. Mater.* **2000**, *215*, 115–117. [CrossRef]
14. Hirakawa, S.I.; Sonoda, T.; Sakai, K.I.; Takeda, K.; Yoshitake, T. Temperature-dependent current-induced magnetization switching in Fe$_3$Si/FeSi$_2$/Fe$_3$Si trilayered films. *Jpn. J. Appl. Phys.* **2011**, *50*, 08JD06. [CrossRef]
15. Takeda, K.; Yoshitake, T.; Nakagauchi, D.; Ogawa, T.; Hara, D.; Itakura, M.; Kuwano, N.; Tomokiyo, Y.; Kajiwara, T.; Nagayama, K. Epitaxy in Fe$_3$Si/FeSi$_2$ superlattices prepared by facing target direct-current sputtering at room temperture. *Jpn. J. Appl. Phys.* **2007**, *46*, 7846. [CrossRef]
16. Sakai, K.I.; Noda, Y.; Daio, T.; Tsumagari, D.; Tominaga, A.; Takeda, K.; Yoshitake, T. Current-induced magnetization switching at low current densities in current-perpendicular-to-plane structural Fe$_3$Si/FeSi$_2$ artificial lattices. *Jpn. J. Appl. Phys.* **2014**, *53*, 02BC15. [CrossRef]
17. Pransilp, P.; Kiatkamjornwong, S.; Bhanthumnavin, W.; Paosawatyanyong, B. Plasma Nano-modification of poly (ethylene terephthalate) fabric for pigment adhesion enhancement. *J. Nanosci. Nanotechnol.* **2012**, *12*, 481–488. [CrossRef]
18. Taweesub, K.; Visuttipitukul, P.; Tungkasmita, S.; Paosawatyanyong, B. Nitridation of Al–6%Cu alloy by rf plasma process. *Surf. Coat. Technol.* **2010**, *204*, 3091–3095. [CrossRef]
19. Berman, J.; Krim, J. Impact of oxygen and argon plasma exposure on the roughness of gold film surfaces. *Thin Solid Films* **2012**, *520*, 6201–6206. [CrossRef]
20. Surdu-Bob, C.C.; Sullivan, J.L.; Saied, S.O.; Layberry, R.; Aflori, M. Surface compositional changes in gaas subjected to argon plasma treatment. *Appl. Surf. Sci.* **2012**, *202*, 183–198. [CrossRef]
21. Prysiazhnyi, V.; Slavicek, P.; Mikmekova, E.; Klima, M. Influence of chemical precleaning on the plasma treatment efficiency of aluminum by rf plasma pencil. *Plasma Sci. Technol.* **2016**, *18*, 430–437. [CrossRef]
22. Ujino, D.; Nishizaki, H.; Higuchi, S.; Komasa, S.; Okazaki, J. Effect of plasma treatment of titanium surface on biocompatibility. *Appl. Sci.* **2019**, *9*, 2257. [CrossRef]
23. Ruand, L.; Jie-Rong, C. Studies on wettability of medical Poly (Vinyl Chloride) by remote argon plasma. *Appl. Surf. Sci.* **2006**, *252*, 5076–5082.
24. Lin, C.; Sun, D.C.; Ming, S.L.; Jiang, E.Y.; Liu, Y.G. Magnetron facing target sputtering system for fabricating single-crystal films. *Thin Solid Films* **1996**, *279*, 49–52. [CrossRef]
25. Odrobina, I.; Kúdela, J.; Kando, M. Characteristics of the planar plasma source sustained by microwave power. *Plasma Sources Sci. Technol.* **1998**, *7*, 238. [CrossRef]
26. Tang, C.B.; Wen, F.R.; Chen, H.X.; Liu, J.J.; Tao, G.Y.; Xu, N.J.; Xue, J.Q. Corrosion characteristics of Fe$_3$Si intermetallic coatings prepared by molten salt infiltration in sulfuric acid solution. *J. Alloys Compd.* **2019**, *778*, 972–981. [CrossRef]
27. Chang-Bin, T.; Hong-Xia, C.; Fu-Rong, W.; Ni-Jun, X.; Wei, H.; Juan-Qin, X. Facile preparation and tribological property of alloyed Fe$_3$Si coatings on stainless steels surface by molten-salt infiltration method. *Surf. Coat. Technol.* **2020**, *397*, 126049. [CrossRef]
28. Sattar, H.; Jielin, S.; Ran, H.; Imran, M.; Ding, W.; Gupta, P.D.; Ding, H. Impact of microstructural properties on hardness of tungsten heavy alloy evaluated by stand-off LIBS after PSI plasma irradiation. *J. Nucl. Mater.* **2020**, *540*, 152389. [CrossRef]
29. Lin, W.T.; Chen, S.H.; Chan, S.H.; Hu, S.C.; Peng, W.X.; Lu, Y.T. Crystallization phase transition in the precursors of CIGS films by Ar-ion plasma etching process. *Vacuum* **2014**, *99*, 1–6. [CrossRef]

30. Fernández, S.; Santos, J.D.; Munuera, C.; García-Hernández, M.; Naranjo, F.B. Effect of Argon Plasma-treated polyethylene terepthalate on ZnO:Al properties for flexible thin film silicon solar cells applications. *Sol. Energy Mater. Sol. Cells* **2015**, *133*, 170–179. [CrossRef]
31. Madera, R.G.B.; Martinez, M.M.; Vasquez, M.R. Effects of rf plasma treatment on spray-pyrolyzed copper oxide films on silicon substrates. *Jpn. J. Appl. Phys.* **2017**, *57*, 01AB05. [CrossRef]
32. Douglas, E.A.; Sheng, J.J.; Verley, J.C.; Carroll, M.S. Argon-Germane In Situ Plasma Clean for Reduced Temperature Ge on Si Epitaxy by High Density Plasma Chemical Vapor Deposition. *J. Vac. Sci. Technol.* **2015**, *33*, 41202. [CrossRef]
33. Chen, G.L.; Peng, J.H.; Xu, W.X. Surface reaction of polycrystalline Fe$_3$Si alloys with oxygen and water vapor. *Intermetallics* **1998**, *6*, 315–322. [CrossRef]
34. Calaway, M.J.; Fries, M.D. Adventitious Carbon on Primary Sample Containment Metal Surfaces. In Proceedings of the 46th Lunar and Planetary Science Conference, The Woodlands Waterway Marriott Hotel and Convention Center, The Woodlands, TX, USA, 16–20 March 2015; p. 1517.
35. Yoshitake, T.; Inokuchi, Y.; Yuri, A.; Nagayama, K. Direct epitaxial growth of semiconducting β-FeSi$_2$ thin films on Si(111) by facing targets direct-current sputtering. *Appl. Phys. Lett.* **2006**, *88*, 182104. [CrossRef]
36. Shaban, M.; Nomoto, K.; Izumi, S.; Yoshitake, T. n-type β-FeSi$_2$/intrinsic-si/p-type si heterojunction photodiodes for near-infrared light detection at room temperature. *Appl. Phys. Lett.* **2009**, *94*, 222113. [CrossRef]
37. Promros, N.; Baba, R.; Takahara, M.; Mostafa, T.M.; Sittimart, P.; Shaban, M.; Yoshitake, T. Epitaxial growth of β-FeSi$_2$ thin films on Si(111) substrates by radio frequency magnetron sputtering and their application to near-infrared photodetection. *Jpn. J. Appl. Phys.* **2016**, *55*, 06HC03. [CrossRef]
38. Abd Jelil, R. A Review of low-temperature plasma treatment of textile materials. *J. Mater. Sci.* **2015**, *50*, 5913–5943. [CrossRef]
39. He, L.P.; Cai, Z.B.; Peng, J.F.; Deng, W.L.; Li, Y.; Yang, L.Y.; Zhu, M.H. Effects of oxidation layer and roughness on the fretting wear behavior of copper under electrical contact. *Mater. Res. Express* **2020**, *6*, 1265e3. [CrossRef]
40. Junior, C.F.L.; de Assis Machado, G.S.; Mendes, P.S.N.; Huguenin, J.A.O.; Ferreira, E.A.; da Silva, L. Analysis of copper and zinc alloy surface by exposure to alcohol aqueous solutions and sugarcane liquor. *J. Mater. Res. Technol.* **2020**, *9*, 2545–2556. [CrossRef]
41. Lavieja, C.; Oriol, L.; Peña, J.I. Creation of superhydrophobic and superhydrophilic surfaces on ABS employing a nanosecond laser. *Materials* **2018**, *11*, 2547. [CrossRef]
42. Charoenyuenyao, P.; Promros, N.; Chaleawpong, R.; Borwornpornmetee, N.; Sittisart, P.; Tanaka, Y.; Yoshitake, T. Impact of annealing temperature and carbon doping on the wetting and surface morphology of semiconducting iron disilicide formed via radio frequency magnetron sputtering. *Thin Solid Films* **2020**, *709*, 138248. [CrossRef]
43. Quéré, D. wetting and roughness. *Annu. Rev. Mater. Res.* **2018**, *38*, 71–99. [CrossRef]
44. Wenzel, R.N. surface roughness and contact angle. *J. Phys. Chem.* **1949**, *53*, 1466–1467. [CrossRef]
45. Cassie, A.B.D.; Baxter, S. Wettability of porous surfaces. *J. Chem. Soc. Faraday Trans.* **1944**, *40*, 546–551. [CrossRef]
46. Fischer-Cripps, A.C. A review of analysis methods for sub-micron indentation testing. *Vacuum* **2000**, *58*, 569–585. [CrossRef]
47. Lu, C.J.; Bogy, D.B. The Effect of tip radius on nano-indentation hardness tests. *Int. J. Solids Struct.* **1995**, *32*, 1759–1770. [CrossRef]
48. Oliver, W.C.; Pharr, G.M. An improved technique for determining hardness and elastic modulus using load and displacement sensing indentation experiments. *J. Mater. Res.* **1992**, *7*, 1564–1583. [CrossRef]
49. Beegan, D.; Chowdhury, S.; Laugier, M.T. Comparison between nanoindentation and scratch test hardness (scratch hardness) values of copper thin films on oxidised silicon substrates. *Surf. Coat. Technol.* **2007**, *201*, 5804–5808. [CrossRef]
50. Murakami, T.; Hibi, Y.; Mano, H.; Matsuzaki, K.; Inui, H. Friction and wear properties of fe-si intermetallic compounds in ethyl alcohol. *Intermetallics* **2012**, *20*, 68–75. [CrossRef]
51. Chen, L.; Ahadi, A.; Zhou, J.; Ståhl, J.-E. Modeling effect of surface roughness on nanoindentation tests. *Procedia CIRP* **2013**, *8*, 334–339. [CrossRef]

Article

Improved Methodology of Cross-Sectional SEM Analysis of Thin-Film Multilayers Prepared by Magnetron Sputtering

Malwina Sikora [1,2], Damian Wojcieszak [1,*], Aleksandra Chudzyńska [3,4] and Aneta Zięba [1,2]

1. Faculty of Electronics, Photonics and Microsystems, Wrocław University of Science and Technology, Janiszewskiego 11/17, 50-372 Wrocław, Poland
2. Nanores Company, Bierutowska 57-59, 51-317 Wroclaw, Poland
3. Compact X Company, Bierutowska 57-59, 51-317 Wroclaw, Poland
4. Institute of Low Temperature and Structure Research, Polish Academy of Sciences, Okólna 2, 50-422 Wroclaw, Poland
* Correspondence: damian.wojcieszak@pwr.edu.pl

Abstract: In this work, an improved methodology of cross-sectional scanning electron microscopy (SEM) analysis of thin-film Ti/V/Ti multilayers was described. Multilayers with various thicknesses of the vanadium middle layer were prepared by magnetron sputtering. The differences in cross sections made by standard fracture, focused ion beam (FIB)/Ga, and plasma focused ion beam (PFIB)/Xe have been compared. For microscopic characterization, the Helios NanoLab 600i microscope and the Helios G4 CXe with the Quanta XFlash 630 energy dispersive spectroscopy detector from Bruker were used. The innovative multi-threaded approach to the SEM preparation itself, which allows us to retain information about the actual microstructure and ensure high material contrast even for elements with similar atomic numbers was proposed. The fracture technique was the most noninvasive for microstructure, whereas FIB/PFIB results in better material contrast (even than EDS). There were only subtle differences in cross sections made by FIB-Ga and PFIB-Xe, but the decrease in local amorphization or slightly better contrast was in favor of Xe plasma. It was found that reliable information about the properties of modern nanomaterials, especially multilayers, can be obtained by analyzing a two-part SEM image, where the first one is a fracture, while the second is a PFIB cross section.

Keywords: cross section; preparation techniques; SEM; FIB/Ga; PFIB/Xe; thin-film materials; multilayer; magnetron sputtering

1. Introduction

Scanning electron microscopy (SEM) allows for the study of a broad range of specimens for scientific and industrial purposes. In addition to metal alloys [1], biological and geological samples [2–8] and polymers [9–11] can also be investigated, as well as electronic devices [12,13] and mechanical components incorporated into larger structures [14]. Advanced imaging techniques require the use of an appropriate method for sample preparation. Nevertheless, preparation for SEM is still not a trivial task, especially when the characteristic dimensions of the objects are at the nanometric level. The choice of the appropriate technique depends on the nature of the sample, including its conductivity, state of aggregation (liquid, solid), and form (powder, bulk material, thin film, etc.). The preparation is closely related to the type of information that should be obtained from the sample with the aid of SEM. The smaller the area of interest, the more demanding sample preparation is, not to mention the need to reduce artifacts and unwanted modification of the sample itself. Selection of the appropriate preparation technique, even from among the already recognized methods, requires knowledge of how each of them changes the actual properties of the sample. Making such a seemingly simple choice is not easy due to the very small number of studies that compare the results of different methods of preparation of the same samples [15,16]. In the case of modern nanomaterials, this is also difficult because the parameters that describe them are often statistical in nature, such as the average size of the crystallites [17]. For this reason, a good research material for a reliable comparison of several methods is multilayer coating, where a well-defined single layer of the desired thickness can be buried at a specific and known depth [18–23]. It should be noted that a series of multilayers with a gradually modified thickness of selected layers allows for better visibility of quite subtle differences and artifacts that could have arisen, especially in the interface area [24–26]. Another issue that makes it difficult to choose the optimal preparation is related to the fact that most publications refer to the advantages and disadvantages of one single method, and unfortunately often without any explanation of the reasons for its choice [27–33]. The third issue to be mentioned is the use of quite contrasting materials (for SEM), which will have a rather poor application for materials consisting of elements with a similar atomic number, such as the titanium and vanadium selected for our research [34,35]. Their use results in a very low electron density contrast at the interface and is an additional difficulty in SEM imaging, but will allow for the elimination of much more subtle artifacts.

An important issue of SEM preparation, especially in the context of imaging advanced nanostructures, is the manufacturing of cross sections. The selection of an appropriate technique, especially in the case of multilayer coatings, is key to the correct visualization of their properties and to limiting the possibility of introducing artifacts. Imaging of advanced electronic components, where characteristic dimensions are at the nanometer level, requires working with cross sections devoid of as many artifacts as possible. Modern electronic or optoelectronic systems usually have a multilayer structure, which results in the need to characterize them at the level of single, nanometric films. For this reason, thin-film materials (in the form of single films or multilayers) are an important group where advanced preparation techniques for SEM analysis are needed. They should include not only unchanged microstructure, surface roughness, or thickness, but should also give true information about the material composition. Nowadays, it is possible to characterize a single film included in the multilayer structure with thicknesses of at least tens of nanometers [36]. Multilayer coatings (used as optical filters), which are a stack of high and low refractive index layers, have thicknesses of individual layers often <10 nm. Similarly, the characteristic dimensions have transistors, which are now produced with the so-called 7 nm technology. Therefore, these factors have led to the development of electron microscopy. Currently, the newest apparatus provides a resolution of ca. 1.4 nm. Therefore, it seems sufficient to visualize various types of advanced electronic, photonic, or optical systems based on nanostructures. However, there are still many problems with the proper preparation of the samples for SEM. It should be noted that while there is no ideal method that would be suitable for every sample, an improved methodology based on the hybrid preparation technique as a multistep approach can be applied as a modern solution.

Improvement of the SEM results requires the use of such cross section preparation techniques that will maintain the real properties of individual layers and interfaces [16,37]. There are a number of methods for manufacturing cross sections of thin films. Among these are the break method [37–39], the pre-cut technique [40], ultramicrotomy [37,41,42], grinding and polishing preceded by resin encapsulation [37], as well as ionic techniques: ion polishing [43–45] and a focused ion beam [46,47]. In each method, there are artifacts that affect the visualization of the properties of the samples, which one needs to be able to dissect. In the literature on thin films, the field of preparation is usually ignored; only SEM images of cross sections are presented. Many works, e.g., [48,49] show results obtained by only one technique, which usually is a standard fracture. There is a lack of summaries comparing different preparation methods with each other, and this paper is an attempt to address this niche. In our opinion, the improved methodology for the manufacturing of thin-film preparations for the purposes of SEM research can be successfully implemented using three methods, which are fracture and focused ion beams with gallium ion source and xenon plasma.

Standard fracture is a very common method. It does not affect the microstructure of the samples, enhances their morphology, and is cheap and fast. However, its main disadvantage is low repetition and high susceptibility to accidental damage. Delicate samples, e.g., coatings deposited on polymer substrates, may require preceding fracture by freezing a sample in liquid nitrogen in order to obtain brittle fracture of the polymer [47].

An ion technique used for the cross section of coatings is the focused ion beam (FIB) [46,47]. The structure and chemical composition of multilayer structures can be best visualized when imaging occurs perpendicularly across the interface [16], hence the ability to perform cross sections without artifacts becomes crucial. FIB enables microstructural characterization of coatings by cross-sectioning and preparation of specimens for scanning purposes, as well as for transmission electron microscopy [48–50]. The most important FIB advantage is the fabrication of the specimen in a selected area of interest. Compared to mechanical polishing, it avoids deformation, streaking of polished layers, and filling of existing cracks [48]. The duration of such a FIB preparation is usually not longer than 15 min. The cross section is formed by multiple passes of a high-energy ion beam (with a current value of ca. 2.7 nA). The penetration depth gradually increases, and at the deepest point of the section, its surface is perpendicular to the surface of the sample. The coarse-picking stage is followed by polishing the cross section with the ion beam obtained at a lower current (about 0.15 nA), after which the sample is ready for imaging. Iterative polishing and imaging of the cross-sectional surface enable a three-dimensional reconstruction of the microstructure [48].

It should be noted that in many works (e.g., [36]) dedicated to SEM studies of thin film materials, especially those prepared by PVD or CVD methods, the methodology for preparation techniques for accurate microscopic visualization is not given or has a residual description. This causes difficulties in the proper interpretation and characterization. Hence, there arises the need to develop a complex methodology. The aim of our work was to develop an improved methodology for the study of thin-film coatings on the basis of known and existing methods, including the aforementioned FIB. This innovative approach does not concern the improvement of standard methods, but it is devoted to a multithreaded approach to the SEM preparation itself, which will allow us to retain information about the actual microstructure and ensure the best possible material contrast (even for elements with similar atomic numbers). The research performed on the example of Ti/V/Ti multilayers, in order to present the sense of using such an improved methodology, showed that the true information about the tested sample can be obtained by assessing its microstructure based on SEM images made using the fracture technique, while the material composition can be well visualized using FIB methods (better than by EDS). The literature review also indicates the lack of data that describe the application of such a methodology to the analysis of multilayer coatings and, in particular, its effect on their structure. Comprehensive studies comparing the use of different techniques for cross section preparation, especially using two different sources of focused ion beam (including focused xenon plasma) are also omitted. Therefore, this work fills that niche.

2. Materials and Methods

2.1. Multilayer Project

Ti/V/Ti multilayer structures were designed for the purpose of the present study. Their construction and elemental composition were chosen for the development of a comprehensive SEM characterization methodology, including both preparation and imaging challenges. The samples were designed as multilayers based on Ti and V, consisting of three single-component metallic layers arranged alternately. Elements with similar atomic numbers (Z_{Ti} = 22 and Z_V = 23) were chosen in order to make a deliberate complication of the analysis, as contrast in SEM microscopy is closely related to atomic number. The top and bottom Ti layers had the same thickness (200 nm), while the thickness of the V middle layer was 100 nm, 50 nm, 30 nm, 20 nm, 10 nm, and 5 nm, respectively. Various thicknesses of the middle layer were supposed to allow the determination of the resolution limit of the microscope as well as the disadvantages of the preparation technique.

2.2. Manufacturing of Multilayer Coatings

Thin-film materials were prepared by pulsed DC magnetron sputtering. We described a detailed description of the applied sputtering method elsewhere [51–59]. Ti/V/Ti multilayers with the desired thickness were obtained by alternately sputtering (with the appropriate power) of targets made of Ti and V. For all prepared multilayers, the deposition processes were carried out in argon plasma at a pressure of 1.2×10^{-2} mbar, which was obtained with an argon flow of approximately 26 mL/min. The sputtered materials had the form of metallic titanium and vanadium targets (diameter: 30 mm, thickness: 3 mm, purity: 99.995%). They were alternately sputtered using two individual magnetrons that were powered with adequate power. The supply parameters were selected to take into account differences in the deposition rate of titanium as compared to vanadium. The distance between the target and the substrates (SiO_2 and Si) on the rotary drum was 90 mm. The deposition time of the bottom and top Ti films was 20 min., while the deposition time of the V middle layer was related to the desired thickness, which was 30 s up to 7 min. for 5 nm and 100 nm, respectively. Detailed data on the technological parameters of the sputtering processes are collected in Table 1. The thickness of individual films was estimated on SEM images with the use of the tools available in the software. The test samples from the deposition of single Ti and V layers allowed us to estimate the sputtering rate of both materials. Therefore, it was possible to accurately determine the time needed to obtain the desired thickness of the individual layers. These results were verified with the aid of an optical profiler (Talysurf CCI from Taylor Hobson).

Table 1. Deposition parameters of Ti/V/Ti multilayers by pulsed DC magnetron sputtering with their thickness.

	Deposition Parameters of Ti/V/Ti Multilayers								
	Bottom Ti Layer -Target Ti			Middle V Layer -Target V			Top Ti Layer -Target Ti		
P_{Ar} [mbar]	Power [W]	Time [min.]	t [nm]	Power [W]	Time [min.]	t [nm]	Power [W]	Time [min.]	t [nm]
1.2×10^{-2}	400	20	200	420	7 / 3.5 / 2 / 1.5 / 1 / 0.5	100 / 50 / 30 / 20 / 10 / 5	400	20	200

Designations: t—thickness, P_{Ar}—pressure of argon during sputtering.

2.3. Preparation Techniques and Details of SEM Measurements

For microscopic visualization of multilayers, three different cross section preparation techniques were used: (1) standard layer fracture, (2) FIB with a gallium ion beam, and (3) PFIB (plasma focused ion beam) with xenon plasma. A diagram of the following stages of the preparation techniques used and their analysis by SEM is shown in Figure 1. In addition to microscopic visualization, the analysis of elemental composition was also performed using EDS. Both the cross sections and their imaging were realized using a dualbeam microscope equipped with an electron column for imaging slides and an ion column for micromachining, respectively. All examined samples were glued to the SEM table using copper tape: (i) double-sided from their bottom and (ii) single-sided along the edge of the sample in order to ensure proper charge dissipation and stable mechanical connection.

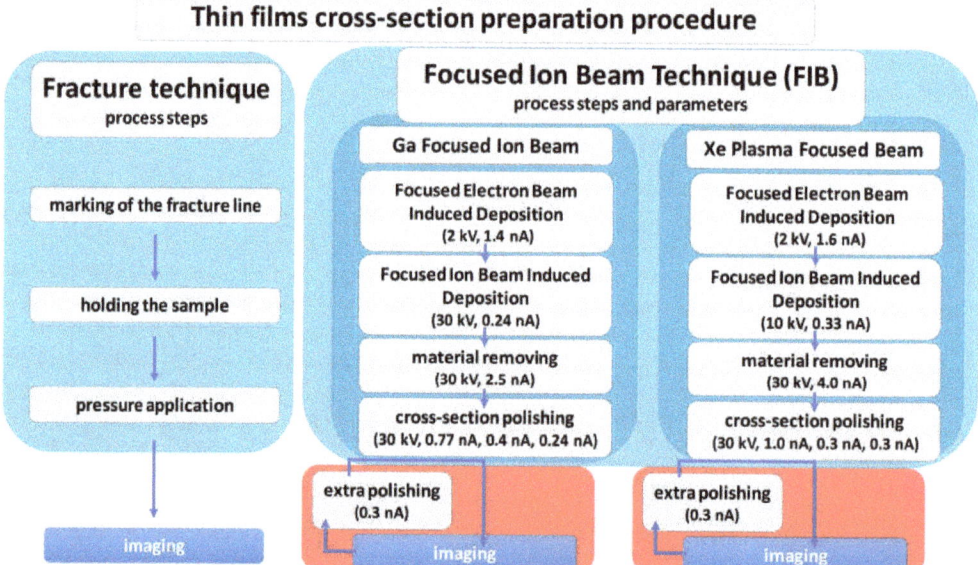

Figure 1. Cross-sectional preparation and SEM imaging procedure of thin-film multilayers.

Fracture of the sample: in this procedure, the line of the breakthrough was marked on the sample (from a side of the silicon substrate) with a diamond stylus. Then, one end of the specimen was held, while the other was pressed with a laboratory spatula (perpendicular to the surface), causing a break along the plotted line. The breakthroughs prepared in this way were placed in an SEM holder in the form of a vise, allowing imaging perpendicular to the plane of the breakthrough, and ensuring proper charge dissipation and mechanical stability. For their microscopic visualization, a Helios NanoLab 600i SEM microscope was used equipped with a Schottky gun, with a claimed resolution of 1.4 nm @ 1 kV. The imaging was carried out in immersion mode, using a TLD detector, at an acceleration voltage of 2 kV and a current of 0.17 nA.

Preparation of cross sections by focused beam techniques:

- **The preparation of FIB/Ga** was carried out with the aid of a Helios NanoLab 600i microscope. Its ion column is equipped in a gallium ion source Ga/LIMS (liquid metal ion source) with the following parameters: (i) current range 0.1–65 nA, (ii) accelerating voltage: 500 V–30 kV. For their microscopic visualization, Helios NanoLab 600i SEM microscope, equipped with a Schottky gun, with a claimed resolution of 1.4 nm @ 1 kV. The imaging was carried out in immersion mode, using a TLD detector, at an acceleration voltage of 2 kV and a current of 0.17 nA.
- **The preparation of PFIB/Xe** was carried out with the aid of the Helios G4 PFIB CXe microscope. The second source is inductively coupled Xe+ plasma with the following parameters: (i) current range: 1 pA–2.5 uA, (ii) accelerating voltage: 2–30 kV. Its electron column contains a Schottky gun with a claimed resolution of 0.6 nm @ 2–15 kV. Imaging was carried out in immersion mode, using a TLD detector, at an accelerating voltage of 2 kV and a current of 0.1 nA.

The first step in both techniques was selection of the area of interest (AOI), which must be protected from the destructive effects of the ion beam. For this purpose, a protective layer of platinum was applied by focused electron beam-induced deposition (FEBID) and focused ion beam-induced deposition (FIBID). In the FIBID process, ion beam bombardment of the surface results in damage to the surface and thus loss of information from the first layers of an examined sample. The use of the FEBID process, as a primary deposition of the Pt layer, offsets this problem. Electrons, compared to ions, have negligible mass, so the Pt application was nondestructive. The dimensions of the Pt layers were (i) $10 \times 1 \times 0.3$ µm (FEBID) and (ii) $10 \times 1 \times 1$ µm (FIBID). In the case of FIB/Ga, the beam parameters were as follows: (i) 2 kV and 1.6 nA, (ii) 30 kV and 0.24 nA, respectively. After the deposition of the protective layers, the next step was to remove the pre-AOI material using FIB. The width of the trench is usually equal to the width of the platinum layer, and the depth depends on the expected thickness of the layers, while the length is chosen so that the deepest layers can be observed. In the case of the tested samples, the trench dimensions were $10 \times 3 \times 1.5$ µm, while the parameters of the FIB/Ga beam were 30 kV, 2.5 nA. The final step in preparing the cross section was polishing its face to obtain a smooth section surface. This step requires several iterations, each with a smaller beam current. In this case, it was 0.77 nA, 0.4 nA, and 0.23 nA, with an acceleration voltage of 30 kV. An analogous procedure was used for a xenon plasma microscope. The differences are in the beam parameters. FEBID platinum was applied with beam parameters at 2 kV and 1.6 nA, while FIBID was deposited at 12 kV and 0.33 nA. The grinding was carried out with a beam-accelerating voltage of 30 kV and a current of 4 nA. Double cross-sectional polishing was performed with 30 kV, while the current was set at 1 nA and 0.3 nA, respectively.

It should also be noted that an important additional step in the developed visualization procedure was additional polishing. We have noticed that for samples with low material contrast, it is necessary to polish after the acquisition of each individual SEM image. It is related to the formation of impurities on the surface of the sample as a result of its interaction with the electron beam. Therefore, each passage of the beam on the cross-sectional surface causes a decrease in contrast. For materials with a similar (especially low) atomic number such as titanium and vanadium, where the material contrast is initially low, the additional reduction in the contrast significantly hinders the differentiation of the layers. Even if only a section of the sample was imaged, the contrast will be reduced over the entire cross section. In the case of examined thin-film multilayers, even a small decrease in contrast significantly affects the ability to distinguish layers. Moreover, changes in contrast mean that individual SEM images cannot be truly compared with each other, especially in the context of distinguishing elements based on their atomic number.

3. Results

In Figure 2, a comparison of the SEM and EDS measurements of the Ti/V/Ti multilayer cross sections prepared by the fracture technique, FIB/Ga and PFIB/Xe, can be seen. The thickness of the middle V layer was 100 nm, 30 nm, and 10 nm, respectively. As can be seen, depending on the preparation method, the visualization of the sample changes significantly. Only in the case of a fracturing procedure has the microstructure of thin films been preserved, and their columnar character can be seen. However, it is difficult to determine the position in which the base and top of the columns are located. Therefore, it is often impossible to determine even the position of the middle vanadium layer. Similar results can be seen in such works as [15,19,28]. In our studies, the columns were sometimes randomly broken off at different heights, regardless of the layered structure of the coating. In the case of the thinnest sample, the individual columns end up where the vanadium layer was located. This effect is similar to the renucleation of a layer after an interrupted deposition process. The column widths for all Ti/V/Ti multilayers are comparable. It is possible to precisely determine their widths of $39 \div 50$ nm for the sample with 100 nm of the middle V layer, as well as $44 \div 56$ nm for the 30 nm of the V layer, and $23 \div 62$ nm for the coating with 10 nm of V layer (Figure 3). The most important limitation of the fracture method is that the cross section does not show the material contrast, and it is impossible to distinguish all individual layers.

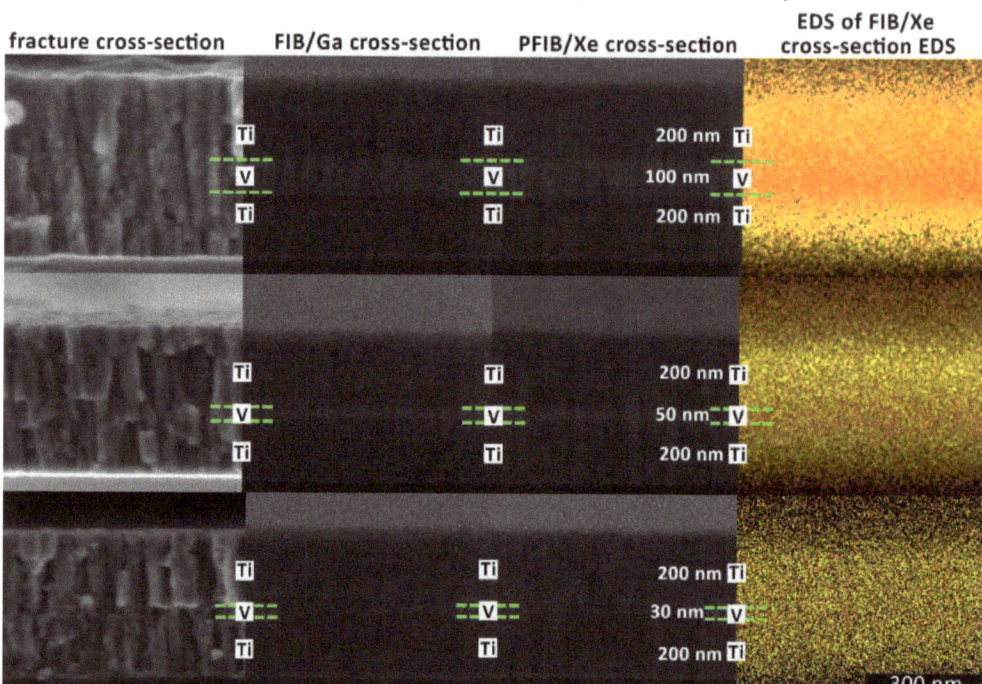

Figure 2. Cross sections of Ti/V/Ti multilayers with various thicknesses of middle V layer (100, 30, and 10 nm) based on SEM and EDS measurements, prepared by fracture technique, FIB/Ga, and PFIB/Xe. Note: The SEM images were recorded at high resolution and their original version can be found in the Supplementary Materials.

The results obtained with a focused ion beam have a different appearance. Both techniques, PFIB/Xe and FIB/Ga, provide high material contrast, and it is possible to distinguish Ti and V layers. However, in both cases, the thickness of the central V layer of 30 nm can be assumed as the limit. For the 10 nm V layer, a slight change in material contrast is apparent, but it is difficult to delineate it precisely. This layer could easily be overlooked, especially when the structure of the multilayer under examination is unknown. It is difficult to relate these results to the literature, as no work has been encountered on FIB cross sections through thin films with similar atomic numbers. In the case of FIB preparation, the main problem is that information about the microstructure of the analyzed coating is mostly lost. Only the outlines of individual columns and the spaces between columns can be identified. Comparison of the structure of the PFIB/Xe and FIB/Ga cross sections indicates a better distinction of the columns based on the PFIB/Xe method. This may be related to the formation of a thinner amorphous layer due to the use of Xe ions compared to Ga ions [1,60–62]. Determining the width of the columns is also possible in the case of both FIB preparation techniques, but it is significantly more difficult than the fracturing technique. For PFIB/Xe, the measured widths were 44 ÷ 46 nm, 35 ÷ 45 nm, and 46 ÷ 45 nm for multilayers with a middle V layer of 100 nm, 30 nm, and 10 nm, respectively (Figure 3). However, for FIB/Ga, these widths were 41 ÷ 49 nm, 46 ÷ 52 nm, and 34 ÷ 44 nm, respectively (Figure 3). These results are in good agreement with those obtained by the fracture method. However, a wider range of measured column widths can be observed in samples prepared by FIB/Ga. Thus, it can be concluded that both cross sections made with a focused ion beam have similar characteristics. The apparent differences related to the preservation of the original microstructure of the sample are relatively insignificant. However, it can be assumed that the best preparation method, burdened with fewer artifacts, is PFIB/Xe.

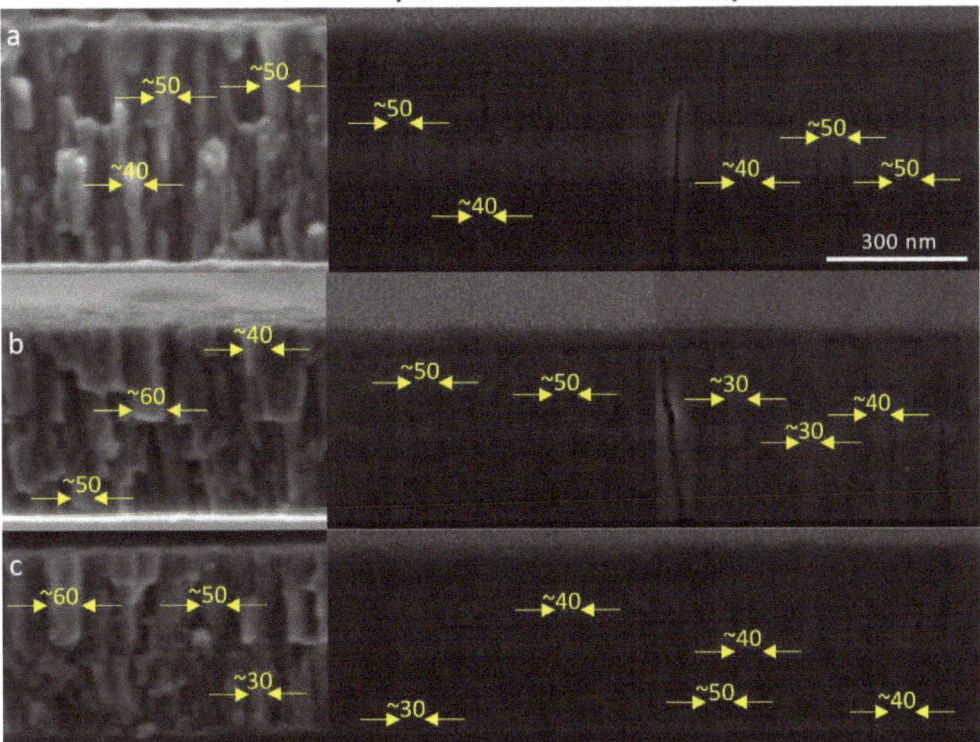

Figure 3. Determination of column width based on cross sections of Ti/V/Ti multilayers (with various thicknesses of the middle V layer: (**a**) 100 nm, (**b**) 30 nm, and (**c**) 10 nm) prepared by fracture technique, FIB/Ga, and PFIB/Xe. Note: The SEM images were recorded at high resolution and their original version can be found in the Supplementary Materials.

It should also be mentioned that the undeniable advantage of the FIB/Ga and PFIB/Xe preparation is the possibility of obtaining a high material contrast, and thus the ability to distinguish the material composition of multilayers with slightly different atomic numbers, i.e., Z_{Ti} = 22 and Z_V = 23. As can be seen in Figure 2, the EDS method does not provide such capabilities. It is impossible to clearly determine the interface between individual layers; thus, the method does not allow one to accurately determine their thickness. Even in the case of a Ti/V/Ti multilayer with the 100 nm middle V layer, where the location of all layers corresponds to their actual arrangement, the exact determination of the vanadium thickness is difficult. The problem worsens as the thickness of the middle layer decreases. For a V thickness of 30 nm, it is possible to speculate with an assumed approximation about the position of the middle layer, but neither its position nor its width corresponds to the SEM images of the cross section. For a V thickness of 10 nm, the signal from both elements is evenly distributed over the cross-sectional area, so it is not possible to observe individual layers. It should be noted that for layers of 10 nm or less, despite the preservation of material contrast, SEM images alone may not be sufficient to define the multilayer structure of the sample. Most of the available literature only reports percentage results of EDS analysis, e.g., [63,64], or elemental maps of the surface, e.g., [65], so it is not possible to confront obtained results with them. In such a case, EDS analysis is used to determine the elements comprising the sample and complements the FIB section imaging.

The next step of the study was to determine the resolution afforded by the preparation of FIB/Ga and PFIB/Xe methods as compared to the fracturing technique. Therefore, cross sections of multilayers with V layer thicknesses of 100, 50, 30, and 5 nm are included (Figure 4). As can be seen, it is difficult to discern where the V layer is on the cross section resulting from conventional fracturing. The growth of both the titanium and vanadium layers during deposition was similar, so their columnar morphology can be distinguished. In some cases, it is even possible to specify only by random breakage of the layers where columns begin with the next layer, but it is not repetitive and is often ambiguous. This is due to the fact that with a small thickness of the films included in the multilayer coating, the effect of reproducing the growth of the previous lower layer occurs (Figure 4a). The situation is completely different in the case of a breakthrough obtained with both FIB techniques, where the 10 nm thick multilayer is the last distinguishable one (Figure 4b,c). Identification of a 5 nm thick layer with an unknown sample structure is impossible. It can be possible only for well-defined and specially designed multilayers, most often as a comparison of SEM images of several samples. The difference in contrast between such thin films is imperceptible. As can be seen in Figure 4, the resolution limit has been reached. Moreover, it can be noticed that the material contrast for cross sections obtained by the PFIB/Xe and FIB/Ga methods is similar. The disclosed microstructures of these two cross sections are convergent. However, the columnar nature of the multilayer was better revealed in the case of PFIB/Xe. Most probably, it is related to the implantation of gallium ions in the structure of the Ti/V/Ti cross sections [1,40,41]. This effect is known [42] and while its influence on tested materials can sometimes be neglected, in the case of the multilayers analyzed, the result of fewer artifacts results from the PFIB/Xe preparation technique.

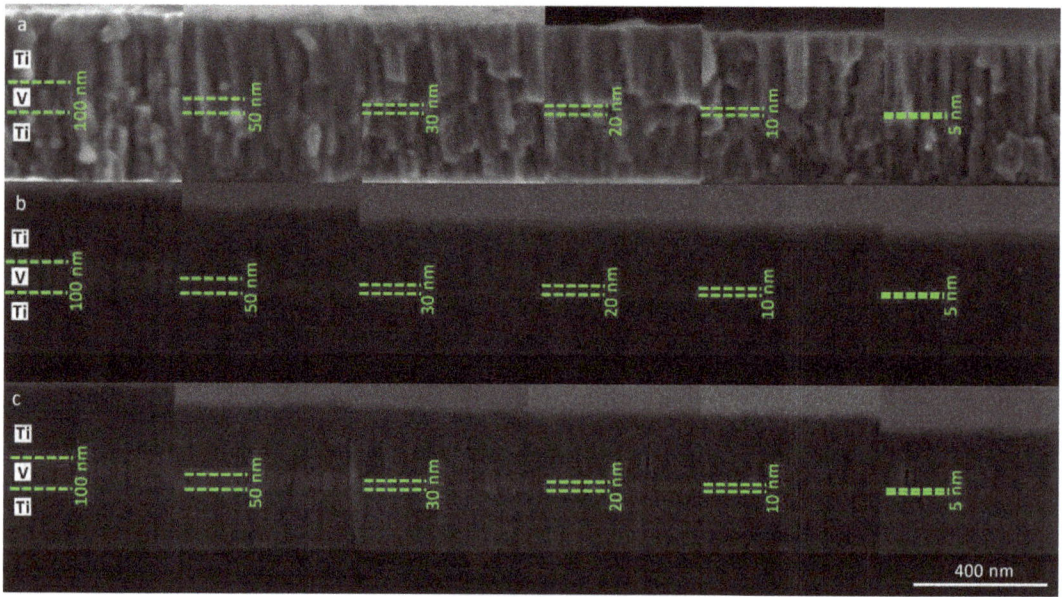

Figure 4. Comparison of the resolution afforded by the preparation of Ti/V/Ti multilayer cross sections by: (**a**) fracturing, (**b**) FIB/Ga, (**c**) PFIB/Xe, where the thickness of the middle V layer was reduced from 100 nm to 5 nm. Note: SEM images were recorded in high resolution, and their original version can be found in Supplementary Materials.

4. Conclusions

An improved methodology for the SEM study of thin-film coatings as an innovative multithreaded approach to the preparation itself was described. It allows information to be retained about the actual microstructure and ensures high material contrast even for elements with similar atomic numbers. It was found that determining the microstructure application of the standard fracture technique is necessary (as with most of the noninvasive methods). However, while the microstructure was preserved, the material contrast remained invisible, making it impossible to distinguish each layer in the construction of a multilayer. EDS analysis was also not sufficient to present their construction and the interfaces were not clearly defined. Even when the thickness of individual V films was around 100 nm, it was difficult to determine by EDS the position of the interfaces between individual layers in a Ti/V/Ti multilayer. Material contrast enhancement occurs only for FIB techniques. Cross-sectional studies showed that the 10 nm mid-V layer was the resolution limit. Moreover, a comparison of the PFIB/Xe and FIB/Ga cross sections revealed that fewer artifacts give the PFIB method, and hence this technique seems to be better for the analysis of multilayer nanostructures. In our opinion, reliable information about the properties of modern nanomaterials, especially multilayers used in electronics, can be obtained by analyzing a two-part SEM image, where the first is a fracture, while the second is a FIB/PFIB cross section. It is worth noting that there were only subtle differences between SEM images of cross sections made by FIB-Ga and PFIB-Xe, but the decrease in local amorphization, the lack of Ga-ions incorporation into the sample, and slightly better contrast are in favor of Xe plasma.

Supplementary Materials: The following supporting information can be downloaded at: https://www.mdpi.com/article/10.3390/coatings13020316/s1, Figure S1: Cross-sectional preparation and SEM imaging procedure of thin-film multilayers.; Figure S2: Cross sections of Ti/V/Ti multilayers with various thicknesses of middle V layer (100, 30, and 10 nm) based on SEM and EDS measurements, prepared by fracture technique, FIB/Ga, and PFIB/Xe. Figure S3: Determination of column width based on cross sections of Ti/V/Ti multilayers (with various thicknesses of the middle V layer: (**a**) 100 nm, (**b**) 30 nm, and (**c**) 10 nm) prepared by fracture technique, FIB/Ga, and PFIB/Xe. Figure S4: Comparison of the resolution afforded by the preparation of Ti/V/Ti multilayer cross sections by: (**a**) fracturing, (**b**) FIB/Ga, (**c**) PFIB/Xe, where the thickness of the middle V layer was reduced from 100 nm to 5 nm.

Author Contributions: Conceptualization, D.W. and M.S.; methodology, M.S. and D.W.; validation, D.W.; investigation, M.S. and A.C.; resources, A.Z.; writing—original draft preparation, M.S. and D.W.; writing—review and editing, M.S. and D.W.; supervision, D.W. All authors have read and agreed to the published version of the manuscript.

Funding: This work was financed from the sources given by the MNiSW in the years 2020–2024 as Implementation Doctorate Programme (project number: DWD/4/5/2020).

Institutional Review Board Statement: Not applicable.

Informed Consent Statement: Not applicable.

Data Availability Statement: The data presented in this study are available on request from the corresponding author.

Conflicts of Interest: The authors declare no conflict of interest.

References

1. Ernst, A.; Wie, M.; Aindow, M. A Comparison of Ga FIB and Xe-Plasma FIB of Complex Al Alloys. *Microsc. Microanal.* **2017**, *23* (Suppl. S1), 288–289. [CrossRef]
2. Lešer, V.; Drobne, D.; Pipan, Ž.; Milani, M.; Tatti, F. Comparison of different preparation methods of biological samples for FIB milling and SEM investigation. *J. Microsc.* **2009**, *233*, 309–319. [CrossRef] [PubMed]
3. Sutton, N.; Hughes, N.; Handley, P. A comparison of conventional SEM techniques, low temperature SEM and the electroscan wet scanning electron microscope to study the structure of a biofilm of Streptococcus crista CR3. *J. Appl. Bacteriol.* **1994**, *76*, 448–454. [CrossRef]

4. Wirth, R. Focused Ion Beam (FIB) combined with SEM and TEM: Advanced analytical tools for studies of chemical composition, microstructure and crystal structure in geomaterials on a nanometre scale. *Chem. Geol.* **2009**, *261*, 217–229. [CrossRef]
5. Kjellsena, K.; Monsøyb, A.; Isachsenb, K.; Detwilerc, R. Preparation of flat-polished specimens for SEM-backscattered electron imaging and X-ray microanalysis—Importance of epoxy impregnation. *Cem. Concr. Res.* **2003**, *33*, 611–616. [CrossRef]
6. Erdman, N.; Campbell, R.; Asahina, S. Precise SEM Cross Section Polishing via Argon Beam Milling. *Microsc. Today* **2006**, *14*, 22–25. [CrossRef]
7. Lee, J.T.Y.; Chow, K. SEM sample preparation for cells on 3D scaffolds by freeze-drying and HMDS. *Scanning* **2012**, *34*, 12–25. [CrossRef]
8. Narbutt, O.; Dąbrowski, H.; Dąbrowska, G. The process of freeze-drying, its wide applications and defense mechanisms against dehydratation. *Eduk. Biol. I Sr.* **2017**, *2*, 20–29.
9. Bassim, N.; De Gregorio, B.; Kilcoyne, A.; Scott, K.; Chou, T.; Wirick, S.; Cody, G.; Stroud, R. Minimizing damage during FIB sample preparation of soft materials. *J. Microsc.* **2012**, *245*, 288–301. [CrossRef]
10. Kim, S.; Park, M.J.; Balsara, N.P.; Liu, G.; Minor, A.M. Minimization of focused ion beam damage in nanostructured polymer thin films. *Ultramicroscopy* **2011**, *111*, 191–199. [CrossRef]
11. Brodusch, N.; Yourdkhani, M.; Hubert, P.; Gauvin, R. Efficient cross-section preparation method for high-resolution imaging of hard polymer composites with a scanning electron microscope. *J. Microsc.* **2015**, *260*, 117–124. [CrossRef] [PubMed]
12. Koh, J.W.; Hwang, G.T.; Hyun, M.S.; Yang, J.; Kim, J.W. Semiconductor layer extraction techniques by SEM. In Proceedings of the 18th IEEE International Symposium on the Physical and Failure Analysis of Integrated Circuits (IPFA), Incheon, Republic of Korea, 4–7 July 2011; pp. 1–3.
13. Zhang, Y.; Popielarski, B.; Davidson, K.; Men, L.; Zhao, W.; Baumann, F. Development of Ultra-thin TEM Lamella Preparation Technique and Its Application in Failure Analysis. *Microsc. Microanal.* **2020**, *26* (Suppl. S2), 1400–1402. [CrossRef]
14. Vazdirvanidis, A.; Pantazopoulos, G.; Louvaris, A. Failure analysis of a hardened and tempered structural steel (42CrMo4) bar for automotive applications. *Eng. Fail. Anal.* **2009**, *16*, 1033–1038. [CrossRef]
15. Strecker, A.; Bäder, U.; Kelsch, M.; Salzberger, U.; Sycha, M.; Gao, M.; Richter, G.; van Benthem, K. Progress in the preparation of cross-sectional TEM specimens by ion-beam thinning. *Int. J. Mater. Res.* **2003**, *94*, 290–297. [CrossRef]
16. Abrahams, M.S.; Buiocchi, C.J. Crosssectional specimens for transmission electron microscopy. *J. Appl. Phys.* **1974**, *45*, 3315–3316. [CrossRef]
17. Naito, M.; Yokoyama, T.; Hosokawa, T.; Nogi, K. *Nanoparticle Technology Handbook*, 3rd ed.; Elsevier: Amsterdam, The Netherlands, 2018.
18. Contreras, E.; Galindez, Y.; Rodas, M.; Bejarano, G.; Gómez, M. CrVN/TiN nanoscale multilayer coatings deposited by DC unbalanced magnetron sputtering. *Surf. Coat. Technol.* **2017**, *332*, 214–222. [CrossRef]
19. Babaei, K.; Fattah-alhosseini, A.; Elmkhah, H.; Ghomi, H. Surface characterization and electrochemical properties of tantalum nitride (TaN) nanostructured coatings produced by reactive DC magnetron sputtering. *Surf. Interfaces* **2020**, *21*, 100685. [CrossRef]
20. Budak, S.; Xiao, Z.; Johnson, B.; Cole, J.; Drabo, M.; Tramble, A.; Casselberry, C. Highly-Efficient Advanced Thermoelectric Devices from Different Multilayer Thin Films. *Am. J. Eng. Appl. Sci.* **2016**, *9*, 356–363. [CrossRef]
21. Rao, S.G.; Shu, R.; Wang, S.; Boyd, R.; Giuliani, F.; le Febvrier, A.; Eklund, P. Thin film growth and mechanical properties of CrFeCoNi/TiNbZrTa multilayers. *Mater. Des.* **2022**, *224*, 111388. [CrossRef]
22. Aulin, C.; Karabulut, E.; Tran, A.; Wågberg, L.; Lindström, T. Transparent Nanocellulosic Multilayer Thin Films on Polylactic Acid with Tunable Gas Barrier Properties. *ACS Appl. Mater. Interfaces* **2013**, *5*, 7352–7359. [CrossRef]
23. Bartosik, M.; Daniel, R.; Mitterer, C.; Matko, I.; Burghammer, M.; Mayrhofer, P.H.; Keckes, J. Cross-sectional X-ray nanobeam diffraction analysis of a compositionally graded CrNx thin film. *Thin Solid Film.* **2013**, *542*, 1–4. [CrossRef]
24. Parreira, N.M.G.; Polcar, T.; Cavaleiro, A. Characterization of W–O coatings deposited by magnetron sputtering with reactive gas pulsing. *Surf. Coat. Technol.* **2007**, *201*, 5481–5486. [CrossRef]
25. Cavaleiro, A.J.; Santos, R.J.; Ramos, A.S.; Vieira, M.T. In-situ thermal evolution of Ni/Ti multilayer thin films. *Intermetallics* **2014**, *51*, 11–17. [CrossRef]
26. Kang, C.; Huang, H. Mechanical load-induced interfacial failure of a thin film multilayer in nanoscratching and diamond lapping. *J. Mater. Process. Technol.* **2016**, *229*, 528–540. [CrossRef]
27. Kang, C.; Huang, H. A comparative study of conventional and high speed grinding characteristics of a thin film multilayer structure. *Precis. Eng.* **2017**, *50*, 222–234. [CrossRef]
28. Bravman, J.C.; Sinclair, R. The preparation of cross-section specimens for transmission electron microscopy. *J. Electron Microsc. Tech.* **1984**, *1*, 53–61. [CrossRef]
29. Song, K.; Kim, S.G.; Kim, H.J.; Kim, S.K.; Kang, W.K.; Lee, J.C.; Yoon, K.H. Preparation of CuIn1−xGaxSe2 thin films by sputtering and selenization process. *Sol. Energy Mater. Sol. Cells* **2003**, *75*, 145–153. [CrossRef]
30. Wang, Y.; Lin, Z.; Cheng, X.; Xiao, H.; Zhang, F.; Zou, S. Study of HfO2 thin films prepared by electron beam evaporation. *Appl. Surf. Sci.* **2004**, *228*, 93–99. [CrossRef]
31. Santana, A.E.; Karimi, A.; Derflinger, V.H.; Schütze, A. Microstructure and mechanical behavior of TiAlCrN multilayer thin films. *Surf. Coat. Technol.* **2004**, *177–178*, 334–340. [CrossRef]
32. Dieterle, L.; Butz, B.; Müller, E. Optimized Ar+-ion milling procedure for TEM cross-section sample preparation. *Ultramicroscopy* **2011**, *111*, 1636–1644. [CrossRef]

33. Stefenelli, M.; Daniel, R.; Ecker, W.; Kiener, D.; Todt, J.; Zeilinger, A.; Mitterer, C.; Burghammer, M.; Keckes, J. X-ray nanodiffraction reveals stress distribution across an indented multilayered CrN–Cr thin film. *Acta Mater.* **2015**, *85*, 24–31. [CrossRef]
34. Lloyd, G.E. Atomic number and crystallographic contrast images with the SEM: A review of backscattered electron techniques. *Mineral. Mag.* **1987**, *51*, 3–19. [CrossRef]
35. Yamashita, S.; Kikkawa, J.; Yanagisawa, K.; Nagai, T.; Ishizuka, K.; Kimoto, K. Atomic number dependence of Z contrast in scanning transmission electron microscopy. *Sci. Rep.* **2018**, *8*, 12325. [CrossRef]
36. Contreras, E.; Bejarano, G.; Gómez, M. Synthesis and microstructural characterization of nanoscale multilayer TiAlN/TaN coatings deposited by DC magnetron sputtering. *Int. J. Adv. Manuf. Technol.* **2019**, *101*, 663–673. [CrossRef]
37. Riester, M.; Lechner, M.D.; Hilgers, H. Physical vapor-deposited thin hard films on polymers: Sample preparation for SEM analysis. *J. Adhes. Sci. Technol.* **1999**, *13*, 963–971. [CrossRef]
38. Gao, W.; Li, Z. ZnO thin films produced by magnetron sputtering. *Ceram. Int.* **2004**, *30*, 1155–1159. [CrossRef]
39. Wang, C.; Brault, P.; Zaepffel, C.; Thiault, J.; Pineau, A.; Sauvage, T. Deposition and structure of W–Cu multilayer coatings by magnetron sputtering. *J. Phys. D: Appl. Phys.* **2003**, *36*, 2709–2713. [CrossRef]
40. Busch, R.; Tielemann, C.; Reinsch, S.; Müller, R.; Patzig, C.; Krause, M.; Höche, T. Sample preparation for analytical scanning electron microscopy using initial notch sectioning. *Micron* **2021**, *150*, 103090. [CrossRef] [PubMed]
41. Marshall, A.F.; Dobbertin, D.C. Cross-sectioning of layered thin films by ultramicrotomy. *Ultramicroscopy* **1986**, *19*, 69–73. [CrossRef]
42. Becker, O.; Bange, K. Ultramicrotomy: An alternative cross section preparation for oxidic thin films on glass. *Ultramicrotomy* **1993**, *52*, 73–84. [CrossRef]
43. Nguyen, T.D.; Gronsky, R.; Kortright, J.B. Cross-sectional transmission electron microscopy of X-ray multilayer thin film structures. *J. Electron Microsc. Tech.* **1991**, *19*, 473–485. [CrossRef] [PubMed]
44. Sunaoshi, T.; Takeuchi, S.; Kamino, A.; Sasajima, M.; Ito, H. Sample preparation technique for the revelation of a semiconductor dopant using an FE-SEM. In Proceedings of the 2016 IEEE 23rd International Symposium on the Physical and Failure Analysis of Integrated Circuits (IPFA), Singapore, 18–21 July 2016; pp. 148–151.
45. Reada, J.C.; Braganca, P.M.; Robertson, N.; Childress, J.R. Magnetic degradation of thin film multilayers during ion milling. *APL Mater.* **2014**, *2*, 046109. [CrossRef]
46. Minor, A.M. FIB sample preparation of thin films and soft materials. *Microsc. Microanal.* **2009**, *15* (Suppl. S2), 1544–1545. [CrossRef]
47. Massl, S.; Thomma, W.; Keckes, J.; Pippan, R. Investigation of fracture properties of magnetron-sputtered TiN films by means of a FIB-based cantilever bending technique. *Acta Mater.* **2009**, *57*, 1768–1776. [CrossRef]
48. Cairney, J.M.; Munroe, P.R.; Hoffman, M. The application of focused ion beam technology to the characterization of coatings. *Surf. Coat. Technol.* **2005**, *198*, 165–168. [CrossRef]
49. Pallecchi, I.; Pellegrino, L.; Bellingeri, E.; Siri, A.S.; Marré, D.; Gazzadi, G.C. Investigation of FIB irradiation damage in La0.7Sr0.3MnO3 thin films. *J. Magn. Magn. Mater.* **2008**, *320*, 1945–1951. [CrossRef]
50. Tanaka, M.; Furuya, K.; Saito, T. TEM observation of FIB induced damages in Ni2Si/Si thin films. *Nucl. Instrum. Methods Phys. Res. B* **1997**, *127–128*, 98–101. [CrossRef]
51. Wojcieszak, D.; Mazur, M.; Kaczmarek, D.; Mazur, P.; Szponar, B.; Domaradzki, J.; Kepinski, L. Influence of the surface properties on bactericidal and fungicidal activity of magnetron sputtered Ti-Ag and Nb-Ag thin films. *Mater. Sci. Eng. C* **2016**, *62*, 86–95. [CrossRef]
52. Mazur, M.; Wojcieszak, D.; Domaradzki, J.; Kaczmarek, D.; Poniedziałek, A.; Domanowski, P. Investigation of microstructure, micro-mechanical and optical properties of HfTiO4 thin films prepared by magnetron co-sputtering. *Mater. Res. Bull.* **2015**, *72*, 116–122. [CrossRef]
53. Mazur, M.; Howind, T.; Gibson, D.; Kaczmarek, D.; Morgiel, J.; Wojcieszak, D.; Zhu, W.; Mazur, P. Modification of various properties of HfO2 thin films obtained by changing magnetron sputtering conditions. *Surf. Coat. Technol.* **2017**, *320*, 426–431. [CrossRef]
54. Mazur, M.; Domaradzki, J.; Wojcieszak, D.; Kaczmarek, D. Investigations of elemental composition and structure evolution in (Ti,Cu)-oxide gradient thin films prepared using (multi)magnetron co-sputtering. *Surf. Coat. Technol.* **2018**, *334*, 150–157. [CrossRef]
55. Adamiak, B.; Wiatrowski, A.; Domaradzki, J.; Kaczmarek, D.; Wojcieszak, D.; Mazur, M. Preparation of multicomponent thin films by magnetron co-sputtering method: The Cu-Ti case study. *Vacuum* **2019**, *161*, 419–428. [CrossRef]
56. Wojcieszak, D.; Mazur, M.; Kaczmarek, D.; Szponar, B.; Grobelny, M.; Kalisz, M.; Pelcarska, A.; Szczygiel, I.; Poniedzialek, A.; Osekowska, M. Structural and surface properties of semitransparent and antibacterial (Cu,Ti,Nb)Ox coating. *Appl. Surf. Sci.* **2016**, *380*, 159–164. [CrossRef]
57. Wojcieszak, D.; Kaczmarek, D.; Antosiak, A.; Mazur, M.; Rybak, Z.; Rusak, A.; Osekowska, M.; Poniedzialek, A.; Gamian, A.; Szponar, B. Influence of Cu-Ti thin film surface properties on antimicrobial activity and viability of living cells. *Mater. Sci. Eng. C* **2015**, *56*, 48–56. [CrossRef]
58. Wojcieszak, D.; Mazur, M.; Kalisz, M.; Grobelny, M. Influence of Cu, Au and Ag on structural and surface properties of bioactive coatings based on titanium. *Mater. Sci. Eng. C* **2017**, *71*, 1115–1121. [CrossRef]

59. Domaradzki, J.; Wiatrowski, A.; Kotwica, T.; Mazur, M. Analysis of electrical properties of forward-to-open (Ti,Cu)Ox memristor rectifier with elemental gradient distribution prepared using (multi)magnetron co-sputtering process. *Mater. Sci. Semicond. Process.* **2019**, *94*, 9–14. [CrossRef]
60. Zhong, X.C.; Wade, A.; Withers, P.J.; Zhou, X.; Cai, C.; Haigh, S.J.; Burke, G.M. Comparing Xe+pFIB and Ga+FIB for TEM sample preparation of Al alloys: Minimising FIB-induced artefacts. *J. Microsc.* **2021**, *282*, 101–112. [CrossRef]
61. Hu, C.; Aindow, M.; Wei, M. Focused ion beam sectioning studies of biomimetic hydroxyapatite coatings on Ti-6Al-4V substrates. *Surf. Coat. Technol.* **2017**, *313*, 255–262. [CrossRef]
62. Burnett, T.L.; Kelley, R.; Winiarski, B.; Contreras, L.; Daly, M.; Gholinia, A.; Burke, M.G.; Withers, P.J. Large volume serial section tomography by Xe Plasma FIB dual beam Microscopy. *Ultramicroscopy* **2016**, *161*, 119–129. [CrossRef]
63. Valdés, M.; Pascual-Winter, M.F.; Bruchhausen, A.; Schreiner, W.; Vázquez, M. Cross-Section Analysis of the Composition of Sprayed Cu2ZnSnS4 Thin Films by XPS, EDS, and Multi-Wavelength Raman Spectroscopy. *Phys. Status Solidi* **2018**, *215*, 1800639. [CrossRef]
64. Giurlani, W.; Enrico Berretti, E.; Lavacchi, A.; Innocenti, M. Thickness determination of metal multilayers by ED-XRF multivariate analysis using Monte Carlo simulated standards. *Anal. Chim. Acta* **2020**, *1130*, 72–79. [CrossRef]
65. Güzelçimen, F. The effect of thickness on surface structure of rf sputtered TiO2 thin films by XPS, SEM/EDS, AFM and SAM. *Vacuum* **2020**, *182*, 109766. [CrossRef]

Disclaimer/Publisher's Note: The statements, opinions and data contained in all publications are solely those of the individual author(s) and contributor(s) and not of MDPI and/or the editor(s). MDPI and/or the editor(s) disclaim responsibility for any injury to people or property resulting from any ideas, methods, instructions or products referred to in the content.

Review

Sputtering Process of Sc$_x$Al$_{1-x}$N Thin Films for Ferroelectric Applications

Jacob M. Wall and Feng Yan *

Department of Metallurgical and Materials Engineering, The University of Alabama, Tuscaloosa, AL 35487, USA
* Correspondence: fyan4@ua.edu

Abstract: Several key sputtering parameters for the deposition of Sc$_x$Al$_{1-x}$N such as target design, sputtering atmosphere, sputtering power, and substrate temperature are reviewed in detail. These parameters serve a crucial role in the ability to deposit satisfactory films, achieve the desired stoichiometry, and meet the required film thickness. Additionally, these qualities directly impact the degree of c-axis orientation, grain size, and surface roughness of the deposited films. It is systematically shown that the electric properties of Sc$_x$Al$_{1-x}$N are dependent on the crystal quality of the film. Although it is not possible to conclusively say what the ideal target design, sputtering atmosphere, sputtering power, and substrate temperature should be for all sputtering processes, the goal of this paper is to analyze the impacts of the various sputtering parameters in detail and provide some overarching themes that arise to assist future researchers in the field in quickly tuning their sputtering processes to achieve optimum results.

Keywords: sputtering; thin film; ScAlN; ferroelectrics

1. Introduction

The use of piezoelectric thin films has seen a surge in recent years due to its various applications such as in microelectromechanical systems (MEMS), bulk acoustic wave (BAW) resonators, and surface acoustic wave (SAW) resonators to name a few [1]. Traditionally, Aluminum Nitride (AlN) was among one of the most used piezoelectric materials for such applications because it is completely semiconductor compatible and has satisfactory piezoelectric properties [2–4]. However, the demand for materials with superior piezoelectric coefficients resulted in the investigation of Sc-doped AlN thin films. In 2009, it was discovered that the partial substitution of aluminum with scandium to form Sc$_x$Al$_{1-x}$N resulted in a substantial increase in the piezoelectric response [5]. The approximately 500% increase in piezoelectric modulus, d$_{33}$, of Sc$_x$Al$_{1-x}$N when x = 0.43 was ascribed to the phase transition from wurtzite to layered hexagonal with increasing scandium content [5]. Furthermore, it was found that the electromechanical coupling factor, k$_t^2$, could be improved from 7% to 10% when x ≤ 0.2 [6]. Due to these excellent piezoelectric characteristics, Sc$_x$Al$_{1-x}$N has garnered substantial interest in the piezoelectric community and is expected to be an ideal candidate for piezoelectric thin-film layers.

More recently in 2019, Fitchner et al., demonstrated the first official instance of ferroelectric switching in Sc$_x$Al$_{1-x}$N [7]. This paramount discovery has generated a resurgence in research related to Sc$_x$Al$_{1-x}$N. High-performance thin-film ferroelectrics exhibiting good technological compatibility with generic semiconductor technology are in urgent demand due to emerging applications based on controlling electrical polarization, multitude of memories, and micro/nano-actuators [7]. Similar to the piezoelectric response observed in Sc$_x$Al$_{1-x}$N, the ferroelectric nature of the material arises as a result of the anisotropic crystal structure that originates from the layered-hexagonal structure when Sc-doping occurs [8]. For this reason, Sc$_x$Al$_{1-x}$N is considered to be a possible candidate for the development of practical two-terminal ferroelectric nonvolatile memory devices (FE-NVMs) [9].

Citation: Wall, J.M.; Yan, F. Sputtering Process of Sc$_x$Al$_{1-x}$N Thin Films for Ferroelectric Applications. *Coatings* **2023**, *13*, 54. https://doi.org/10.3390/coatings13010054

Academic Editor: Rafal Chodun

Received: 21 November 2022
Revised: 10 December 2022
Accepted: 20 December 2022
Published: 28 December 2022

Copyright: © 2022 by the authors. Licensee MDPI, Basel, Switzerland. This article is an open access article distributed under the terms and conditions of the Creative Commons Attribution (CC BY) license (https://creativecommons.org/licenses/by/4.0/).

To date, there have been two primary deposition routes used when creating $Sc_xAl_{1-x}N$ thin films. These include both molecular beam epitaxy (MBE) and RF or magnetron sputtering. Although there are benefits to MBE deposition such as better control over thickness, crystallinity, and stoichiometry, sputtering represents a more viable deposition process because of its relatively low cost, reproducibility, good adhesion with the substrate, ability to deposit stoichiometry of target material, and potential for large-scale manufacturing. However, a systematic review of the sputtering parameters for thin film $Sc_xAl_{1-x}N$ is absent from the literature.

In this review, the evolution and optimization of the various sputtering parameters for $Sc_xAl_{1-x}N$ is reported. There are several key parameters during sputtering that have a critical outcome on film properties. For instance, in this review target design, sputtering atmosphere, sputtering power, and substrate temperature will be examined in detail. Deposition parameters such as these will directly affect the deposited film quality, including the thickness and morphology. Many reports have analyzed the effects of one or more of these deposition parameters on such quality indicators as film thickness, crystallinity, or final composition. Moreover, various reports have then connected those film quality indicators to the eventual electric properties of the resulting film. The type of substrate can also play a role in controlling the final crystal quality of the $Sc_xAl_{1-x}N$ films. Li et al., compared ScAlN films prepared on both silicon and C276 alloy substrates and found that the films prepared on C276 alloy substrates resulted in non-ideal, a-axis oriented ScAlN films. However, films prepared on Si resulted in all c-axis oriented crystals. The non-ideal behavior was attributed to the superior lattice mismatch of the Si substrate [10]. For this reason, most literature reports depositing ScAlN on c-axis oriented substrates such as silicon or sapphire. Thus, since the literature comparing the impacts of different substrates (i.e., amorphous silicon) are scarce this review chooses not to focus on this topic. Specifically, in this review, the goal is to interconnect the findings of all existing reports and generate a better understanding of the sputtering process for $Sc_xAl_{1-x}N$.

2. Structure and Properties of $Sc_xAl_{1-x}N$

2.1. Structure of $Sc_xAl_{1-x}N$

It has long been known that the III-nitride materials, such as AlN, GaN, and InN, possess a wurtzite-type crystal structure (space group P63mc) [11]. Moreover, there is a spontaneous polarization along the c-axis of these III-V semiconductors, which leads to the separation of group-III and nitrogen atoms in individual planes [7]. This phenomenon leads to the piezoelectric response in this class of materials. Moreover, it was discovered that the piezoelectric response of AlN can be significantly increased by forming solid solutions with ScN [5]. While pure ScN has a stable cubic rock salt crystal structure, it also maintains a highly metastable, nearly fivefold coordinated layered-hexagonal phase (space group P63/mmc) [12]. Therefore, the belief in the research community is that there is a transition from pure wurtzite to a more layered-hexagonal crystal structure with increasing scandium content as shown in Figure 1.

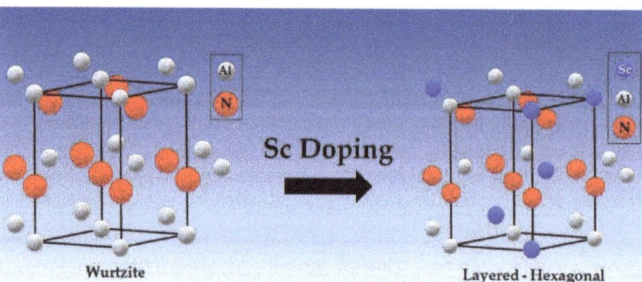

Figure 1. Crystal structure transition of AlN from wurtzite to layered-hexagonal with Sc doping to form $Sc_xAl_{1-x}N$.

Moreover, the metastable hexagonal phase works to flatten the ionic potential landscape, which causes the wurtzite basal plane and the length of metal-nitrogen bonds parallel to the c-axis relative to the lattice parameter, c, to increase [13,14]. In other words, the distance between the (0001) planes that hold both the nitrogen atoms and the metal atoms increases with increasing Sc-content. On account of this, the energy barrier is set to decrease with increasing Sc content, leading to an overall reduction of the energy barrier between the two polarization states of the wurtzite structure and creating an avenue for ferroelectric switching. Furthermore, the wurtzite to layered-hexagonal transition is responsible for both the reciprocal increase in piezoelectric response with increasing Sc content and the ability of $Sc_xAl_{1-x}N$ films to exhibit ferroelectric switching [7].

2.2. Properties of $Sc_xAl_{1-x}N$

A summary of some of the structural, optical, electrical, thermal, piezoelectric, and ferroelectric properties of $Sc_xAl_{1-x}N$ are reported in Table 1. Since $Sc_xAl_{1-x}N$ is a ternary alloy with properties that are highly dependent on the Sc content, most of the reported values are presented as a range. Additionally, the equations used to generate the values with respect to Sc content are provided as well.

Table 1. $Sc_xAl_{1-x}N$ properties.

	Property	Reported Value	[Ref]
Structural Properties	Density (g/cm^3)	3.255–3.456; $\rho(x) = 3.806x + 3.255(1-x) - 0.298x(1-x)$	[15]
	Elastic Modulus (GPa)	535–269 (for x = 0–0.41)	[16]
	Elastic constant C_{11} (GPa)	396.00–280.96; $C_{11}(x) = 285.12x + 396(1-x) - 238.39x(1-x)$	[15]
	Elastic constant C_{12} (GPa)	137.00–161.59; $C_{12}(x) = 180.57x + 137(1-x) + 11.23x(1-x)$	[15]
	Elastic constant C_{13} (GPa)	108.00–137.84; $C_{13}(x) = 141.70x + 108(1-x) + 51.95x(1-x)$	[15]
	Poisson's ratio	$\nu_{21} = 0.343$ (when x = 0.5)	[15]
	Crystal structure	wurtzite/layered-hexagonal	[15]
	Lattice constant (Å)	a = 3.0997; c = 4.59569	[17]
Optical Properties	Effective electron mass	0.46 m_0 (for x = 0.18)	[18]
	Refractive index (visible to IR)	2.05	[19]
Electrical Properties	Breakdown field (MV/cm)	12.44 (for x = 0.18)	[18]
	Mobility of electrons/holes (cm^2/V-s)	147–205 (for x = 0.18)	[18]
	Dielectric constant	10.31–34.52; $\varepsilon_{33} = 89.93x + 10.31(1-x) - 62.48x(1-x)$	[15]
	Energy band gap (eV)	4.29–6.15; $E_g(x) = 6.15 - 9.32x$ (for x ≤ 0.2)	[20]
	Resistivity (10^{12} Ω-cm)	1.0–3.5	[10]
Thermal Properties	Thermal conductivity (W/m-K)	3.0–8.0 (for x = 0–0.20)	[21]
	Coefficient of thermal expansion ($\times 10^{-6}$/K)	4.29–4.65 (for x = 0–0.41)	[16]
	Debye temperature (K)	933 (for x = 0.18)/737 (for x = 0.25)	[18,19]
Piezoelectric Properties	Piezoelectric coeff. e_{15} (C/m^2)	−0.313–−0.135; $e_{15} = 0.308x - 0.313(1-x) - 0.528x(1-x)$	[15]
	Piezoelectric coeff. e_{31} (C/m^2)	−0.593–−0.829; $e_{31} = -1.353x - 0.593(1-x) + 0.576x(1-x)$	[15]
	Piezoelectric coeff. e_{33} (C/m^2)	1.471–3.642; $e_{33} = 9.125x + 1.471(1-x) - 6.625x(1-x)$	[15]
	Relative permittivity coeff. ε_{33}	9.37–13.06 (for x = 0–0.26)	[22]
Ferroelectric Properties	Ferroelectric switching (μC/cm^2)	~80–153	[7,8,23]
	Coercive field (MV/cm)	2–5 (for x = 0.27–0.43)	[7]

It is clear from the excellent piezoelectric and ferroelectric properties of $Sc_xAl_{1-x}N$ that it has tremendous potential for use in various power electronics. Moreover, the excellent piezoelectric properties of $Sc_xAl_{1-x}N$ and low processing temperature make it a suitable choice for power devices such as surface and bulk acoustic wave resonators [24]. Additionally, the recent discovery of ferroelectricity in low-temperature processed $Sc_xAl_{1-x}N$ provides substantial opportunities for direct memory integration with logic transistors, providing the possibility for the back-end of the line (BEOL) integration on silicon logic. Thus, taking advantage of high ferroelectric switching and coercive fields, ferroelectric field-effect transistors (FE-FET) can be fabricated [25]. However, in order to maximize the piezoelectric and ferroelectric properties in ScAlN thin films, high-quality thin films must be deposited, which is only possible by utilizing optimum sputtering conditions.

3. Sputtering Process for Scandium Aluminum Nitride

3.1. Deposition Parameters

Reactive sputtering is a physical vapor deposition (PVD) process that utilizes charged ions from a mixture of argon and reactive gases to bombard a target causing the ejection of surface atoms from that target and the eventual deposition of those target atoms onto a substrate following the reaction with the reactive gas. In the case of $Sc_xAl_{1-x}N$, the reactive gas is nitrogen (N_2). Many different parameters can be altered during the reactive sputtering process to achieve desired deposition outcomes. These include parameters such as target design, sputtering atmosphere, sputtering power/power density, substrate temperature, sputtering time, and target-to-substrate distance. However, the primary parameters that have been focused on and investigated thus far in the literature are target design, sputtering atmosphere, sputtering power/power density, and substrate temperature [6,24,26,27]. Table 2 lists the published articles that deposited c-axis ScAlN films on top of several substrates using a variety of sputtering equipment and processes.

3.2. Target Design

Target design is a crucial step in the sputtering process because it will directly impact both the final film's stoichiometry and uniformity. Three types of target designs have been widely used for the sputtering of $Sc_xAl_{1-x}N$ thin films. including alloy sputtering targets, pure metal sputtering targets (used in conjunction with dual co-sputtering), and segmented targets [24]. A schematic representation of the three different target types is illustrated in Figure 2. The final thickness and composition of the $Sc_xAl_{1-x}N$ film deposited can be controlled in different ways depending on the target chosen. For example, for a dual co-sputtering target the final film composition can be varied by selectively adjusting the power applied to the Sc and Al sources, respectively. Correspondingly, for a Sc-Al alloy target the final film composition can be varied by carefully tuning the alloy composition of the target. Lastly, for Sc-Al segmented targets the distribution, size, and quantity of various Al and Sc segments allows for facile and precise tuning of film composition and homogeneity [24].

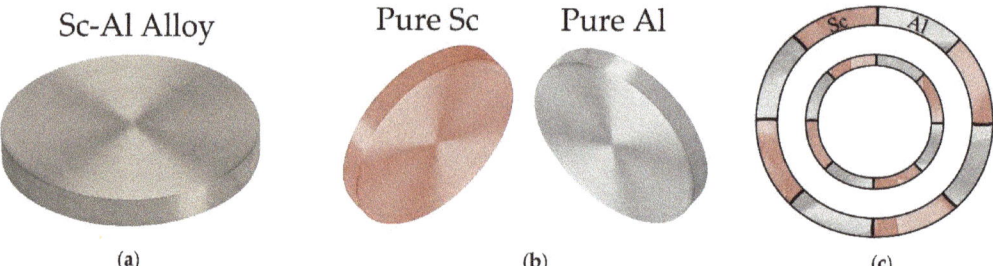

Figure 2. Schematic representation of different target designs with (**a**) Sc-Al Alloy target, (**b**) pure Sc and pure Al targets for dual co-sputtering, and (**c**) ring-shaped segmented target with alternating Sc and Al tiles.

Table 2. Summary of sputtering deposited ScAlN growth conditions.

Author [Ref] (Year of Publication)	Substrate	Sputtering Type	Power (W)/Power Density (W/cm^2)	Substrate Temperature (°C)	Sputtering Pressure (Pa)	Base Pressure (Pa)	Gas Composition Ratio [N2:Ar] [%N$_2$]	Target to Substrate Distance [mm]	Final Composition (Sc$_x$Al$_{1-x}$N)	FWHM (°)	Deposition Rate (nm/min)	Film Thickness (μm)	Surface Roughness (nm)
Tominaga et al. [28] (2022)	(100) Si	RF magnetron sputtering	200/3.98	300	0.6	3.00 × 10^{-4}	1:2/50%	25	x = 0.3	3.2–6.0	NA	4.0–4.5	NA
Tominaga et al. [29] (2021)	(100) Si	RF magnetron sputtering	200/3.98	300	0.14, 0.25, 0.35, 0.45, 0.56	2.00 × 10^{-4}	1:2/50%	25	NA	2.3–4.7	NA	1–2	NA
Rassay et al. [8] (2021)	NA	DC magnetron sputtering	2000, 3500, 5500/8.06, 4.80, 7.54	280	NA	NA	15:6, 20:3, 28:1.6/NA	NA	x = 0.22, 0.25, 0.30	2–2.8	NA	0.025–0.250	NA
Liu et al. [25] (2021)	Pt/(100) Si	Pulsed DC magnetron co-sputtering	Al-target: 1000/12.34; Sc-target: 450/5.55	350	NA	NA	NA	NA	x = 0.29	NA	NA	0.100	NA
Liu et al. [9] (2021)	Pt/(100) Si	Pulsed DC magnetron co-sputtering	Al-target: 1000/12.34; Sc-target: 655/8.08	350	NA	NA	NA	NA	x = 0.36	NA	NA	0.200	NA
Zhang et al. [17] (2021)	Mo/SiO$_2$/AlN/SOI	DC magnetron sputtering	7500/10.28	300	0.347	NA	1:3/25%	70	x = 0.29	4.13	NA	0.780	NA
Wang et al. [30] (2020)	Pt(111)/Ti/SiO$_2$/si	Pulsed DC magnetron co-sputtering	Al-target: 1000/12.34; Sc-target (x = 0.32): 555/6.85; Sc-target (x = 0.36): 655/8.08	350	NA	8.30 × 10^{-2}	20:80/20%	33	x = 0.32, 0.36	2.7–2.8	15.6–16.8	0.2	NA
Dong et al. [31] (2019)	Pt	DC magnetron sputtering	140–190/NA	24	0.3	NA	13:17/43.3%	NA	x = 0.175	0.38–0.29	23.3	0.7	NA
Felmetsger et al. [24] (2019)	(100) Si	AC magnetron reactive Sputtering with segmented target	2000–5000/NA	24	NA	NA	NA	NA	x = 0.3	1.6	NA	0.500–2.00	2.3
Fichtner et al. [7] (2019)	NA	DC reactive magnetron sputter deposition	600/NA	400	NA	NA	15:7.5/NA	NA	x = 0.36	NA	NA	0.600	NA
Tabaru et al. [32] (2019)	(100) p-Si	RF reactive magnetron sputtering	400/8.78	207	0.4, 1.0	5.00 × 10^{-5}	4:6/NA	70	x = 0.4	4.6, 8.5	NA	2.3, 2.6	NA

Table 2. Cont.

Author [Ref] (Year of Publication)	Substrate	Sputtering Type	Power (W)/Power Density (W/cm^2)	Substrate Temperature (°C)	Sputtering Pressure (Pa)	Base Pressure (Pa)	Gas Composition Ratio [N$_2$:Ar] /%N$_2$	Target to Substrate Distance [mm]	Final Composition (Sc$_x$Al$_{1-x}$N)	FWHM (°)	Deposition Rate (nm/min)	Film Thickness (um)	Surface Roughness (nm)
Henry et al. [33] (2018)	(100) Si	Pulsed DC magnetron sputtering	80, 90, 100, 110, 120/0.116, 0.127, 0.140, 0.156, 0.170	350	NA	NA	1:3, 1:4, 1:5/25%, 20%, 16.7%	NA	x = 0.12	1.884	NA	0.750	NA
Lozano et al. [34] (2018)	(100) As-doped Si & (100) B-doped Si	DC reactive balanced magnetron sputtering	300, 500, 700/3.70, 6.17, 8.64	24	0.53, 0.79, 1.06	1.00 × 10^{-2}	1:3/25%	45	x = 0.26	2–5	24–90	1.00	NA
Mertin et al. [35] (2018)	NA	Pulsed DC magnetron sputtering/co-sputtering	30.4 cm-Target: 7500/10.34; 10 cm-Target: 200–1000/2.55–12.74	300–350	NA	1.00 × 10^{-5}	1:2/33.3%	NA	x = 0, 0.1, 0.31, 0.42	1.2–2.0	12–60	NA	NA
Perez-Campos et al. [36] (2017)	(100) As-doped Si & (100) B-doped Si	DC reactive balanced magnetron sputtering	300, 500, 700, 900/3.70, 6.17, 8.64, 11.11	24	0.26, 0.53, 0.79, 1.06	9.99 × 10^{-5}	3:9/25%	45	x~0.23–0.26	2.5–10	24–110	1.00	NA
Tang et al. [26] (2017)	PT/Ti/Si	RF reactive magnetron sputtering	100, 120, 135, 145, 160/1.05, 1.26, 1.42, 1.52, 1.68	600	0.47	1.50 × 10^{-4}	3:4.7/32.7%	120	x = 0.15	2.38–6.55	NA	1.00	3.25–10.34
Felmetsger et al. [37] (2017)	(100) Si	AC powered S-gun sputtering	2000/NA	350	NA	NA	9:3.5/NA	NA	x = 0.07	1.55	NA	1.00	3.3
Fichtner et al. [38] (2017)	(100) c-Si	Pulsed DC reactive co-sputtering	1000/NA	300	0.21	5.00 × 10^{-5}	15:5.3/NA	NA	x = 0.27, 0.29	1.7	NA	0.4–2	NA
Li et al. [10] (2016)	(100) p-Si & Ni-Cr-Mo (Hastelloy)	DC reactive magnetron sputtering	NA/1.16–2.10	600	0.45	2.00 × 10^{-4}	3:3.7	NA	x = 0.43	1.5–11	NA	1.1–2.0	2.0–4.9
Tang et al. [6] (2016)	(100) p-Si	DC reactive magnetron sputtering	130/1.37	600	0.4, 0.8	2.00 × 10^{-4}	30:70, 35:65, 40:60, 50:50, 60–40/30%–60%	100	NA	1.7	16.6–21.0	1.50	3–21
Zhang et al. [39] (2014)	(0001) Sapphire	DC reactive magnetron sputtering	130/1.37	650	0.3–0.7	4.00 × 10^{-4}	3.1:7–3.6:7/30.7%–34%	NA	NA	2.6	16.67	1.50	2.65
Akiyama et al. [40] (2013)	(100) n-Si	Dual RF magnetron reactive co-sputtering	NA	NA	NA	1.20 × 10^{-6}	NA	NA	x = 0.41	1.8–7.9	NA	0.500–1.10	NA

Table 2. Cont.

Author [Ref] (Year of Publication)	Substrate	Sputtering Type	Power (W)/Power Density (W/cm²)	Substrate Temperature (°C)	Sputtering Pressure (Pa)	Base Pressure (Pa)	Gas Composition Ratio [N2:Ar] /%N2	Target to Substrate Distance [mm]	Final Composition ($Sc_xAl_{1-x}N$)	FWHM (°)	Deposition Rate (nm/min)	Film Thickness (um)	Surface Roughness (nm)
Zukauskaite et al. [41] (2012)	TiN(111)/Al$_2$O$_3$ (0001)-100–200 nm	Magnetically unbalanced reactive DC magnetron sputtering	150/7.64	400, 600, 800	0.17	6.00×10^{-7}	19.8:30/NA	NA	x = 0, 0.1, 0.2, 0.3	1.0–2.0	NA	0.25	NA
Akiyama et al. [29] (2010)	(100) n-Si-600 um	RF reactive magnetron sputtering	300/6.58	200	0.3	5.00×10^{-5}	3:7/30%	NA	x = 0.38	2.3	NA	0.500–1.20	NA
Hoglund et al. [42] (2010)	ScN(111)/ MgO(111)	Magnetron sputter epitaxy	Al-Target: 250, 230, 180, 130, 80; Sc-Target: 0, 20, 70, 120, 170	800	0.46	1.33×10^{-6}	NA	NA	x = 0.4, 0.32, 0.26, 0.22	NA	4.2	0.080	NA
Hoglund et al. [43] (2010)	ScN(111)/ MgO(111)	Magnetron sputter epitaxy	Al-Target: 0, 20, 60, 100, 140, 180, 200; Sc-Target: 200, 180, 140, 100, 60, 20, 0	600	1.2	1.33×10^{-6}	0.13:1.07/NA	NA	x = 0, 0.1, 0.27, 0.49, 0.71, 0.86, 1	NA	5.4	0.05–0.06	NA
Akiyama et al. [27] (2009)	(100) n-Si	Dual RF magnetron reactive co-sputtering	0–200/0–9.87	27–580	0.25	1.20×10^{-6}	NA/40%	NA	x = 0–0.43	2.3–7.5	NA	0.5–1.1	0.3–2.7
Akiyama et al. [5] (2009)	(100) n-Si	Dual RF magnetron reactive co-sputtering	0–200/0–9.87	580	0.25	1.20×10^{-6}	NA/40%	NA	x = 0–0.43	1.8–7.9	NA	0.5–1.1	NA

NA (Not Available) is used to indicate when specific parameters were not provided in the given literature.

Early in the research days of sputtering $Sc_xAl_{1-x}N$, the primary sputtering technique was dual co-sputtering [5,27,42–44]. However, there was a clear transition from dual co-sputtering, using two pure targets, to the use of Sc-Al alloy targets for sputtering. In fact, approximately 67% of the literature cited the use of alloy targets during the period of 2010 to 2018 [6,7,10,17,24,32–35,39,45]. The primary reason for this transition was that during this period dual co-sputtering was thought to make it difficult to maintain a consistent scandium concentration in $Sc_xAl_{1-x}N$ films over large substrates due to the difference in scandium and aluminum sputtering yields [45]. Additionally, single-target sputtering is more attractive for industrial high-volume production because it has a higher deposition rate [35].

Although single-alloy target sputtering has several advantages, it is confined by its lack of tunability of the Sc concentration. Thus, as interest grew to develop $Sc_xAl_{1-x}N$ films with ever increasing Sc content, the preferred deposition mode again pivoted towards dual co-sputtering. Dual co-sputtering is unique in that the Sc concentration can be set directly by tuning the two powers of the sputtering targets [35]. In Table 3, the transition between different sputtering targets is made especially clear, as is the correlation with the desired $Sc_xAl_{1-x}N$ composition. In general, it can be stated that single-alloy-target sputtering is best suited for the deposition of $Sc_xAl_{1-x}N$ films with $x \leq 0.3$ and dual co-sputtering is better suited for handling deposition of compositions where $x > 0.3$ and precise control over chemistry is required.

Table 3. $Sc_xAl_{1-x}N$ films produced using various target designs.

$Sc_xAl_{1-x}N$ Composition	Target Composition	Ref
Single-Alloy-Target		
x = 0.38	Sc:Al = 42:58	[45]
NA	Sc:Al = 0.1:0.9	[39]
x = 0.43	Sc: Al = 0.1:0.9	[10]
NA	Sc:Al = 1:9	[6]
x = 0.23–0.26	Sc:Al = 0.4:0.6	[36]
x = 0.15	Sc:Al = 0.15:0.85	[26]
x = 0.12	Sc:Al = 12.5:87.5	[33]
x = 0.26	Sc:Al = 0.4:0.6	[34]
x = 0.1, 0.31	Sc:Al = 6:9.5, 15:28	[35]
x = 0.3	Sc:Al = 8:92	[24]
x = 0.36	Sc:Al = 43:57	[7]
x = 0.4	Sc:Al = 43:57	[32]
x = 0.29	Sc:Al = 0.3:0.7	[17]
Dual Co-Sputtering Target		
x = 0, 0.1, 0.27, 0.49, 0.71, 0.86, 1	Al (pure) and Sc (pure)	[42]
x = 0.36	Al (pure) and Sc (pure)	[5]
NA	Al (pure) and Sc (pure)	[44]
x = 0.4, 0.32, 0.26, 0.22	Al (pure) and Sc (pure)	[43]
x = 0, 0.1, 0.2, 0.3	Al (pure) and Sc (pure)	[41]
x = 0.41	Al (pure) and Sc (pure)	[40]
x = 0.27, 0.29	Al (pure) and Sc (pure)	[38]
x = 0.42	Al (pure) and Sc (pure)	[35]
x = 0.175	Al (pure) and Sc (pure)	[31]
x = 0.32, 0.36	Al (pure) and Sc (pure)	[30]
x = 0.29	Al (pure) and Sc (pure)	[9]
x = 0.36	Al(pure) and Sc (pure)	[25]
x = 0.3	Al(pure) and Sc (pure)	[29]
x = 0.43	Al(pure) and Sc (pure)	[28]
Segmented Target		
x = 0.3	Segmented Al-Sc	[24]
x = 0.22, 0.25, 0.30	Segmented Al-Sc	[8]

NA (Not Available) is used to indicate when specific parameters were not provided in the given literature.

In recent years, there has also been the introduction of a third class of sputtering targets known as a segmented targeted. The primary advantages and disadvantages of Sc-Al alloy targets and dual co-sputtering targets are compared in table IV. Sc-Al alloy targets allow for higher deposition rates, which make them well suited for use in industrial applications. Additionally, the alloy target provides consistent Sc content across large substrate areas. However, this target design is very expensive to produce due to the complicated metallurgy required to produce it. Moreover, the alloy design makes quick tuning of the final films Sc content very difficult. On the other hand, using two pure Al and Sc targets in a dual co-sputtering setup allows for facile tuning of Sc content by tuning the power on the respective targets. Additionally, this setup significantly reduces the cost of the targets due to its simple nature. Yet, the dual co-sputtering target design only allows for residual stress tailoring by altering the reactive gas flow, which ultimately degrades film uniformity and crystallinity. Additionally, this design makes it difficult to obtain final films with tensile stress. Thus, Sc-Al segmented targets were developed to overcome the disadvantages of the two prior methods while maintaining as many of the advantages of them as possible. A comparison of the various target designs is presented in Table 4. By combining alternating segments of Sc and Al as shown in Figure 2c, it was possible to get a more uniform distribution of Sc in the $Sc_xAl_{1-x}N$ films [24]. Additionally, the segmented target design allowed for a significant reduction in film thickness while maintaining excellent control over Sc content and residual film stress [8]. Furthermore, achieving thin films of such thicknesses was especially beneficial for use in ferroelectrics.

Table 4. Comparison of various target designs.

Sc-Al Alloy Target	Dual Co-Sputtering Target	Sc-Al Segmented Target
• Higher deposition rate • Applicable for industrial applications • Constant Sc content across large substrates	• Easy Sc content tuning by adjusting target power • Reduced cost of target	• Enables thickness reduction • Large tunability of Sc content • Large tailorability of residual film stress (independent of gas flows)
• Expensive targets due to complicated metallurgy • Difficult tuning of Sc content	• Residual stress tailoring only through altering reactive gas flow, which degrades film uniformity and crystallinity • Difficult to achieve tensile stress	• Lower deposition rate • Less applicable for industrial applications

Although few papers have been published that explicitly compare $Sc_xAl_{1-x}N$ deposition with different targets, the target design remains one of the most crucial sputtering parameters [24,35]. In fact, there remains more opportunities to investigate novel modifications to target design and setup. For instance, Posadowski et al., explored the critical role that the placement of substrates versus the target axis, i.e., the so called on- or off-axis mode, had on the properties of deposited films of Al_xZn_yO when using a two element segmented target [46]. During the engineering of the electric properties of piezoelectric and ferroelectric materials such as $Sc_xAl_{1-x}N$, it is vital to achieving desired material stoichiometry [40]. Furthermore, choosing the correct target to achieve the desired chemistry and electric properties is of paramount importance and must not be understated.

3.3. Sputtering Atmosphere

A second critical reactive sputtering parameter that must be taken into consideration is the sputtering atmosphere. Sputtering atmosphere consists of both the chamber pressure and the proportion of argon to nitrogen gas [6]. There have been several studies conducted that examine how the variation in sputtering atmosphere affects parameters such as crystal quality, sputtering rate, residual stress, and the electric properties of the film [6,28,34,36,39]. Because sputtering atmosphere is one of the parameters that directly impacts the energy for adatoms to bombard the surface of the substrate, it is especially pertinent to optimize the sputtering atmosphere to achieve desired film quality and the subsequent electric properties [47].

3.3.1. Sputtering Pressure

(a) Effect on Crystal Quality: Sputtering working pressure is well known to have a significant impact on the crystal quality of deposited $Sc_xAl_{1-x}N$ films. Numerous researchers have demonstrated that the highest level of c-axis orientation of $Sc_xAl_{1-x}N$ corresponds to the lowest full width half maximum (FWHM) of the (0002) peak [36]. Furthermore, when comparing the FWHM values of the rocking curve vs. pressure from several sources, as shown in Figure 3, there appears to be a relative congruence in the ideal pressure for $Sc_xAl_{1-x}N$ sputtering. Most of the papers reported the lowest FWHM value at an approximate pressure between 0.4 Pa and 0.6 Pa. Additionally, these studies also highlight that the preferred c-axis texture deteriorates at both extremely high and low pressures. In 2022, Tominaga et al., proposed a possible explanation for the deterioration of crystallinity and even electric properties at low-pressure depositions. In short, the report claims that negative-ion bombardment will increase both in quantity and energy at lower pressures, which will negatively impact $Sc_xAl_{1-x}N$ crystal growth and cause substantial film quality reduction [29].

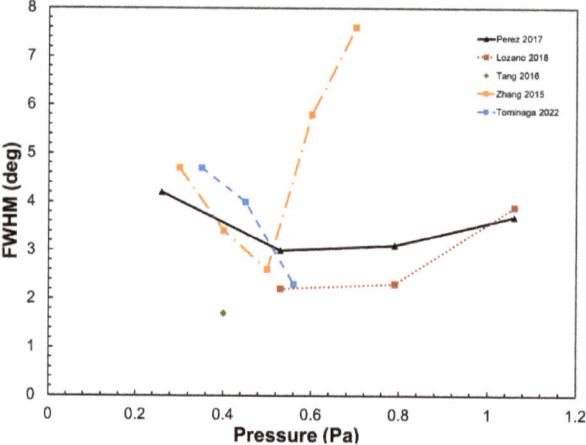

Figure 3. Rocking curve FWHM of the (0002) reflections of ScAlN thin films under various working pressure conditions.

(b) Effect on Sputter Deposition Rate: The impact of sputtering pressure on the sputter deposition rate is partially obfuscated when comparing reports from existing literature related to the reactive sputtering of $Sc_xAl_{1-x}N$. For instance, Tang et al., reported that there was a decrease in the sputtering rate with an increase in pressure from 0.4 Pa to 0.8 Pa which was attributed to more scattering between the ejected atoms and the gas atoms, which reduced the mean free path of the ejected atoms [6]. On the other hand, Perez and Lozano reported, respectively, that during deposition at low pressures the rate decreases due to the lower quantity of ions available in the plasma to eject the target material [34,36]. Regardless of these competing theories, the actual deviation of the sputtering rate reported due to the change in pressure is rather minor when compared to the deviation of the sputtering rate due to other parameters such as the sputtering power or N_2 proportion. For this reason, more focus should be placed on tuning parameters such as those when seeking control over the sputtering rate of $Sc_xAl_{1-x}N$.

(c) Effect on Electric Properties: Resistivity is an important benchmark when comparing piezoelectric thin films. It stands as a crucial indicator of a piezoelectric thin-films ability to achieve both reduced dielectric losses and lower insertion losses when used in electronic devices such as SAW resonators [6,39]. It is well known that a film's crystallinity directly impacts its dielectric properties. Thus, the pressure that provides the optimum c-axis orientation in the case of $Sc_xAl_{1-x}N$ should also be the pressure that generates the greatest resistivity. In reality, this theory stands true as is supported by the literature. Tang et al., claim that this occurs because non-ideal sputtering pressure leads to incomplete crystallization, which will ultimately lead to a reduction in resistivity of $Sc_xAl_{1-x}N$ thin-films [6].

Similar to the resistivity of piezoelectric thin films, the piezoelectric response, d_{33}, is directly impacted by the film's crystal quality. The trends for rocking curve FWHM values for different sputtering pressures coincide with the trends for the piezoelectric constant (d_{33}) at different sputtering pressures [34]. For this reason, there is a consensus in the literature that the ideal piezoelectric response is achieved when the greatest degree of c-axis texturing occurs [6,28,34,36,39]. Furthermore, by optimizing the sputtering pressure to reach the maximum c-axis orientation, the greatest piezoelectric response can be achieved, paving the way for the creation of piezoelectric sensors and SAW devices [39].

Currently, limited published investigations regarding the effect of sputtering pressure on the ferroelectric response of $Sc_xAl_{1-x}N$. However, it can be assumed that the ferroelectric properties of $Sc_xAl_{1-x}N$ would be tied to the film crystallinity just as the piezoelectric properties were. Since most ferroelectric $Sc_xAl_{1-x}N$ films are significantly thinner than their piezoelectric counterparts, there is a potential for differences to exist in the ideal sputtering pressure in each case. Regardless, this knowledge gap provides an opportunity for additional research and discovery and warrants further investigation in the future.

3.3.2. Gas Flow Ratio

The gas flow ratio is the ratio of inert gas (Ar) to reactive gas (N_2) that is in the sputtering atmosphere during reactive sputtering. The gas flow ratio plays a crucial role during the reactive sputtering process because it controls the amount of reactive gas that is involved during thin film deposition. As shown in Figure 4, deposition parameters such as sputtering rate, structural properties such as crystallinity and surface morphology, and electronic properties such as the resistivity, the dielectric constant, and the piezoelectric response are all greatly impacted and controlled according to the ratio of gas flows [6]. In other words, tuning the concentration of N_2 gas present during sputtering is a crucial parameter that must be considered during the reactive sputtering of $Sc_xAl_{1-x}N$.

(a) Effect on Sputtering Rate: The effects of gas flow ratio on the sputtering rate of $Sc_xAl_{1-x}N$ have been shown to exhibit similar trends to those of the AlN system, where sputtering rate decreases with an increasing proportion of N_2 gas [48]. The deleterious effect of the N_2 proportion on the sputtering rate is clearly demonstrated in Figure 4a. Moreover, Tang et al., explained that there are two possible explanations for this phenomenon. The first explanation is that pure Ar^+ ions in the working gas have a higher sputtering yield than N^+ or N_2^+ ions due to their higher mass. The second explanation is related to target poisoning, which occurs with an increasing proportion of N_2 [6]. For these reasons, it is very important to be aware of the impacts of the gas flow ratio on the sputtering rate of $Sc_xAl_{1-x}N$.

(b) Effect on Crystal Quality: It has been established that excellent c-axis orientation is a prerequisite for achieving maximum piezoelectric or ferroelectric properties in $Sc_xAl_{1-x}N$ thin-films [38]. Thus, when developing the reactive sputtering procedures for $Sc_xAl_{1-x}N$ thin-films, it was vital to screen the gas flow ratios to determine the ideal percentage of N_2 gas necessary to achieve maximum c-axis orientation. Moreover, it was determined from the literature that the XRD intensity and FWHM of the rocking curves for the (0002) plane are both closely tied to the concentration of N_2 [6,39]. At approximately 32%–35% N_2, both the XRD intensity for the (0002) plane and the FWHM were found to be optimized as shown in Figure 4b. There is a clear consensus that an atmosphere oversaturated with N_2 will lead to "target poisoning" due to excessive N_2 reacting on top of the target [39]. On the other hand, an atmosphere undersaturated with N_2 will develop poor crystallinity and film quality due to the formation of N-vacancies and Al-interstitials [6]. For this reason, there must be strict control and optimization of the gas flow ratios to ensure proper film quality and crystallographic orientation during the reactive sputtering $Sc_xAl_{1-x}N$.

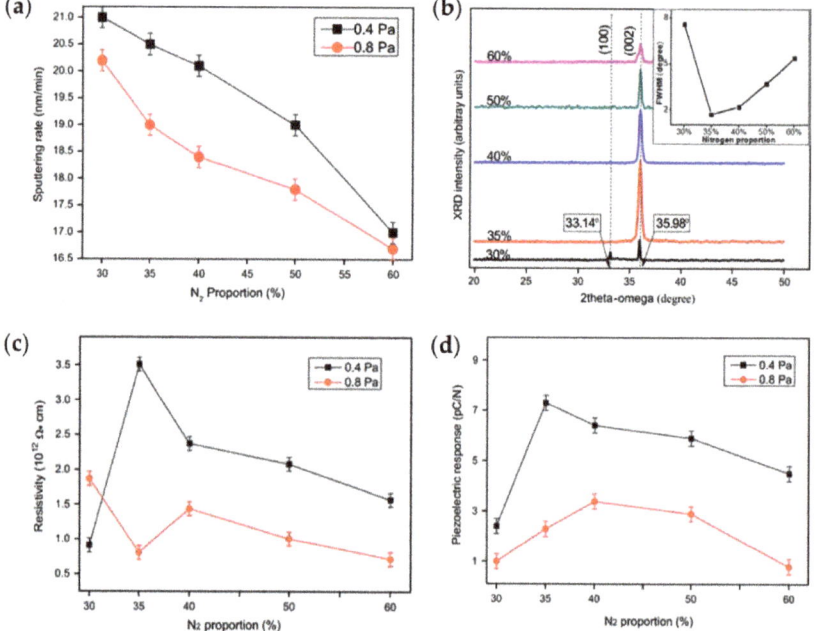

Figure 4. Effect of sputtering atmosphere on (**a**) sputtering rate, (**b**) crystal quality, (**c**) resistivity, and (**d**) piezoelectric response. Reproduced with permission from Ref. [6].

(c) Effect on Surface Roughness: Surface roughness becomes especially important when considering the applications of $Sc_xAl_{1-x}N$ thin-films such as in SAW devices. Surface acoustic waves are propagated only on the surface, and all the energy is approximately contained in a wavelength from surface to inside [49,50]. Furthermore, SAWs will be critically hindered when the surface roughness of the film is greater than the wavelength. Therefore, it is beneficial to minimize the RMS of $Sc_xAl_{1-x}N$ films to achieve reduced insertion losses for SAW devices [51]. It has been discovered that the gas flow ratio also has an important impact on the surface roughness of $Sc_xAl_{1-x}N$ [39]. By optimizing the gas flow ratio to achieve the desired c-axis orientation, it has been shown to minimize the surface roughness [6,39].

(d) Effect on Electric Properties: As mentioned in the previous sections regarding the resistivity and piezoelectric response of $Sc_xAl_{1-x}N$ films with respect to the sputtering pressure, both parameters are directly dictated by the degree of c-axis orientation of the film. Thus, by optimizing the gas flow ratio to achieve the maximum c-axis orientation, both the highest resistivity and largest piezoelectric response can be achieved. As is shown in Figure 4c,d, the resistivity, piezoelectric response, and dielectric constant are all shown to improve initially with the increasing N_2 concentration. However, they show significant degradation after reaching a saturation point [6]. Thus, a harmonious proportion of N_2 is required to achieve films with uniform size and minimized defects, which will ultimately lead to the greatest improvement of electric properties.

3.4. Sputtering Power Density

The sputtering power/power density is an additional parameter that is critical to the thin-film deposition of $Sc_xAl_{1-x}N$. At the beginning, it was established that the sputtering power significantly influences the deposition of AlN films, affecting both its crystallinity and its electric properties [52]. Similarly, it was discovered that sputtering power plays an analogous role in the thin-film deposition of $Sc_xAl_{1-x}N$ as is shown in Figure 5 [26].

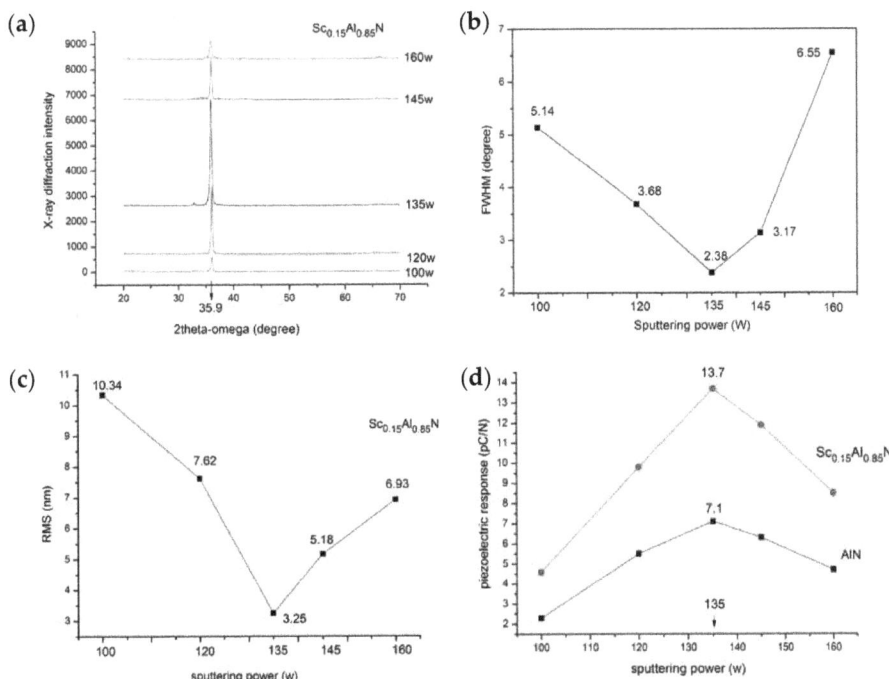

Figure 5. Effect of sputtering power on (**a**) XRD intensity, (**b**) FWHM, (**c**) RMS, and (**d**) piezoelectric response. Reproduced with permission from Ref. [26].

(a) Effect on Sputtering Rate and Thickness: Sputtering power density becomes especially important during the process of reactive dual co-sputtering of $Sc_xAl_{1-x}N$ because the variation in power density between the Sc and Al targets allows for the precise control over the concentration of Sc in the alloy [42]. By maintaining a constant power density on the Al target, the Sc concentration can be increased linearly by gradually increasing power density applied to the Sc target [31]. On account of this, most papers employing a dual co-sputtering setup for $Sc_xAl_{1-x}N$ tend to modulate the power density only as a means to dictate the final stoichiometry of the film. The mechanism governing the linear increase in Sc content with increasing power density is closely related to the mechanism governing the effect of the power density on both the sputtering rate and the film thickness.

Moreover, increased power density generally results in an increase in the sputtering rate of $Sc_xAl_{1-x}N$. This can be mainly attributed to the increase in the kinetic energy of the adatoms arriving at the substrate [36]. However, there have been reports of exceedingly high power densities leading to a reduction in sputtering rate due to the occurrence of re-sputtering [10]. Furthermore, since the thickness is a rate-dependent property, it follows the same trend as the sputtering rate for a constant sputtering time. This means that the thickness of sputtered $Sc_xAl_{1-x}N$ can be systematically tailored by increasing or decreasing the sputtering power for a given sputtering time. Thickness tuning via power density modulation is primarily reserved for cases in which Sc-Al alloys are utilized as the target.

(b) Effect on Crystal Quality: As previously discussed, $Sc_xAl_{1-x}N$ has a wurtzite structure. As a result of this crystal structure, the (0002) plane has the lowest surface energy because it is the closest packed plane [10]. Additionally, it has been established that with increasing power density, there is an increase in the kinetic energy of Sc adatoms. This helps to promote the rearrangement of atoms to align according to (0002) crystal orientation on the substrate surface, which will contribute to a reduction in surface energy [26]. However, at exceedingly high discharge powers/power densities, the adatoms can harbor an overabundance of kinetic energy that forces the B_2 bonds in wurtzite to disassociate, causing crystal quality deterioration [31]. Correspondingly, it has been stated that at higher discharge powers the increase of adatom mobility not only improves the crystal quality, but it is also beneficial in preventing the incorporation of impurities from the background gas [34]. Indeed, these phenomena are widely observable in the data available in the literature where the XRD patterns of $Sc_xAl_{1-x}N$ and plots of the rocking curve FWHM are shown to improve initially with increasing power density, and then degrade at higher power densities [10,26,31]. An example of this is shown in Figure 5a,b. Moreover, Henry et al., showed that increasing RF power from 80 to 120 W leads to a significant reduction in both compressive stress and the inclusion number, where the inclusion number is the number of non-c-axis orientated grains [33].

Additionally, the surface morphology of $Sc_xAl_{1-x}N$ films is similarly tied to the sputtering power used during film deposition as the surface properties are improved with increasing c-axis orientation due to the increase in sputtering power. With insufficient power, the film is unable to properly form the desired texture. On the contrary, with increasing power, the mobility of the atoms improves, providing the opportunity to eliminate defects such as inclusions [53]. However, Tang et al., warn that when the power density is increased toward the extreme, the surface of the film is more susceptible to cracking due to thermal stresses [26]. This is reflected in the measured RMS as shown in Figure 5c.

(c) Effect on Electric Properties: The electric properties of $Sc_xAl_{1-x}N$ such as resistivity, leakage current, and piezoelectric response are closely tied to the crystal quality of the deposited film. With better crystal quality, there are fewer defects in the lattice, which serves an advantageous role in the enhancement of electrical properties [10]. Moreover, when the optimum power condition is chosen such that there is a coincidence between the best crystal quality and surface morphology, the maximum piezoelectric response is achieved [26]. When this occurs, the literature has shown that $Sc_xAl_{1-x}N$ thin-films are capable of achieving a piezoelectric response at least 250% greater than that of AlN [31].

3.5. Sputtering Substrate Temperature

The substrate temperature is a very important parameter to consider during the sputtering of $Sc_xAl_{1-x}N$ films as it impacts the crystal quality, grain size, and even the final electric properties of the film as shown in Figure 6 [27]. Additionally, the substrate temperature can directly limit what applications the film can be used for. For example, a perquisite for using $Sc_xAl_{1-x}N$ thin-films in complementary metal-oxide-semiconductor (CMOS)-integrated devices such as ferroelectric field effect transistors (FE-FETs) and random-access memories (RAMs) is a low-temperature (T < 400 °C) deposition [25,30]. Moreover, the low temperature deposition of $Sc_xAl_{1-x}N$ thin films could alleviate the process integration failures commonly associated with PZT FE-RAMs that require deposition temperatures exceeding 600 °C, which are detrimental for CMOS transistors [8,23].

(a) Effect on Crystal Quality, Grain Size, and Film Stress: It has long been acknowledged that the crystal quality and grain size of $Sc_xAl_{1-x}N$ films share a dependency with the substrate or growth temperature [5,27,41]. In 2009, Akiyama et al., showed that there was a significant decrease in the crystal quality of films deposited above 400 °C as shown in Figure 6 [27]. Additionally, this decrease in crystal quality was proposed to be the result of a drastic increase in the size and disorder of grains when the growth temperature exceeded 400 °C. Moreover, Zukauskaite et al., showed that at high growth temperatures above 400 °C and higher Sc concentrations, there was a structural degradation into Al-rich and Sc-rich domains, which most likely contributed to the reduction in crystal quality [41]. Likewise, it was found that the number of non-c-axis oriented grains decreases with increasing substrate temperature up to 375 °C before increasing again with subsequent temperature increase [33]. On account of these investigations, the vast majority of literature utilizes a sputtering temperature that is less than or equal to 400 °C.

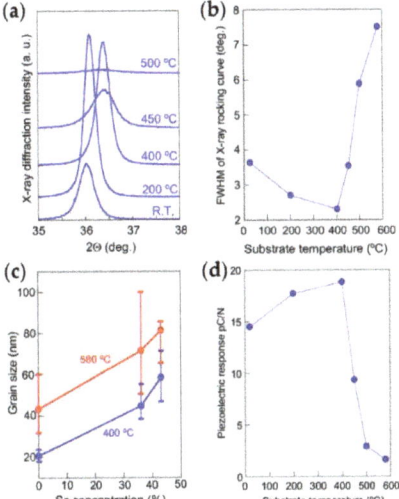

Figure 6. Effect of substrate temperature on (**a**) XRD intensity, (**b**) FWHM, (**c**) grain size, and (**d**) piezoelectric response. Reproduced with permission from Ref. [27].

(b) Effect on Electric Properties: As mentioned in the previous section, the crystal quality of $Sc_xAl_{1-x}N$ films deteriorates significantly when the substrate temperature exceeds 400 °C. Since the electrical properties of the film are dependent on the crystal quality, this means that they too should degrade after surpassing the temperature threshold. Indeed, this is supported by the data presented in the literature. When $T_{substrate} > 400$ °C, there is a significant increase in the leakage current [41]. Additionally, the piezoelectric response was shown to drop off drastically when $T_{substrate} > 400$ °C as shown in Figure 6d [27]. The degradation of the electric properties such as piezoelectric response could also be linked to the formation of Al-rich and Sc-rich clusters, which occurs at higher temperatures [22]. To conclude, it is vital to minimize the substrate temperature during the deposition of thin-film $Sc_xAl_{1-x}N$ because exceeding the threshold temperature of approximately 400 °C leads to a significant reduction in crystal quality and electric properties while also excluding the film from being easily integrated into CMOS devices.

4. Conclusions

In summary, key sputtering parameters for the deposition of $Sc_xAl_{1-x}N$ such as target design, sputtering atmosphere, sputtering power, and substrate temperature are vital for depositing high-quality thin films, achieving desired stoichiometry, and meeting the required film thickness. Moreover, this review has shown that these qualities directly impact the degree of c-axis orientation, grain size, and surface roughness of the deposited films. Additionally, the electric properties of $Sc_xAl_{1-x}N$ films share a clear dependence on the crystal quality of the film. It should be stated that there is no one set of sputtering parameters that is ideal for all applications. Therefore, it is impossible to say definitively what the ideal target design, sputtering atmosphere, sputtering power, and substrate temperature should be. However, some overarching truths can be gleaned from the review of current literature on the sputtering of $Sc_xAl_{1-x}N$. These are as follows:

(1) Sputtering target design is essential to ensure the Sc and Al composition in the $Sc_xAl_{1-x}N$ films. In general, single-alloy targets appear to be better when depositing films with Sc concentrations less than 30%, whereas dual co-sputtering targets are better suited for applications where concentrations exceed 30% and when the precise control of Sc content is necessary. Lastly, segmented targets are interesting in the ability to combine the advantages of both single-alloy targets and dual co-sputtering targets, but they are less applicable for industrial applications.

(2) In general, sputtering pressures should be kept between 0.4–0.6 Pa to avoid the issues associated with extremely low or high pressures. Moreover, the gas flow ratio should be kept such that there is between 30% and 35% N_2 present.
(3) Increased sputtering power can benefit the crystal quality and electric properties of $Sc_xAl_{1-x}N$. However, there exists a maximum power density whereupon further increase the film quality will become damaged.
(4) Substrate temperature should not exceed 400 °C during the deposition of $Sc_xAl_{1-x}N$.

These insights and this comprehensive analysis can serve as reference points for any novel research project utilizing the sputtering of $Sc_xAl_{1-x}N$ thin films. Moreover, by utilizing this review future researchers should be able to quickly tune and adjust their sputtering processes to rapidly obtain exceptional results.

Author Contributions: J.M.W. writing—original draft preparation, F.Y.; writing—review and editing. All authors have read and agreed to the published version of the manuscript.

Funding: This work is supported by National Science Foundation under contracts No. ECCS-1944374, CMMI-2019473 and DMR-2127640, National Aeronautics and Space Administration, Alabama EPSCoR International Space Station Flight Opportunity program (contract# 80NSSC20M0141), and USDA National Institute of Food and Agriculture, AFRI project award (contract# 2020-67022-31376).

Institutional Review Board Statement: Not applicable.

Informed Consent Statement: Not applicable.

Data Availability Statement: Not applicable.

Conflicts of Interest: The authors declare no conflict of interest.

References

1. Piazza, G.; Felmetsger, V.; Muralt, P.; Olsson, R.H., III; Ruby, R. Piezoelectric aluminum nitride thin films for microelectromechanical systems. *MRS Bull.* **2012**, *37*, 1051–1061. [CrossRef]
2. Elfrink, R.; Kamel, T.M.; Goedbloed, M.; Matova, S.; Hohlfeld, D.; Van Andel, Y.; Van Schaijk, R. Vibration energy harvesting with aluminum nitride-based piezoelectric devices. *J. Micromech. Microeng.* **2009**, *19*, 094005. [CrossRef]
3. Stoppel, F.; Schröder, C.; Senger, F.; Wagner, B.; Benecke, W. AlN-based piezoelectric micropower generator for low ambient vibration energy harvesting. *Procedia Eng.* **2011**, *25*, 721–724. [CrossRef]
4. Lu, Y.; Heidari, A.; Horsley, D. A High Fill-Factor Annular Array of High Frequency Piezoelectric Micromachined Ultrasonic Transducers. *J. Microelectromech. Syst.* **2015**, *24*, 904–913. [CrossRef]
5. Akiyama, M.; Kamohara, T.; Kano, K.; Teshigahara, A.; Takeuchi, Y.; Kawahara, N. Enhancement of Piezoelectric Response in Scandium Aluminum Nitride Alloy Thin Films Prepared by Dual Reactive Cosputtering. *Adv. Mater.* **2009**, *21*, 593–596. [CrossRef] [PubMed]
6. Tang, J.; Niu, D.; Yang, Y.; Zhou, D.; Yang, C. Preparation of ScAlN films as a function of sputtering atmosphere. *J. Mater. Sci. Mater. Electron.* **2016**, *27*, 4788–4793. [CrossRef]
7. Fichtner, S.; Wolff, N.; Lofink, F.; Kienle, L.; Wagner, B. AlScN: A III–V semiconductor based ferroelectric. *J. Appl. Phys.* **2019**, *125*, 114103. [CrossRef]
8. Rassay, S.; Hakim, F.; Li, C.; Forgey, C.; Choudhary, N.; Tabrizian, R. A Segmented-Target Sputtering Process for Growth of Sub-50 nm Ferroelectric Scandium–Aluminum–Nitride Films with Composition and Stress Tuning. *Phys. Status Solidi (RRL) Rapid Res. Lett.* **2021**, *15*, 2100087. [CrossRef]
9. Liu, X.; Zheng, J.; Wang, D.; Musavigharavi, P.; Stach, E.A.; Olsson, R., III; Jariwala, D. Aluminum scandium nitride-based metal–ferroelectric–metal diode memory devices with high on/off ratios. *Appl. Phys. Lett.* **2021**, *118*, 202901. [CrossRef]
10. Li, X.; Yang, Y.; Zhou, D.; Yang, C.; Feng, F.; Yang, J.; Hu, Q. Preparation of ScAlN films as a function of power density on Si and flexible substrate by dc reactive magnetron sputtering. *J. Mater. Sci. Mater. Electron.* **2016**, *27*, 171–176. [CrossRef]
11. Zoroddu, A.; Bernardini, F.; Ruggerone, P.; Fiorentini, V. First-principles prediction of structure, energetics, formation enthalpy, elastic constants, polarization, and piezoelectric constants of AlN, GaN, and InN: Comparison of local and gradient-corrected density-functional theory. *Phys. Rev. B* **2001**, *64*, 045208. [CrossRef]
12. Tasnádi, F.; Alling, B.; Höglund, C.; Wingqvist, G.; Birch, J.; Hultman, L.; Abrikosov, I.A. Origin of the anomalous piezoelectric response in wurtzite $Sc_xAl_{1-x}N$ alloys. *Phys. Rev. Lett.* **2010**, *104*, 137601. [CrossRef] [PubMed]
13. Zhang, S.; Holec, D.; Fu, W.Y.; Humphreys, C.J.; Moram, M.A. Tunable optoelectronic and ferroelectric properties in Sc-based III-nitrides. *J. Appl. Phys.* **2013**, *114*, 133510. [CrossRef]
14. Deng, R.; Jiang, K.; Gall, D. Optical phonon modes in $Al_{1-x}Sc_xN$. *J. Appl. Phys.* **2014**, *115*, 013506. [CrossRef]

15. Ambacher, O.; Christian, B.; Feil, N.; Urban, D.F.; Elsässer, C.; Prescher, M.; Kirste, L. Wurtzite ScAlN, InAlN, and GaAlN crystals, a comparison of structural, elastic, dielectric, and piezoelectric properties. *J. Appl. Phys.* **2021**, *130*, 045102. [CrossRef]
16. Lu, Y.; Reusch, M.; Kurz, N.; Ding, A.; Christoph, T.; Prescher, M.; Kirste, L.; Ambacher, O.; Žukauskaitė, A. Elastic modulus and coefficient of thermal expansion of piezoelectric $Al_{1-x}Sc_xN$ (up to x = 0.41) thin films. *APL Mater.* **2018**, *6*, 076105. [CrossRef]
17. Zhang, Q.; Chen, M.; Liu, H.; Zhao, X.; Qin, X.; Wang, F.; Tang, Y.; Yeoh, K.H.; Chew, K.H.; Sun, X. Deposition, Characterization, and Modeling of Scandium-Doped Aluminum Nitride Thin Film for Piezoelectric Devices. *Materials* **2021**, *14*, 6437. [CrossRef] [PubMed]
18. Fu, H.; Goodrich, J.C.; Ogidi-Ekoko, O.; Tansu, N. Power electronics figure-of-merit of ScAlN. *Appl. Phys. Lett.* **2021**, *119*, 072101. [CrossRef]
19. Saada, S.; Lakel, S.; Almi, K. Optical, electronic and elastic properties of ScAlN alloys in WZ and ZB phases: Prospective material for optoelectronics and solar cell applications. *Superlattices Microstruct.* **2017**, *109*, 915–926. [CrossRef]
20. Deng, R.; Evans, S.; Gall, D. Bandgap in $Al_{1-x}Sc_xN$. *Appl. Phys. Lett.* **2013**, *102*, 112103. [CrossRef]
21. Song, Y.; Perez, C.; Esteves, G.; Lundh, J.S.; Saltonstall, C.B.; Beechem, T.E.; Yang, J.I.; Ferri, K.; Brown, J.E.; Tang, Z.; et al. Thermal Conductivity of Aluminum Scandium Nitride for 5G Mobile Applications and Beyond. *ACS Appl. Mater. Interfaces* **2021**, *13*, 19031–19041. [CrossRef] [PubMed]
22. Caro, M.A.; Zhang, S.; Riekkinen, T.; Ylilammi, M.; Moram, M.A.; Lopez-Acevedo, O.; Molarius, J.; Laurila, T. Piezoelectric coefficients and spontaneous polarization of ScAlN. *J. Phys. Condens. Matter* **2015**, *27*, 245901. [CrossRef] [PubMed]
23. Olsson, R.H.; Tang, Z.; Agati, M.D. Doping of Aluminum Nitride and the Impact on Thin Film Piezoelectric and Ferroelectric Device Performance. In Proceedings of the 2020 IEEE Custom Integrated Circuits Conference (CICC), Boston, MA, USA, 22–25 March 2020.
24. Felmetsger, V.; Mikhov, M.; Ramezani, M.; Tabrizian, R. Sputter Process Optimization for $Al_{0.7}Sc_{0.3}N$ Piezoelectric Films. In Proceedings of the 2019 IEEE International Ultrasonics Symposium (IUS), Glasgow, UK, 6–9 October 2019.
25. Liu, X.; Wang, D.; Kim, K.H.; Katti, K.; Zheng, J.; Musavigharavi, P.; Miao, J.; Stach, E.A.; Olsson, R.H., III; Jariwala, D. Post-CMOS Compatible Aluminum Scandium Nitride/2D Channel Ferroelectric Field-Effect-Transistor Memory. *Nano Lett.* **2021**, *21*, 3753–3761. [CrossRef] [PubMed]
26. Tang, J.; Niu, D.; Tai, Z.; Hu, X. Deposition of highly c-axis-oriented ScAlN thin films at different sputtering power. *J. Mater. Sci. Mater. Electron.* **2017**, *28*, 5512–5517. [CrossRef]
27. Akiyama, M.; Kano, K.; Teshigahara, A. Influence of growth temperature and scandium concentration on piezoelectric response of scandium aluminum nitride alloy thin films. *Appl. Phys. Lett.* **2009**, *95*, 162107. [CrossRef]
28. Tominaga, T.; Takayanagi, S.; Yanagitani, T. c-Axis-tilted ScAlN films grown on silicon substrates for surface acoustic wave devices. *Jpn. J. Appl. Phys.* **2022**, *61*, SG1054. [CrossRef]
29. Tominaga, T.; Takayanagi, S.; Yanagitani, T. Negative-ion bombardment increases during low-pressure sputtering deposition and their effects on the crystallinities and piezoelectric properties of scandium aluminum nitride films. *J. Phys. D Appl. Phys.* **2021**, *55*, 105306. [CrossRef]
30. Wang, D.; Zheng, J.; Musavigharavi, P.; Zhu, W.; Foucher, A.C.; Trolier-McKinstry, S.E.; Stach, E.A.; Olsson, R.H. Ferroelectric Switching in Sub-20 nm Aluminum Scandium Nitride Thin Films. *IEEE Electron Device Lett.* **2020**, *41*, 1774–1777. [CrossRef]
31. Dong, K.; Wang, F.; Deng, M.; Yan, S.; Zhang, K. Piezoelectric performance improvement of ScAlN film and two-port SAW resonator application. *Electron. Lett.* **2019**, *55*, 1355–1357. [CrossRef]
32. Tabaru, T.; Akiyama, M. Residual stress reduction in piezoelectric $Sc_{0.4}Al_{0.6}N$ films by variable-pressure sputtering from 0.4 to 1.0 Pa. *Thin Solid Film.* **2019**, *692*, 137625. [CrossRef]
33. Henry, M.D.; Young, T.R.; Douglas, E.A.; Griffin, B.A. Reactive sputter deposition of piezoelectric $Sc_{0.12}Al_{0.88}N$ for contour mode resonators. *J. Vac. Sci. Technol. B* **2018**, *36*, 03E104. [CrossRef]
34. Lozano, M.S.; Pérez-Campos, A.; Reusch, M.; Kirste, L.; Fuchs, T.; Žukauskaitė, A.; Chen, Z.; Iriarte, G.F. Piezoelectric characterization of $Sc_{0.26}Al_{0.74}N$ layers on Si (001) substrates. *Mater. Res. Express* **2018**, *5*, 036407. [CrossRef]
35. Mertin, S.; Heinz, B.; Rattunde, O.; Christmann, G.; Dubois, M.A.; Nicolay, S.; Muralt, P. Piezoelectric and structural properties of c-axis textured aluminium scandium nitride thin films up to high scandium content. *Surf. Coat. Technol.* **2018**, *343*, 2–6. [CrossRef]
36. Pérez-Campos, A.; Sinusía Lozano, M.; Garcia-Garcia, F.J.; Chen, Z.; Iriarte, G.F. Synthesis of ScAlN thin films on Si (100) substrates at room temperature. *Microsyst. Technol.* **2018**, *24*, 2711–2718. [CrossRef]
37. Felmetsger, V.V. Sputter Technique for Deposition of AlN, ScAlN, and Bragg Reflector Thin Films in Mass Production. In Proceedings of the 2017 IEEE International Ultrasonics Symposium (IUS), Washington, DC, USA, 6–9 September 2017.
38. Fichtner, S.; Wolff, N.; Krishnamurthy, G.; Petraru, A.; Bohse, S.; Lofink, F.; Chemnitz, S.; Kohlstedt, H.; Kienle, L.; Wagner, B. Identifying and overcoming the interface originating c-axis instability in highly Sc enhanced AlN for piezoelectric microelectromechanical systems. *J. Appl. Phys.* **2017**, *122*, 035301. [CrossRef]
39. Zhang, Y.; Zhu, W.; Zhou, D.; Yang, Y.; Yang, C. Effects of sputtering atmosphere on the properties of c-plane ScAlN thin films prepared on sapphire substrate. *J. Mater. Sci. Mater. Electron.* **2014**, *26*, 472–478. [CrossRef]
40. Akiyama, M.; Umeda, K.; Honda, A.; Nagase, T. Influence of scandium concentration on power generation figure of merit of scandium aluminum nitride thin films. *Appl. Phys. Lett.* **2013**, *102*, 021915. [CrossRef]

41. Zukauskaite, A.; Wingqvist, G.; Palisaitis, J.; Jensen, J.; Persson, P.O.; Matloub, R.; Muralt, P.; Kim, Y.; Birch, J.; Hultman, L. Microstructure and dielectric properties of piezoelectric magnetron sputtered w-Sc$_x$Al$_{1-x}$N thin films. *J. Appl. Phys.* **2012**, *111*, 093527. [CrossRef]
42. Höglund, C.; Bareno, J.; Birch, J.; Alling, B.; Czigany, Z.; Hultman, L. Cubic Sc$_{1-x}$Al$_x$N solid solution thin films deposited by reactive magnetron sputter epitaxy onto ScN (111). *J. Appl. Phys.* **2009**, *105*, 113517. [CrossRef]
43. Höglund, C.; Birch, J.; Alling, B.; Bareño, J.; Czigány, Z.; Persson, P.O.; Wingqvist, G.; Zukauskaite, A.; Hultman, L. Wurtzite structure Sc$_{1-x}$Al$_x$N solid solution films grown by reactive magnetron sputter epitaxy: Structural characterization and first-principles calculations. *J. Appl. Phys.* **2010**, *107*, 123515. [CrossRef]
44. Lei, W.W.; Liu, D.; Zhu, P.W.; Chen, X.H.; Zhao, Q.; Wen, G.H.; Cui, Q.L.; Zou, G.T. Ferromagnetic Sc-doped AlN sixfold-symmetrical hierarchical nanostructures. *Appl. Phys. Lett.* **2009**, *95*, 162501. [CrossRef]
45. Akiyama, M.; Tabaru, T.; Nishikubo, K.; Teshigahara, A.; Kano, K. Preparation of scandium aluminum nitride thin films by using scandium aluminum alloy sputtering target and design of experiments. *J. Ceram. Soc. Jpn.* **2010**, *118*, 1166–1169. [CrossRef]
46. Posadowski, W.; Wiatrowski, A.; Domaradzki, J.; Mazur, M. Selected properties of Al$_x$Zn$_y$O thin films prepared by reactive pulsed magnetron sputtering using a two-element Zn/Al target. *Beilstein J. Nanotechnol.* **2022**, *13*, 344–354. [CrossRef] [PubMed]
47. Berg, S.; Nyberg, T. Fundamental understanding and modeling of reactive sputtering processes. *Thin Solid Film.* **2005**, *476*, 215–230. [CrossRef]
48. Moreira, M.A.; Doi, I.; Souza, J.F.; Diniz, J.A. Electrical characterization and morphological properties of AlN films prepared by dc reactive magnetron sputtering. *Microelectron. Eng.* **2011**, *88*, 802–806. [CrossRef]
49. Xu, X.-H.; Wu, H.S.; Zhang, C.J.; Jin, Z.H. Morphological properties of AlN piezoelectric thin films deposited by DC reactive magnetron sputtering. *Thin Solid Film.* **2001**, *388*, 62–67. [CrossRef]
50. Rodríguez-Madrid, J.G.; Iriarte, G.F.; Williams, O.A.; Calle, F. High precision pressure sensors based on SAW devices in the GHz range. *Sens. Actuators A Phys.* **2013**, *189*, 364–369. [CrossRef]
51. Fujii, S.; Odawara, T.; Yamada, H.; Omori, T.; Hashimoto, K.Y.; Torii, H.; Umezawa, H.; Shikata, S. Low Propagation Loss in a One-Port SAW Resonator Fabricated on Single-Crystal Diamond for Super-High-Frequency Applications. *IEEE Trans. Ultrason. Ferroelectr. Freq. Control.* **2013**, *60*, 986–992. [CrossRef]
52. Ababneh, A.; Schmid, U.; Hernando, J.; Sánchez-Rojas, J.L.; Seidel, H. The influence of sputter deposition parameters on piezoelectric and mechanical properties of AlN thin films. *Mater. Sci. Eng. B* **2010**, *172*, 253–258. [CrossRef]
53. Yang, Y.; Zhou, D.; Yang, C.; Feng, F.; Yang, J.; Hu, Q. Preparation of ScAlN film on Hastelloy alloys under different sputtering power. *Mater. Lett.* **2015**, *161*, 26–28. [CrossRef]

Disclaimer/Publisher's Note: The statements, opinions and data contained in all publications are solely those of the individual author(s) and contributor(s) and not of MDPI and/or the editor(s). MDPI and/or the editor(s) disclaim responsibility for any injury to people or property resulting from any ideas, methods, instructions or products referred to in the content.

Article

Properties of RF Magnetron-Sputtered Copper Gallium Oxide (CuGa$_2$O$_4$) Thin Films

Ashwin Kumar Saikumar *, Sreeram Sundaresh, Shraddha Dhanraj Nehate and Kalpathy B. Sundaram

Department of Electrical and Computer Engineering, University of Central Florida, Orlando, FL 32816, USA; sreeram.sundaresh@knights.ucf.edu (S.S.); shraddha.nehate@knights.ucf.edu (S.D.N.); kalpathy.sundaram@ucf.edu (K.B.S.)
* Correspondence: saikumarashwin1991@knights.ucf.edu

Abstract: Thin films of CuGa$_2$O$_4$ were deposited using an RF magnetron-sputtering technique for the first time. The sputtered CuGa$_2$O$_4$ thin films were post-deposition annealed at temperatures varying from 100 to 900 °C in a constant O$_2$ ambience for 1.5 h. Structural and morphological studies were performed on the films using X-ray diffraction analysis (XRD) and a Field Emission Scanning Electron Microscope (FESEM). The presence of CuGa$_2$O$_4$ phases along with the CuO phases was confirmed from the XRD analysis. The minimum critical temperature required to promote the crystal growth in the films was identified to be 500 °C using XRD analysis. The FESEM images showed an increase in the grain size with an increase in the annealing temperature. The resistivity values of the films were calculated to range between 6.47×10^3 and 2.5×10^8 Ωcm. Optical studies were performed on all of the films using a UV-Vis spectrophotometer. The optical transmission in the 200–800 nm wavelength region was noted to decrease with an increase in the annealing temperature. The optical bandgap value was recorded to range between 3.59 and 4.5 eV and showed an increasing trend with an increase in the annealing temperature.

Keywords: CuGa$_2$O$_4$; cubic spinel; annealing studies; optical characteristics; XRD; electrical characteristics

1. Introduction

Wide-bandgap semiconductors such as Ga$_2$O$_3$, ZnO, indium tin oxide, HfO$_2$, Y$_2$O$_3$, ZrO$_2$, AlGa$_2$O$_3$, IGZO exhibit many attractive properties beyond the capabilities of Si, and hence find applications in electronics, optical, optoelectronics, photonics and magneto-electronic devices [1–7]. As a result, the research and development of metal oxides with versatile properties are imperative. Among the various metal-oxide materials, cubic spinels have attracted great attention due to their chemical structure consisting of tetrahedral and octahedral sites [8]. Cubic spinels with the formula AB$_2$O$_4$ have cations distributed randomly among one octahedral site and two tetrahedral sites. B^{3+} ions occupy half of the octahedral holes, and A^{2+} ions occupy one eighth of the tetrahedral holes [9]. One such cubic spinel material exhibiting a wide bandgap like Ga$_2$O$_3$ is CuGa$_2$O$_4$. The copper and gallium ions in CuGa$_2$O$_4$ are distributed randomly in the A and B sublattices. This pseudo-binary system consisting of CuO-Ga$_2$O$_3$ phases displays a high potential for optoelectronic applications. Furthermore, CuGa$_2$O$_4$ has distinguished physical and chemical stability and exhibits a catalytic property. The fundamental requirement to promote catalytic and photocatalytic properties demands control over the thin films regarding the particle size, surface area and crystallinity. Due to its high sensitivity and rapid response to reducing/oxidizing, CuGa$_2$O$_4$ is a noteworthy candidate material for gas sensing towards H$_2$, liquefied petroleum and NH$_3$ [10].

Due to its unique structure, CuGa$_2$O$_4$ finds applications in supercapacitors [11], and as an active catalyst for a hydrogen gas source [12], a photocatalyst for solar hydrogen production [13], anode materials for sodium-ion batteries [14], and in organic photovoltaic

devices [15]. Previous studies have synthesized $CuGa_2O_4$ using techniques such as chemical vapor deposition (CVD), aerosol-assisted CVD [9], solid-state reaction [16], thermal decomposition [17], laser molecular beam epitaxy (L-MBE) [18] and hydrothermal techniques [19]. Synthesizing spinels requires multiple steps, high processing temperatures and a longer synthesis time [12,20]. Additionally, conventional CVD techniques involve volatile precursors, and the processes can be limited due to the formation of toxic byproducts. Among all of the thin-film deposition techniques, the magnetron sputtering technique is an established method of depositing thin films with high uniformity and homogeneity [1]. The magnetron sputtering techniques provide the flexibility to choose target materials with a wide range of melting points. This feature provides the capability to regulate the film thickness and deposition rate. Moreover, the magnetron sputtering technique does not demand the use of toxic or specialized precursors, which are required for CVD, and it offers great adhesion over a large surface area.

There have been no reported attempts to deposit $CuGa_2O_4$ thin films using the RF magnetron sputtering technique. This work addresses this inadequacy and focuses on the investigation of the properties of $CuGa_2O_4$ thin films synthesized using RF magnetron sputtering. Thin films of $CuGa_2O_4$ were deposited using a stoichiometric target mixture of Cu_2O and Ga_2O_3 in the presence of argon gas. The influence of annealing the films in the presence of oxygen gas at different temperatures was evaluated. The chemical structure, morphological properties, optical properties and electrical properties of $CuGa_2O_4$ thin films were investigated.

2. Experimental

2.1. Deposition of the $CuGa_2O_4$ Thin Films

$CuGa_2O_4$ films were deposited using an ultra-high vacuum 3 gun sputtering system (AJA international, Scituate, MA, USA). Fused quartz substrates were used for the annealing of the films at temperatures above 500 °C, and regular glass substrates were used for the annealing of the films at temperatures lower than 500 °C. The substrates were cleaned using acetone, methanol and DI water, followed by drying using nitrogen gas. The $CuGa_2O_4$ films were deposited using a 3-inch powder pressed sputtering target with stochiometric proportions of Cu_2O and Ga_2O_3 (99.99% purity). A base pressure of 3×10^{-7} Torr was achieved before every deposition. The sputter depositions were performed at the RF frequency (13.56 MHz) and a constant power of 200 W, using ultra-pure grade Ar as the sputtering gas. The Ar flow was kept constant at 10 sccm, and the deposition pressure was maintained constant at 10 mTorr for all of the depositions. All of the $CuGa_2O_4$ films reported in this research had a uniform thickness of 2000 Å. The substrate holder was rotated at a speed of 20 rpm in order to obtain a uniform thin-film thickness. All of the film depositions were performed at room temperature. The films were then annealed, post deposition, for 1.5 h at temperatures varying from 100 to 900 °C in O_2 ambience. The O_2 flow into the annealing furnace was maintained at 100 sccm for all of the film annealing.

2.2. Film Characterization

A Veeco Dektak 150 profilometer (Veeco, NY, USA) was used to measure the thickness of the films. The XRD measurements were performed using a PANalytical Empyrean XRD system (Malvern Panalytical, Westborough, MA USA), using radiation from a Cu source at 45 kV and 40 mA. The diffraction patterns were recorded between 2θ angles of 25–70°, and the phase information was analyzed using HighScore Plus software (Malvern Panalytical, Westborough, MA USA). The surface morphology of the film was identified using field emission scanning electron microscope Zeiss ULTRA-55 FEG SEM, (Zeiss Microscopy, White Plains, NY, USA). The optical transmission studies were performed using a Cary 100 UV-Vis spectrophotometer (Varian Analytical Instruments, Walnut creek, CA, USA). In order to perform the optical transmission studies, a wavelength range of 200–800 nm was used for the film deposited on the quartz substrates, and the wavelength range of 300–800 nm was used for the films deposited on the glass substrates. The resistivity of the film was measured by patterning parallel Al contact pads on the annealed CuGa$_2$O$_4$ films. In order to pattern the Al contact pads, a liftoff process was used. Once the contact pad windows were patterned, Al metal was deposited by adopting thermal evaporation. A schematic representation of the configuration used for the electrical measurements, along with the dimensions of the contact pads, is shown in Figure 1. A Keithley 2450 source meter (Tektronix Inc, Beaverton, OR, USA) unit was used to measure the I-V characteristics. From the measured I-V characteristics data, the resistance R was calculated. From the calculated R value, the resistivity (ρ) of the film was identified using the formula mentioned below:

$$\rho = R \frac{A}{L} \qquad (1)$$

where R is the resistance, L is the length and A is the area of cross section.

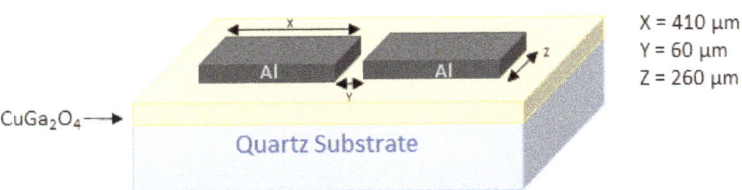

Figure 1. Schematic representation of the configuration used for the electrical measurements.

3. Results and Discussions

3.1. XRD Analysis

Thin-film XRD diffractograms of the as-deposited CuGa$_2$O$_4$ thin films, as well as films annealed at temperatures varying from 500–900 °C, are shown in Figure 2. The XRD pattern did not display any distinguished peaks for the as-deposited CuGa$_2$O$_4$ thin films or the thin films annealed at temperatures lower than 500 °C. However, the CuGa$_2$O$_4$ thin films annealed at temperatures of 500–900 °C displayed distinguishable peaks. The films annealed at 500 °C showed only two distinct peaks with comparatively lower intensity. They are the (311) peak related to the CuGa$_2$O$_4$ and (111) related to CuO. This suggests that the as-deposited films and the films annealed at temperatures less than 500 °C were majorly amorphous; therefore, for the sake of comparison, the XRD patterns of films annealed at 300 °C and 400 °C are not shown in Figure 2. It is likely that the films require a minimum annealing temperature of 500 °C in order to crystallize. The diffraction peaks identified from the XRD pattern were indexed well with the peaks pertaining to CuGa$_2$O$_4$ (JCPDS PDF # 44-0183). The peaks pertaining to CuO were also identified, suggesting the presence of mixed phases in the deposited thin films. Such a similar presence of CuO peaks along with CuGa$_2$O$_4$ has been previously reported in earlier research [9,20]. The major peaks identified at 2θ of 35.73, 37.68 and 63.59 were indexed to the (311), (222) and (440) phases of CuGa$_2$O$_4$. No peaks pertaining to Ga$_2$O$_3$ were identified. From Figure 2, a steady improvement in the peak intensity and peak sharpness can be noticed with the increase in the annealing temperature, thereby denoting a gradual improvement in the crystallinity. The peak intensity of the strongest peak (311) of CuGa$_2$O$_4$ was plotted as a function of the annealing temperature in Figure 3. A steady increase in the (311) peak intensity with an increase in the annealing temperature is evident from Figure 3.

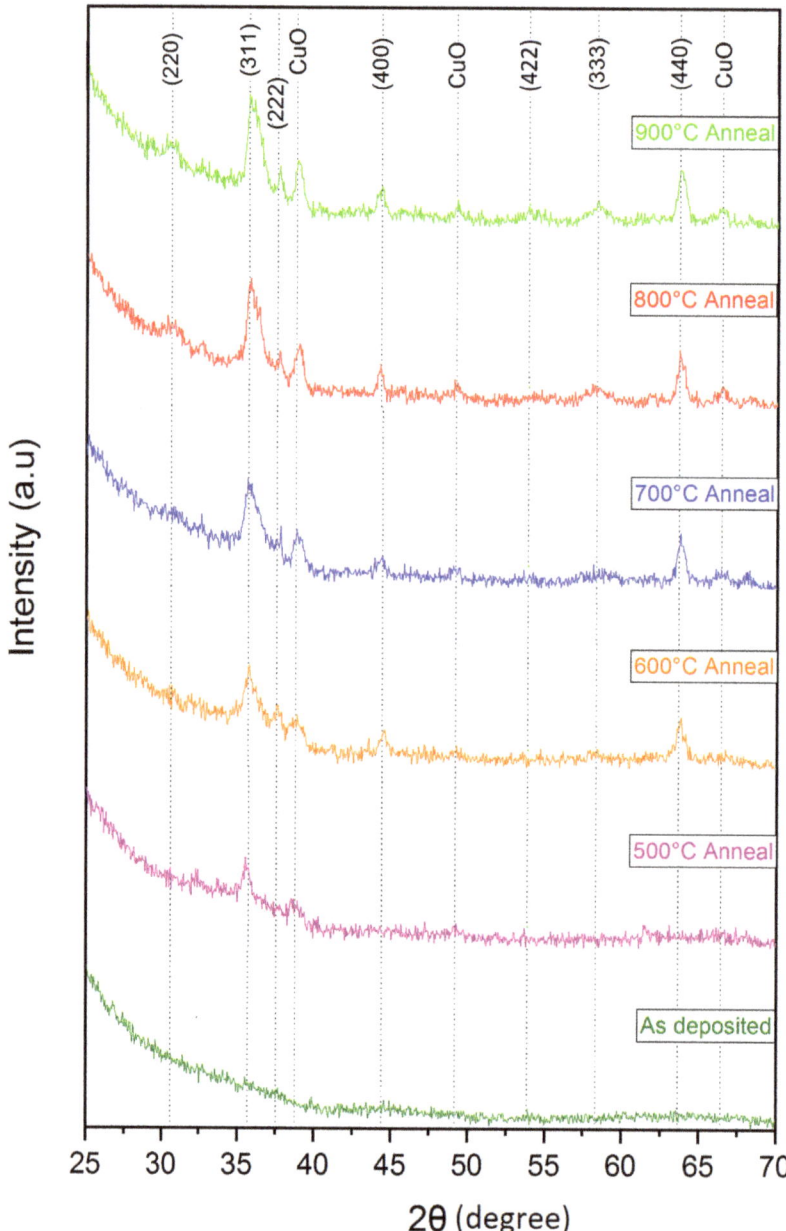

Figure 2. X-ray diffractions of the as-deposited $CuGa_2O_4$ thin film and the $CuGa_2O_4$ thin films annealed at 500–900 °C in O_2 ambience for 1.5 h.

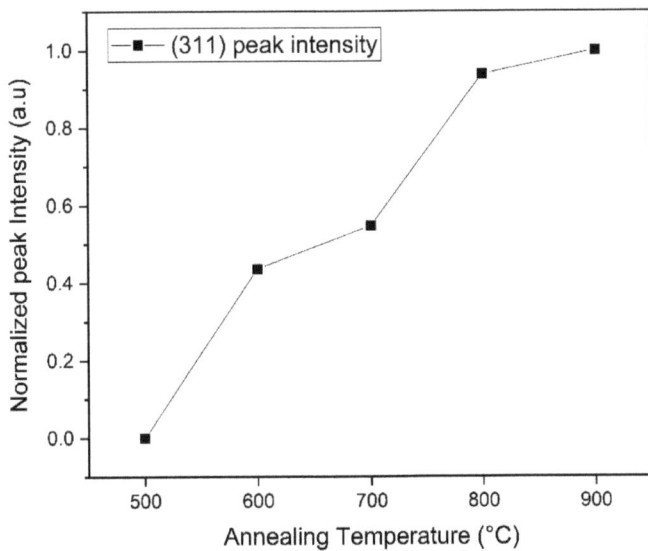

Figure 3. (311) peak intensity identified in the CuGa$_2$O$_4$ films as a function of the annealing temperature.

3.2. Morphology Studies

Figure 4 shows the FESEM images of the CuGa$_2$O$_4$ films. The morphological changes in both the as-deposited CuGa$_2$O$_4$ film (Figure 4a) and the CuGa$_2$O$_4$ films annealed at 300–900 °C (Figure 4b–h) can be seen clearly. Although the as-deposited film and films annealed at 300 °C and 400 °C displayed a tiny grain/particle presence in Figure 4a–c, they remained majorly amorphous and did not reveal any evidence of a peak presence in the XRD analysis. However, the film annealed at 500 °C displayed higher evidence of a grain presence and grain boundaries compared to the films annealed at 300 °C and 400 °C. This corroborates with the results from the XRD analysis in which the first appearance of diffraction peaks was observed in the film annealed at 500 °C. Deriving from the SEM findings and the appearance of diffraction peaks in the XRD analysis, it can be concluded that 500 °C is the minimum critical temperature required to promote nanocrystalline growth. The increase in the grain size with the increase in the annealing temperature can be evidently noted from Figure 4b–h. The small grains coalesce together and produce bigger grains when there is an increase in the annealing temperature [21]. This implies that the increment in the annealing temperature has an incremental influence on the CuGa$_2$O$_4$ film grain size. The film annealed at 900 °C was noted to have the largest-sized grains of 89 nm when compared to the other films. Figure 5 reiterates the steady increase in the grain size with the increase in the annealing temperature. The elemental analysis of all of the samples was performed using the EDAX incorporated in the FESEM. Table 1 shows the composition present in all of the films.

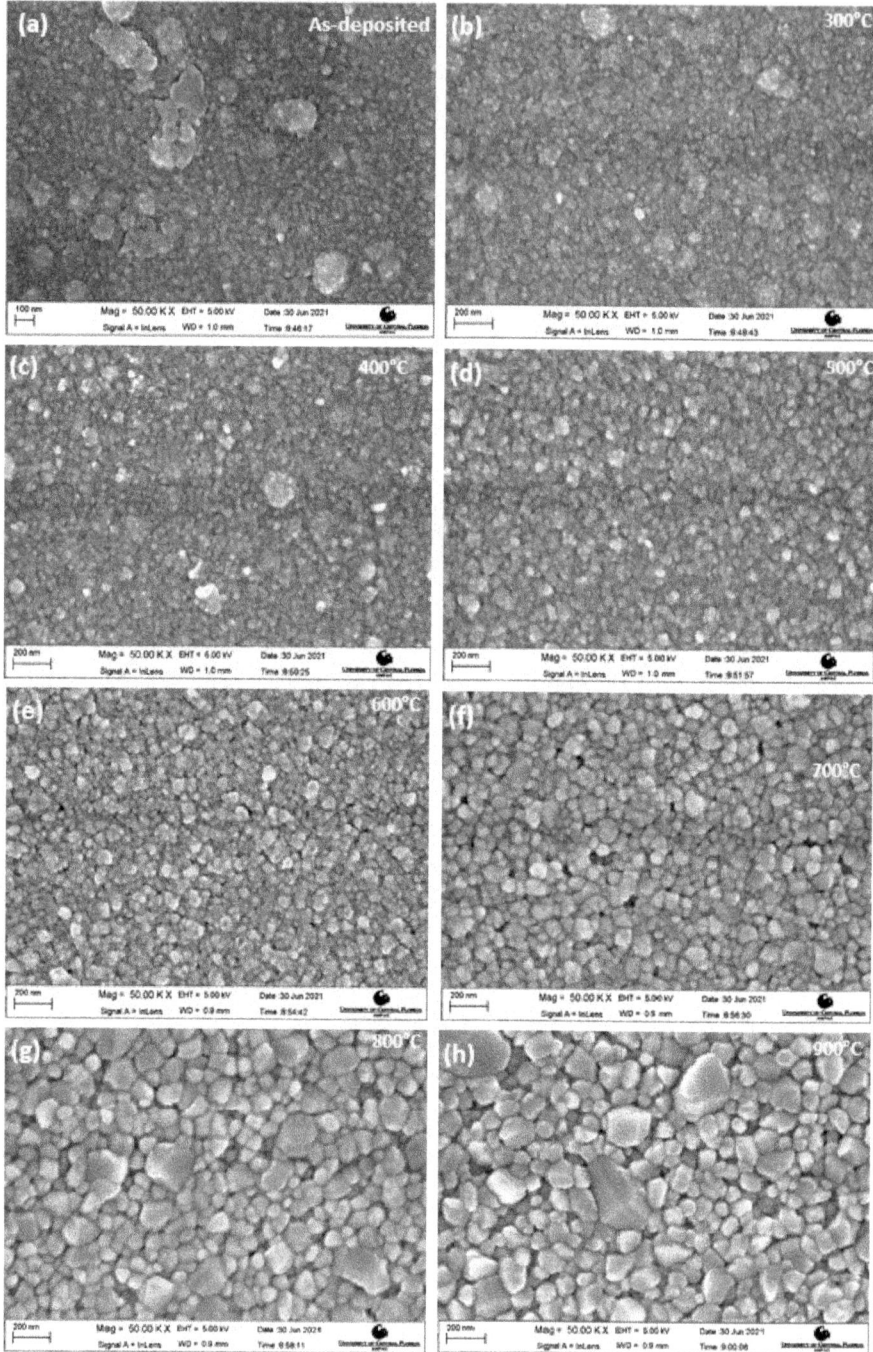

Figure 4. FESEM images of (**a**) the as-deposited $CuGa_2O_4$ films and the $CuGa_2O_4$ films annealed at (**b**) 300 °C, (**c**) 400 °C, (**d**) 500 °C, (**e**) 600 °C, (**f**) 700 °C, (**g**) 800 °C and (**h**) 900 °C (SEM magnification were constant at 50 KX. All scales in the image are in nm range).

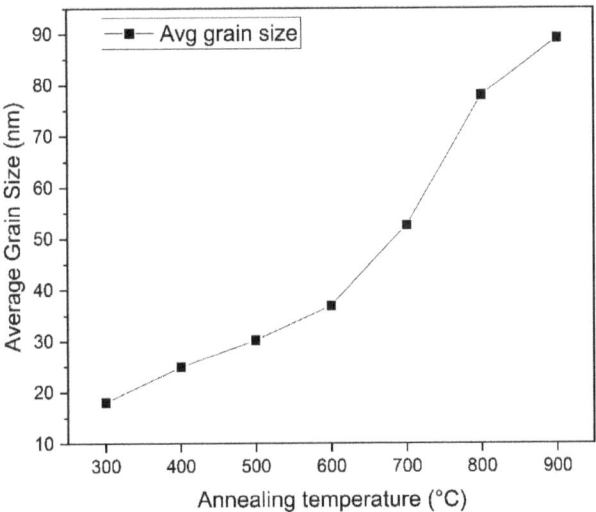

Figure 5. Average grain size as a function of the annealing temperature.

Table 1. $CuGa_2O_4$ film composition as a function of the annealing temperature, measured using EDAX.

Film	Cu atm%	Ga atm%	O atm%
As deposited film	13.64	25.88	60.48
Annealed at 300 °C	14.92	24.90	60.18
Annealed at 400 °C	14.78	24.07	61.15
Annealed at 500 °C	13.93	20.21	65.86
Annealed at 600 °C	14.10	20.79	65.11
Annealed at 700 °C	14.56	20.43	65.01
Annealed at 800 °C	13.88	19.92	66.19
Annealed at 900 °C	13.12	19.71	67.16

3.3. Optical Studies

3.3.1. Optical Transmission

The optical studies were performed on $CuGa_2O_4$ thin films deposited on quartz and glass substrates. Figure 6 shows the % transmission values recorded using a UV–Visible spectrophotometer in the wavelength range of 200–800 nm. The as-deposited $CuGa_2O_4$ thin films exhibited an optical transmission of ~60% at 500 nm. The increase in the annealing temperature resulted in the reduction of the optical transmission. The films annealed at 300 °C, 400 °C and 500 °C showed reduced transmission values between 30 and 40% at 500 nm. However, the higher annealing temperature further reduced the transmission from 60% for the as-deposited film to 30% for the film annealed at 600 °C, at the wavelength of 500 nm. This reduction in the optical transmission of $CuGa_2O_4$ thin films was attributed to a change in the grain size, as seen from the XRD analysis and FESEM images. The films with a low grain size were more transparent, while the optical transmission reduces with the annealing temperatures, as there is an increase in the grain size and crystallinity [22,23]. A further increase in the annealing temperature from 600 °C to 900 °C did not show much variation in the optical transmission of the $CuGa_2O_4$ thin films.

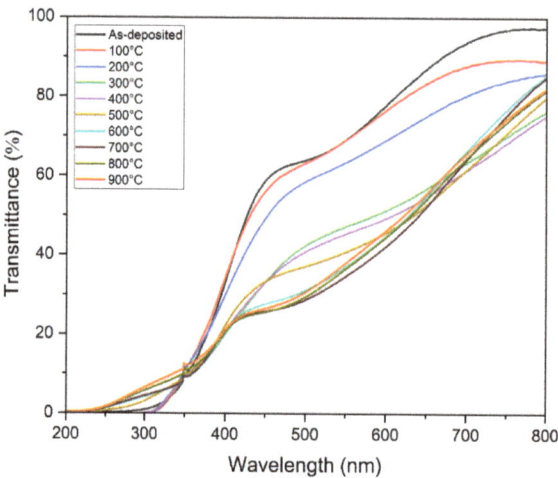

Figure 6. Optical transmission spectra of the CuGa$_2$O$_4$ thin films annealed at different temperatures.

3.3.2. Optical Bandgap

The optical transmission data was used to calculate the optical bandgap of the CuGa$_2$O$_4$ thin films. The absorption coefficient (α) was calculated using the following equation:

$$\alpha = \left(\frac{-2.303}{t}\right) \log_{10}(\%T) \quad (2)$$

where t is the thickness of the CuGa$_2$O$_4$ thin films, and T is the transmission. The optical bandgap (Eg) was estimated [24] using the following equation:

$$(\alpha h\nu)^{\frac{1}{n}} = B\left(h\nu - E_g\right) \quad (3)$$

where hν is the photon energy, B is a constant and Eg is the optical bandgap.

Figure 7 shows the Tauc plot generated using Equations (2) and (3). The linear region of the curves was extrapolated to the x-axis in order to achieve the Eg value. The optical bandgap values displayed an increasing trend with an increase in the annealing temperature. The bandgap of as-deposited film was 3.59 eV, which increased to 4.5 eV for the CuGa$_2$O$_4$ thin film annealed at 900 °C. The bandgap values obtained in this study are comparable to the bandgap values reported for CuGa$_2$O$_4$ thin films deposited using the laser MBE technique [18]. Table 2 shows the variation in the optical bandgap as a function of the annealing temperature. The unique structure and wide bandgap make CuGa$_2$O$_4$ a potential candidate material for ultraviolet optoelectronic device applications [18].

3.4. Electrical Studies

Figure 8 shows the electrical resistivity of the CuGa$_2$O$_4$ films. The electrical resistivities were calculated using the electrical resistance values obtained from the I–V curves. From Figure 8, it is evident that the resistivity showed a decreasing trend with increase in the annealing temperature up until 700 °C (6.4 × 10^3 Ωcm). This reduction in resistivity could be ascribed to the crystallization of the films. However, the resistivity increased in the films annealed at higher temperatures, with the film annealed at 900 °C showing a higher resistivity of 3.89 × 10^5 Ωcm. Such similar trends of decreasing resistivity up until a particular annealing temperature, followed by a subsequent increase in the resistivity at higher temperatures have been reported for other compounds belonging to the cubic spinel family [25]. Because there are no observed changes in the structural diffraction phases of the annealed films, the effect of annealing on the resistivity is primarily related to the grain density and grain boundaries. A similar increase in resistivity at very high annealing temperatures, despite with steady increase in the diffraction peak intensity (XRD peaks), has been previously reported in compounds belonging to the cubic spinel family [26,27]. One of the two plausible reasonings behind the increase in resistivity at higher temperatures is the decrease in the actual number of grains per volume, as noted from the FESEM image, thereby resulting in reduced carrier transitions [21,28]. This would potentially result in a resistivity increase. The other reason could be the presence of CuO phases in the film. Research conducted by Valladares et al. showed a sudden increase of resistivity in CuO films annealed at a temperature of 800 °C and above [29]. The authors attribute the increase in resistivity to the potential defects in the film. Similarly, the $CuGa_2O_4$ films sputtered in this research also showed an increase in resistivity in the films annealed at temperatures 800 °C and 900 °C.

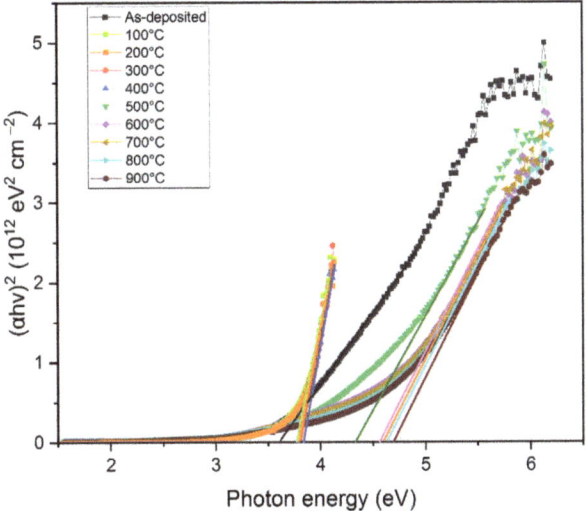

Figure 7. Tauc plot of the $CuGa_2O_4$ thin films annealed at different temperatures.

Table 2. Optical bandgap values obtained for $CuGa_2O_4$ thin films.

Annealing Temperature	Eg, Bandgap (eV)
As deposited	3.59
100 °C	3.72
200 °C	3.76
300 °C	3.8
400 °C	3.84
500 °C	4.12
600 °C	4.36
700 °C	4.4
800 °C	4.44
900 °C	4.5

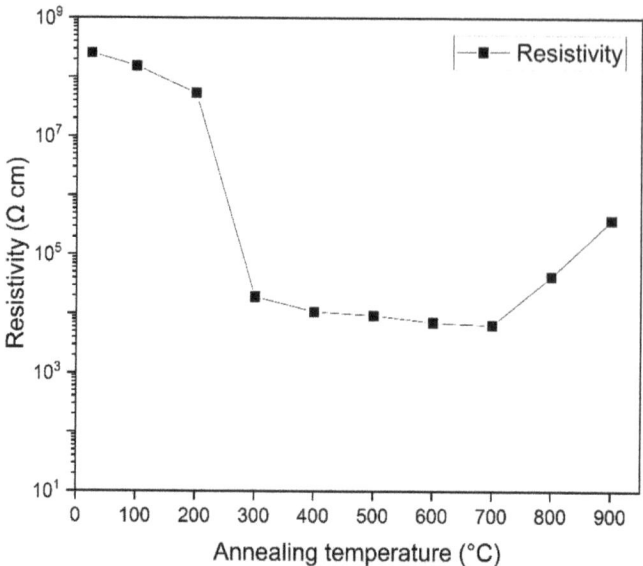

Figure 8. Influence of the annealing temperature on the electrical resistivity.

4. Conclusions

In this study, thin films of copper gallium oxide spinel, $CuGa_2O_4$, were successfully deposited using a sputtering target with a stochiometric composition of Cu_2O and Ga_2O_3 by an RF sputtering technique. The structural, morphological, optical and electrical effects due to post-deposition annealing of the $CuGa_2O_4$ thin films at temperatures varying from 100 to 900 °C in constant O_2 ambience were examined using XRD, FESEM, UV-Vis spectrophotometer and I–V curves. The XRD peaks pertaining to both $CuGa_2O_4$ and CuO were identified from the XRD studies. It was found that a minimum critical temperature of 500 °C was required to promote the crystallization of the films. The crystallinity of the films was noted to improve with increase in the annealing temperature. The average grain size was recognized to increase from 30 nm to 89 nm when the annealing temperature was increased from 500 °C to 900 °C. The optical bandgap recorded an increments from 3.59 eV to 4.5 eV in the non-annealed films and the films annealed at 900 °C, respectively. The tunability of the electrical resistivity of the $CuGa_2O_4$ thin films as a function of the annealing temperature was demonstrated. The $CuGa_2O_4$ films showed a stark decrease in resistivity from the highest value of 2.5×10^8 Ωcm in the as-sputtered films to the lowest resistivity of 6.47×10^3 Ωcm in the films annealed at 700 °C. Due to its high bandgap and cubic structure, $CuGa_2O_4$ could potentially be used extensively in UV optoelectronic device applications.

Author Contributions: Conceptualization, A.K.S. and K.B.S.; methodology, A.K.S.; validation, A.K.S. and K.B.S.; investigation, A.K.S.; writing—original draft preparation, A.K.S.; writing—review and editing, A.K.S., S.S., S.D.N., K.B.S.; visualization, A.K.S.; supervision, K.B.S. All authors have read and agreed to the published version of the manuscript.

Funding: This research received no external funding.

Institutional Review Board Statement: Not applicable.

Informed Consent Statement: Not applicable.

Data Availability Statement: Data available on request.

Conflicts of Interest: The authors declare no conflict of interest.

References

1. Saikumar, A.K.; Nehate, S.D.; Sundaram, K.B. RF sputtered films of Ga2O3. *ECS J. Solid State Sci. Technol.* **2019**, *8*, Q3064. [CrossRef]
2. Saikumar, A.K.; Skaria, G.; Sundaram, K.B. ZnO gate based MOSFETs for sensor applications. *ECS Trans.* **2014**, *61*, 65. [CrossRef]
3. Nehate, S.D.; Prakash, A.; Mani, P.D.; Sundaram, K.B. Work Function Extraction of Indium Tin Oxide Films from MOSFET Devices. *ECS J. Solid State Sci. Technol.* **2018**, *7*, P87–P90. [CrossRef]
4. Mudavakkat, V.; Atuchin, V.; Kruchinin, V.; Kayani, A.; Ramana, C. Structure, morphology and optical properties of nanocrystalline yttrium oxide (Y2O3) thin films. *Opt. Mater.* **2012**, *34*, 893–900. [CrossRef]
5. Pearton, S.; Abernathy, C.; Norton, D.; Hebard, A.; Park, Y.; Boatner, L.; Budai, J. Advances in wide bandgap materials for semiconductor spintronics. *Mater. Sci. Eng. R Rep.* **2003**, *40*, 137–168. [CrossRef]
6. Saikumar, A.K.; Nehate, S.D.; Sundaram, K.B. A review of recent developments in aluminum gallium oxide thin films and devices. *Rev. Solid State Mater. Sci.* **2021**, 1–32.
7. Sanctis, S.; Hoffmann, R.C.; Koslowski, N.; Foro, S.; Bruns, M.; Schneider, J.J. Aqueous Solution Processing of Combustible Precursor Compounds into Amorphous Indium Gallium Zinc Oxide (IGZO) Semiconductors for Thin Film Transistor Applications. *Chem. Asian J.* **2018**, *13*, 3912–3919. [CrossRef] [PubMed]
8. Petrakovskii, G.A.; Aleksandrov, K.S.; Bezmaternikh, L.N.; Aplesnin, S.S.; Roessli, B.; Semadeni, F.; Amato, A.; Baines, C.; Bartolomé, J.; Evangelisti, M. Spin-glass state in $CuGa_2O_4$. *Phys. Rev. B* **2001**, *63*, 184425. [CrossRef]
9. Knapp, C.E.; Prassides, I.D.; Sathasivam, S.; Parkin, I.P.; Carmalt, C.J. Aerosol-Assisted Chemical Vapour Deposition of a Copper Gallium Oxide Spinel. *ChemPlusChem* **2013**, *79*, 122–127. [CrossRef] [PubMed]
10. Shi, J.; Liang, H.; Xia, X.; Li, Z.; Long, Z.; Zhang, H.; Liu, Y. Preparation of high-quality $CuGa_2O_4$ film via annealing process of Cu/β-Ga_2O_3. *J. Mater. Sci.* **2019**, *54*, 11111–11116. [CrossRef]
11. Zardkhoshoui, A.M.; Davarani, S.S.H. Designing a flexible all-solid-state supercapacitor based on $CuGa_2O_4$ and FeP-rGO electrodes. *J. Alloy. Compd.* **2019**, *773*, 527–536. [CrossRef]
12. Faungnawakij, K.; Shimoda, N.; Fukunaga, T.; Kikuchi, R.; Eguchi, K. Cu-based spinel catalysts CuB_2O_4 (B= Fe, Mn, Cr, Ga, Al, Fe0. 75Mn0. 25) for steam reforming of dimethyl ether. *Appl. Catal. A Gen.* **2008**, *341*, 139–145. [CrossRef]
13. Gurunathan, K.; Baeg, J.-O.; Lee, S.M.; Subramanian, E.; Moon, S.-J.; Kong, K.-J. Visible light active pristine and Fe^{3+} doped $CuGa_2O_4$ spinel photocatalysts for solar hydrogen production. *Int. J. Hydrog. Energy* **2008**, *33*, 2646–2652. [CrossRef]
14. Pilliadugula, R.; Nithya, C.; Krishnan, N.G. Influence of Ga_2O_3, $CuGa_2O_4$ and Cu_4O_3 phases on the sodium-ion storage behaviour of CuO and its gallium composites. *Nanoscale Adv.* **2020**, *2*, 1269–1281. [CrossRef]
15. Wang, J.; Ibarra, V.; Barrera, D.; Xu, L.; Lee, Y.-J.; Hsu, J.W.P. Solution Synthesized p-Type Copper Gallium Oxide Nanoplates as Hole Transport Layer for Organic Photovoltaic Devices. *J. Phys. Chem. Lett.* **2015**, *6*, 1071–1075. [CrossRef] [PubMed]
16. Fenner, L.A.; Wills, A.S.; Bramwell, S.T.; Dahlberg, M.; Schiffer, P. Zero-point entropy of the spinel spin glasses $CuGa_2O_4$ and $CuAl_2O_4$. *J. Phys. Conf. Ser.* **2009**, *145*. [CrossRef]
17. Biswas, S.K.; Sarkar, A.; Pathak, A.; Pramanik, P. Studies on the sensing behaviour of nanocrystalline $CuGa_2O_4$ towards hydrogen, liquefied petroleum gas and ammonia. *Talanta* **2010**, *81*, 1607–1612. [CrossRef] [PubMed]
18. Wei, H.; Chen, Z.; Wu, Z.; Cui, W.; Huang, Y.; Tang, W. Epitaxial growth and characterization of $CuGa_2O_4$ films by laser molecular beam epitaxy. *AIP Adv.* **2017**, *7*, 115216. [CrossRef]
19. Yin, H.; Shi, Y.; Dong, Y.; Chu, X. Synthesis of spinel-type $CuGa_2O_4$ nanoparticles as a sensitive non-enzymatic electrochemical sensor for hydrogen peroxide and glucose detection. *J. Electroanal. Chem.* **2021**, *885*, 115100. [CrossRef]
20. Gingasu, D.; Mindru, I.; Patron, L.; Marinescu, G.; Tuna, F.; Preda, S.; Calderon-Moreno, J.M.; Andronescu, C. Synthesis of $CuGa_2O_4$ nanoparticles by precursor and self-propagating combustion methods. *Ceram. Int.* **2012**, *38*, 6739–6751. [CrossRef]
21. Ahmadipour, M.; Ain, M.F.; Ahmad, Z.A. Effects of annealing temperature on the structural, morphology, optical properties and resistivity of sputtered CCTO thin film. *J. Mater. Sci. Mater. Electron.* **2017**, *28*, 12458–12466. [CrossRef]
22. Shakti, N.; Gupta, P.S. Structural and optical properties of sol-gel prepared ZnO thin film. *Appl. Phys. Res.* **2010**, *2*, 19. [CrossRef]
23. Ramana, C.V.; Smith, R.J.; Hussain, O.M. Grain size effects on the optical characteristics of pulsed-laser deposited vanadium oxide thin films. *Phys. Status Solidi* **2003**, *199*, R4–R6. [CrossRef]
24. Tauc, J.; Grigorovici, R.; Vancu, A. Optical Properties and Electronic Structure of Amorphous Germanium. *Phys. Status Solidi* **1966**, *15*, 627–637. [CrossRef]
25. Li, R.; Fu, Q.; Zou, X.; Zheng, Z.; Luo, W.; Yan, L. Mn-Co-Ni-O thin films prepared by sputtering with alloy target. *J. Adv. Ceram.* **2020**, *9*, 64–71. [CrossRef]
26. Schulze, H.; Li, J.; Dickey, E.C.; Trolier-McKinstry, S. Synthesis, Phase Characterization, and Properties of Chemical Solution-Deposited Nickel Manganite Thermistor Thin Films. *J. Am. Ceram. Soc.* **2009**, *92*, 738–744. [CrossRef]
27. Kukuruznyak, D.A.; Bulkley, S.A.; Omland, K.A.; Ohuchi, F.S.; Gregg, M.C. Preparation and properties of thermistor thin-films by metal organic decomposition. *Thin Solid Films* **2001**, *385*, 89–95. [CrossRef]
28. Ohyama, M.; Kozuka, H.; Yoko, T. Sol-Gel Preparation of Transparent and Conductive Aluminum-Doped Zinc Oxide Films with Highly Preferential Crystal Orientation. *J. Am. Ceram. Soc.* **2005**, *81*, 1622–1632. [CrossRef]
29. Valladares, L.D.L.S.; Salinas, D.H.; Dominguez, A.B.; Najarro, D.A.; Khondaker, S.; Mitrelias, T.; Barnes, C.; Aguiar, J.A.; Majima, Y. Crystallization and electrical resistivity of Cu_2O and CuO obtained by thermal oxidation of Cu thin films on SiO_2/Si substrates. *Thin Solid Films* **2012**, *520*, 6368–6374. [CrossRef]

Article

Microstructure and Properties of MAO-Cu/Cu-(HEA)N Composite Coatings on Titanium Alloy

Zhao Wang [1], Nan Lan [2,*], Yong Zhang [3] and Wanrong Deng [3]

1. Shaanxi Office of Science, Technology and Industry for National Defense, Xi'an 710061, China
2. Xi'an Saitesimai Titanium Industry Co., Ltd., Xi'an 710201, China
3. School of Materials Science and Chemical Engineering, Xi'an Technological University, Xi'an 710032, China
* Correspondence: 15082383495@163.com

Abstract: In this paper, MAO-Cu/Cu-(HEA)N composite coatings on TC4 titanium alloy were prepared by combining micro arc oxidation (MAO) with magnetron sputtering (MS) to enhance the wear resistance and antibacterial ability of the substrate in simulated seawater. The number of micropores on the surface of the composite coatings decreased with increasing $CuSO_4$ concentration in the electrolyte, causing the surfaces to be flat and smooth. XPS and EDS analyses revealed that the MAO-Cu/Cu-(HEA)N composite coatings predominately contained TiO_2, Cu_2O, and (HEA)N. Moreover, the addition of $CuSO_4$ increased the growth rate of the MAO coatings. Comparatively, the MAO-Cu/Cu-(HEA)N composite coating with 5 g/L $CuSO_4$ showed superior wear resistance, reduced friction coefficient (approximately 0.2), and shallow and narrow grinding cracks were observed compared to the other coatings. Antibacterial experiments showed that the MAO-Cu/Cu-(HEA)N composite coatings had better bacterial killing effects than the TC4 substrate, which is of great significance to the antifouling abilities of titanium alloys in marine applications.

Keywords: TC4 titanium alloy; micro arc oxidation; magnetron sputtering; HEA film; microstructure and properties

1. Introduction

Titanium alloys are widely used in the marine and biomedical field owing to their light weights, high specific strengths, and many other advantages [1–6]. Titanium alloys are extensively used as biomedical materials in various artificial joints, bone fixings, dental implants, heart stents, etc., [5–8]. The use of titanium alloys in the marine industry can effectively reduce the weight of equipment, thus improving their payloads, which ultimately improves their reliability and reduces their maintenance costs. They are therefore ideal materials for marine equipment. However, titanium alloys face extremely harsh service conditions in marine environments, such as abrasion resulting from the erosion and collision of seawater and stones, corrosion of other alloy parts after connection, and adhesion and fouling of marine bacteria and microorganisms. Although titanium alloys exhibit excellent corrosion resistance when used as marine materials, they have poor wear resistance compared with stainless steel and other alloys [6,9,10]. In addition, the loss of titanium alloys from biological fouling is also a problem requiring urgent solutions. Micro arc oxidation (MAO) is a promising surface engineering technology which is used for the surface modification of various metals [11–17]. Some studies have shown that metallurgically bonded MAO layers on titanium alloy substrates can effectively improve the corrosion resistance, wear resistance, and hardness of the alloy [18–20]. It has also been reported that the addition of various soluble chemicals into the electrolyte used for the fabrication of MAO coatings can modify the structure of the coatings, which in turn affects the performance of titanium substrates [9,21–23]. The addition of particles into the electrolyte can also affect the discharge channel in the MAO process, change the structures and surface

Citation: Wang, Z.; Lan, N.; Zhang, Y.; Deng, W. Microstructure and Properties of MAO-Cu/Cu-(HEA)N Composite Coatings on Titanium Alloy. *Coatings* **2022**, *12*, 1877. https://doi.org/10.3390/coatings12121877

Academic Editor: Alexander D. Modestov

Received: 4 October 2022
Accepted: 11 November 2022
Published: 3 December 2022

Publisher's Note: MDPI stays neutral with regard to jurisdictional claims in published maps and institutional affiliations.

Copyright: © 2022 by the authors. Licensee MDPI, Basel, Switzerland. This article is an open access article distributed under the terms and conditions of the Creative Commons Attribution (CC BY) license (https://creativecommons.org/licenses/by/4.0/).

morphologies of the films, and affect the protective performances of titanium alloys [24]. However, the porous oxide coating formed on the surface of titanium alloys provides more possibilities for the adhesion of bacteria and microorganisms due to its usually large surface area. The addition of antibacterial elements into the coating is one of the methods adopted to improve the antibacterial properties of the alloy [25]. Noting that Cu is a common antibacterial element [26–28], V. Stranak et al. [29] found that the Ti-Cu films prepared by the dual-HiPIMS technique had good antibacterial ability against *Staphylococcus epidermidis* and *Staphylococcus aureus*. Zhang et al. [30] found that Cu exists in the form of Cu_2O and CuO for MAO coating doped with Cu. The addition of Cu significantly improved the antifouling performance of the studied metal substrate. Despite having excellent antibacterial properties, Cu has relatively poor corrosion and wear resistance [31,32]. Therefore, using a single Cu film as a protective coating for titanium alloy in marine environments will pose some protective limitations. Recently, high-entropy alloy nitride ((HEA)N) films have gained popularity due to their high hardness, improved oxidation and corrosion resistance, and high temperature stability [33,34], and a comprehensive literature survey proves that there is scant information existing for the addition of Cu to (HEA)N films.

In this paper, the morphology and structure of MAO coatings were adjusted by adding different concentrations of $CuSO_4$ in a phosphate system, followed by the deposition of Cu and high-entropy alloy nitrides (Cu-(HEA)N) film by magnetron sputtering. Cu and (HEA)N films were deposited on the surface of the MAO layer to provide a feasible approach for fabricating anti-wear and antibacterial composite coatings. The MAO-Cu/Cu-(HEA)N composite coating prepared by this method is expected to improve the wider application of titanium alloys in the marine field.

2. Experimental

2.1. Preparation of MAO-Cu/Cu-(HEA)N Composite Coatings

TC4 titanium alloy (Baoji Titanium Industry Co., Ltd, Xi'an, China) circular disk samples with dimension of ϕ 20 mm × 4 mm were used as the substrate material in the experiments. The chemical composition of TC4 is displayed in Table 1. The TC4 substrates were polished by SiC sandpaper from 200# to 2000#, ultrasonically cleaned with alcohol for 10 min to remove surface stains, and dried by a blower. The MAO equipment was independently developed by Xi'an University of technology, with TC4 substrates as anode and stainless-steel barrel as cathode. In the micro-arc oxidation process, DC pulsed constant current mode is used. The basic electrolyte contained sodium hexametaphosphate (($NaP_2O_5)_6$, 20 g/L), potassium hydroxide (KOH, 4 g/L), sodium tungstate ($Na_2WO_4 \cdot 2H_2O$, 3 g/L) and potassium fluoride ($KF \cdot 2H_2O$, 3 g/L). Varying concentrations of $CuSO_4$ solution (5, 10, and 15 g/L) were added to the base electrolyte with a MAO process of 10 min at constant voltage of 450 V, frequency of 800 Hz, and a duty cycle of 3%.

Table 1. The composition of the simulated seawater solution.

Composition	NaCl	$MgCl_2$	Na_2SO_4	$CaCl_2$	KCl	$NaHCO_3$	KBr	HBO_3	SrCl	NaF
g/L	24.53	11.11	4.09	1.16	0.685	0.201	0.101	0.027	0.028	0.003

The MAO coatings with different $CuSO_4$ concentrations were ultrasonically cleaned with alcohol for 10 min to remove the electrolyte on the surface of the coatings. Copper alloy target and AlSiTiCrNbV high entropy alloy target prepared by an arc melting furnace were cut into 50 mm × 2 mm disc. Cu-(HEA)N films with thickness of 1 μm were deposited on the surface of the MAO coatings by a double target magnetron sputtering deposition system. Pretreatment of the substrate cavities was carried out before deposition. The surface impurities on the sleeve of the fixed target were removed by a sandblasting machine (Taseken Trading Co., Shanghai, China) and the substrates were ultrasonically cleaned with acetone and alcohol at room temperature for 10 min. All the substrates were wiped clean with dry non-woven cloths, and residual surface impurities or oxides were removed by setting a negative bias of 400 V for 20 min before deposition. During the deposition, the powers of the copper alloy target and AlSiTiCrNbV high entropy alloy target were 20 W and 200 W, respectively. The negative bias voltage was 200 V and deposition time was 120 min. The preparation process of the MAO-Cu/Cu-(HEA)N composite coatings are shown in Figure 1.

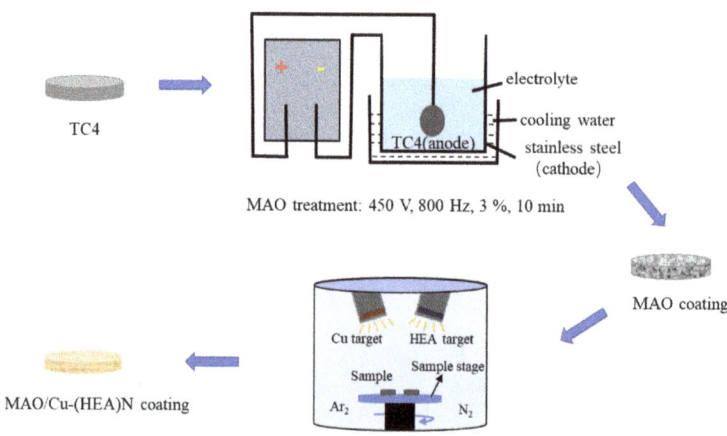

Figure 1. Preparation process of MAO-Cu/Cu-(HEA)N composite coatings.

2.2. Coating Characterization

The surface morphologies and elemental compositions of the coatings were analyzed by VEGA3-SBH scanning electron microscope (Taseken Trading Co., Shanghai, China) and energy dispersive spectrometer with a working voltage of 10 KV and a working distance of 3 mm (Taseken Trading Co., Shanghai, China). The composition and chemical states of coatings were analyzed by XPS (Shimadzu-Kratos Co., Hadano, Japan). The thicknesses of the MAO coatings were measured with a coating thickness gauge, and the average thickness after fifteen random measurements at different points was taken as the final thickness value. The friction coefficients of the coatings in simulated seawater were obtained using a HT-1000 high temperature friction and wear tester (Zhongke Kaihua Technology Development Co., Lanzhou, China). The chemical composition of the simulated seawater is shown in Table 1. A GCr15 bearing steel ball with a diameter of 6 mm was used as the wear material, and the wear time in the wear resistance test was set at 20 min. The *Staphylococcus aureus* (*S. aureus*) used for antimicrobial testing was cultured in LB medium and added into the coatings with simulated seawater by a pipette gun, and then cultured for 3 days at room temperature. The optical density of the bacterial solution was measured by a microplate reader, and bacterial adhesion was observed and measured based on the drying of the coating. Then, the bacteria liquid was coated on the surface of solid medium for 24 h, and colony growth was observed.

3. Results and Discussion
3.1. Analysis of Surface Morphology and Composition

The surface morphologies and compositions of the MAO-Cu/Cu-(HEA)N composite coatings with different concentrations of $CuSO_4$ are shown in Figure 2 and Table 2. It can be seen that the four composite coatings exhibited typical crater-like morphologies with uniform pore distributions. The pores on the surface of the MAO coatings were not completely covered by the Cu-(HEA)N films, maintaining the typical feature of microarc oxidation coatings [35–37]. Figure 2 shows that the nanoscale representation of the deposited Cu-(HEA)N films on micron-scale MAO layers had no obvious effect on the surface morphology of the MAO coating. The composite coating prepared from the base solution had a considerable number of protrusions and pores in various sizes, as observed in Figure 2a. However, with the addition of $CuSuO_4$ into electrolyte, the number of pores and protrusions on the coating surface decreased, as shown in Figure 2b. It is noteworthy that small amounts of $CuSO_4$ increased the conductivity of the solution, thereby promoting the growth of the MAO coating. The conductivity of the solution and its reaction rate were continuously enhanced with increasing $CuSO_4$ concentration. More melt was produced at high temperatures and high pressures during the micro arc oxidation [31], which solidified and accumulated on the surface of the TC4 substrate after expulsion through the discharge channel. Therefore, the increase in concentration to 10 g/L and 15 g/L $CuSO_4$ did not reduce the number of micropores and protrusions on the surface of the composite coatings, as displayed in Figure 2c,d. Table 2 shows that the composition of the MAO-Cu/Cu-(HEA)N composite coatings with different $CuSO_4$ concentrations exhibited a certain trend. The coatings were mainly composed of nitrides formed by the reaction of the elements in the (HEA)N target with N_2 and oxides of Cu and Ti. It was observed that the Cu content in the coatings increased with increasing $CuSO_4$ concentration in the electrolyte.

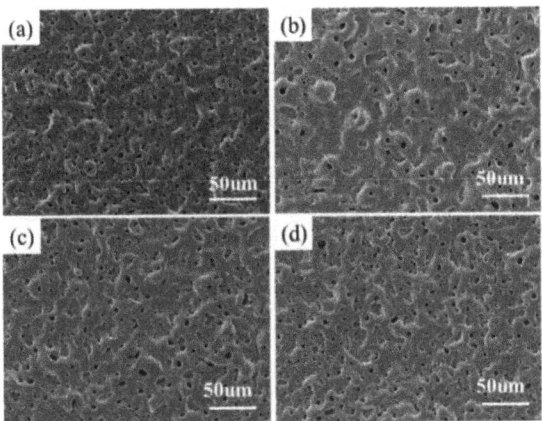

Figure 2. The surface morphologies of MAO-Cu/Cu-(HEA)N composite coatings with different $CuSO_4$ concentrations:(**a**) 0 g/L; (**b**) 5 g/L; (**c**) 10 g/L; and (**d**) 15 g/L.

Table 2. The composition of MAO-Cu/Cu-(HEA)N composite coatings with different $CuSO_4$ concentrations.

Samples	N	O	Al	Si	Ti	Cr	Nb	V	Cu
0 g/L	24.0	24.0	4.5	2.8	10.1	3.7	4.1	4.1	21.6
5 g/L	24.5	24.5	4.3	2.7	10.3	3.6	4.2	4.1	20.6
10 g/L	21.7	21.7	4.5	3.0	11.8	3.8	4.6	4.6	22.7
15 g/L	22.9	22.9	2.3	1.5	12.6	3.8	4.6	3.4	25.0

3.2. XPS Analysis

The surface characteristics of the MAO-Cu/Cu-(HEA)N samples were analyzed by XPS and displayed in Figure 3. The XPS fit was carried out using advantage software (version 5), and the peak of C1s was used to calibrate the peak of each element. The results show that the MAO-Cu/Cu-(HEA)N composite coatings surfaces were mainly composed of Ti, Cu, and O elements. The binding energy of Ti 2p was located at 458.17 eV and 464.48 eV, and assigned to Ti $2p_{3/2}$ and Ti $2p_{1/2}$ of Ti^{4+}, as shown in Figure 3a, indicating that O element mainly existed in the MAO-Cu/Cu-(HEA)N composite coatings in the form of TiN and TiO_2. The Cu 2p two main peaks were observed at 932.71 eV and 952.78 eV, and assigned to Cu $2p_{3/2}$ and Cu $2p_{1/2}$ of Cu^+, but no obvious peaks were detected around 944.0–941.5 eV according to Figure 3b, signifying that the Cu^+ of Cu_2O rather than the Cu^{2+} of CuO was the existing form of Cu in the MAO-Cu/Cu-(HEA)N composite coating [38–40]. The peak value of the O 1s peak was detected at 530.45 eV and 532.09 eV in Figure 3c, confirming the existence of TiO_2 and Cu_2O [41,42]. In the micro-arc oxidation process, the arc discharge generates instantaneous high temperature and high pressure. $CuSO_4$ decomposes at high temperatures to produce CuO, which in turn decomposes to produce Cu_2O.

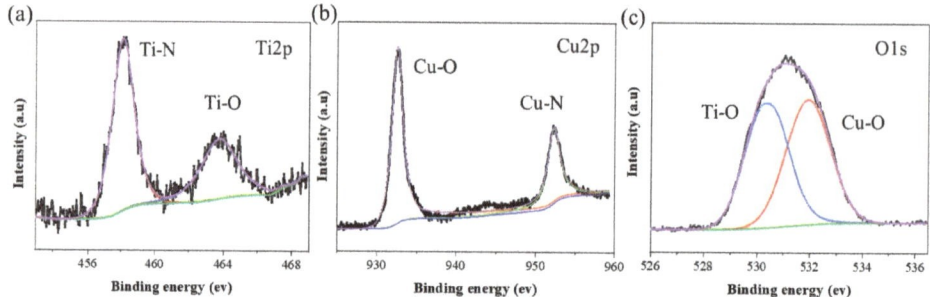

Figure 3. XPS spectra analysis (black line) and XPS fitting (purple line) on the surface of the MAO-Cu/Cu-(HEA)N composite coating: (**a**) Ti 2p (red line is Ti-N, blue line is Ti-O and green line is base line), (**b**) Cu 2p (red line is Cu-O, green line is Cu-N and blue line is base line), (**c**) O 1s (red line is Ti-O, red line is Cu-O and green line is base line).

3.3. Thickness

The thicknesses of the MAO coatings are shown in Figure 4. It can be seen from the figure that the thickness values of the four coatings were between 20 and 30 μm. It is possible that the Cu^{2+} in $CuSO_4$ solution was responsible for the enhancement of the conductivity of the electrolyte and the increase in current which significantly increased the generation and dissolution rate of the surface melt, which improved the growth rate of the MAO coatings. Compared to the coating prepared by the base solution, the thickness of the MAO coating increased by more than 40% with the addition of 5 g/L $CuSO_4$ solution. However, the effect of the MAO treatment was limited at higher $CuSO_4$ concentrations because electric breakdown was more difficult in the thicker coatings [43]. Hence, the thickness of the composite coating did not necessarily increase at $CuSO_4$ concentrations above 5 g/L.

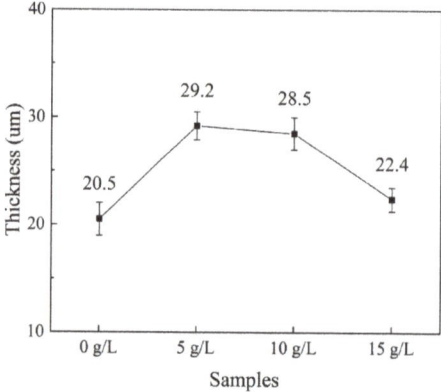

Figure 4. The thickness of MAO coatings with different $CuSO_4$ concentrations.

3.4. Wear Resistance

Friction coefficient is an important parameter employed to evaluate the wear resistances of materials [44–46]. Figure 5 depicts the friction coefficient (CoF) vs. sliding time for the TC4 material and the four MAO-Cu/Cu-(HEA)N composite coatings at an applied load of 2 N in simulated seawater. The friction coefficient of the TC4 material fluctuated significantly in simulated seawater with values ranging from 0.3 to 0.8. The TC4 material had a relatively low hardness and high viscosity [37]. With continuous friction of the grinding ball and the TC4 surface, the material viscosity and wear contact surface increased together with the increasing frictional force. Therefore, TC4 material had large friction coefficient and obvious fluctuations. The friction coefficient curve of the composite coating with the base solution exhibited a heavy fluctuation, and the friction coefficient gradually increased with increasing wear time. Specifically, the friction coefficient of the MAO-Cu/Cu-(HEA)N composite coating with 5 g/L $CuSO_4$ was about 0.21 with the extension of wear time, and the friction coefficient curve was relatively stable. However, the friction coefficient of the MAO-Cu/Cu-(HEA)N composite coatings with 10 g/L and 15 g/L $CuSO_4$ increased to 0.3 and 0.4, respectively. The addition of $CuSO_4$ to the electrolyte reduced the number of micropores in the MAO layers, making the coatings flat and smooth [47–49], and the Cu-(HEA)N films covered parts of the tiny pores in the MAO layers. The low friction of the Cu element in the MAO layers decreased the friction coefficient, thereby improving the wear resistance. In addition, the simulated seawater stored in the MAO layers formed a liquid film during wear, which further decreased the friction coefficient. The weightlessness percentages of the four composite coatings calculated after wear are shown in Table 3. As can be seen from the table, the weightlessness percentages of the four composite coatings were lower than that of the TC4 material. The lowest weightlessness percent of the MAO-Cu/Cu-(HEA)N composite coating was recorded for the coating with 5 g/L $CuSO_4$, indicating that it had the best wear resistance of the four coatings. Overall, the MAO-Cu/Cu-(HEA)N composite coatings improved the wear damage usually associated with TC4 materials.

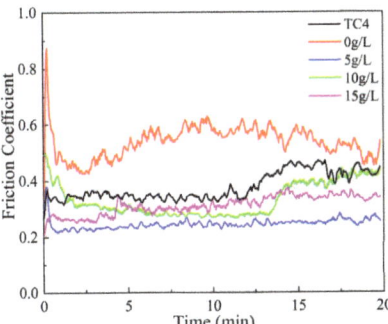

Figure 5. Friction coefficient curves of TC4 and MAO-Cu/Cu-(HEA)N composite coatings with different $CuSO_4$ concentrations.

Table 3. Percent of weight loss of TC4 and MAO-Cu/Cu-(HEA)N composite coatings with different $CuSO_4$ concentrations.

Samples	Percent of Weightloss (%)
TC4	1.310 ± 0.004
0 g/L	0.016 ± 0.001
5 g/L	0.011 ± 0.001
10 g/L	0.016 ± 0.001
15 g/L	0.014 ± 0.001

The surface morphologies by SEM examination and EDS micrographs of the wear scars are shown in Figure 6. Additionally, the widths of the wear scars were marked in the SEM micrographs. After wear, the surface of the TC4 material had the widest wear mark width of 551.45 μm with obvious furrow morphology, indicating a typical abrasive wear characteristic (Figure 6a) [43]. The rough surface of the GCr15 grinding ball played a vital role in ploughing the surface of the TC4 alloy, leaving ploughing grooves on the worn surface. Compared to the TC4 material, the wear marks on the surface of the MAO-Cu/Cu-(HEA)N composite coatings were shallow without obvious grooves, as displayed in Figure 6b–e. The four MAO-Cu/Cu-(HEA)N composite coatings had narrow wear scar widths, which first increased and then decreased with increasing $CuSO_4$ concentrations. The composite coating with 5 g/L $CuSO_4$ exhibited the least wear mark width and friction coefficient. The surface of the composite coating without the addition of $CuSO_4$ has more micropores and larger bumps, with a large degree of wear on the grinding balls. The introduction of $CuSO_4$ improves the structure of the coating, the number of micropores is reduced and the surface of the coating is smooth, so the wear on the grinding balls and the coating is slight. The TiO_2 layer of microporous sodium storage seawater formed a liquid film to reduce the degree of wear on the grinding ball and the coating. As can be seen from the EDS diagram, Fe element did not only exist at the wear mark locations, but also on the whole surface of the TC4 material, as shown in Figure 6a. Obviously, the Fe element was largely characterized on the wear marks of the composite coatings. This was due to compaction of the wear debris produced by repeated friction and wear on the grinding balls in the numerous micropores on the surface of the composite coatings (Figure 6b–e). The GCr15 grinding ball only wore out the surface of the four composite coatings but did not damage the TC4 substrate, indicating that the four composite coatings played effective roles in protecting the TC4 substrate.

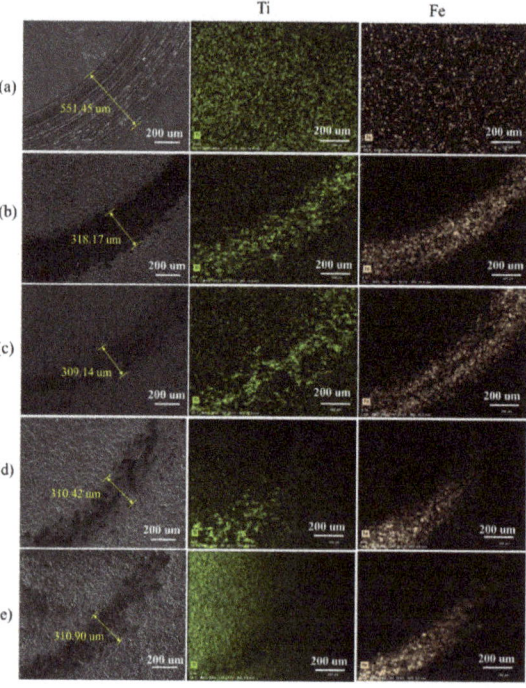

Figure 6. The surface SEM and EDS of TC4 and MAO-Cu/Cu-(HEA)N composite coatings with different $CuSO_4$ concentrations: (**a**) TC4; (**b**) 0 g/L; (**c**) 5 g/L; (**d**) 10 g/L; and (**e**) 15 g/L.

3.5. Antibacterial Ability

Figure 7 shows the macrophotographs of *Staphylococcus aureus* (*S. aureus*) colonies cultured on TC4 and MAO-Cu/Cu-(HEA)N composite coatings with different $CuSO_4$ concentrations attached to the surface of LB solid medium. The bacteria cultured on the surface of TC4 survived well on the surface of LB solid medium and they proliferated and grew in large numbers, as shown in Figure 7a. With increasing $CuSO_4$ concentrations, the MAO-Cu/Cu-(HEA)N composite coatings demonstrated excellent antibacterial abilities. For the coating containing 5 g/L $CuSO_4$, only few *S. aureus* existed on the surface of the solid medium after culture, indicating that the increasing $CuSO_4$ concentration was beneficial to the improvement of the antibacterial properties of the MAO-Cu/Cu-(HEA)N composite coatings (Figure 7b). It was reported that the presence of Cu_2O nanoparticles will completely inactivate bacterial growth through the release of Cu ions which generates reactive oxygen species [50]. In this study, the inactivation of *S. aureus* on the surface of the solid medium is also attributed to the presence of Cu ions in the MAO-Cu/Cu-(HEA)N composite coatings.

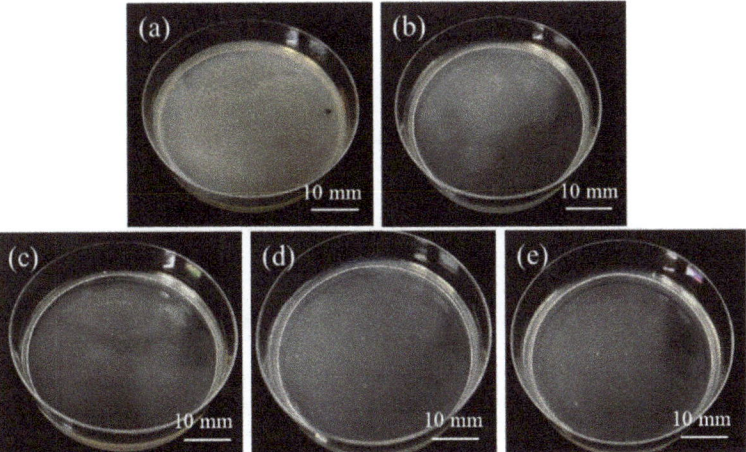

Figure 7. The macrophotographs of *Staphylococcus aureus* colonies cultured by TC4 and MAO-Cu/Cu-(HEA)N composite coatings with different $CuSO_4$ concentrations attached to the surface of LB solid medium: (**a**)TC4; (**b**) 0 g/L; (**c**) 5 g/L; (**d**) 10 g/L; and (**e**) 15 g/L.

Figure 8 shows the optical density of *S. aureus* increment in the simulated seawater. Optical density reflects the antibacterial ability of samples to *S. aureus* [51], and a low optical density represents poor livability of *S. aureus* and an excellent antibacterial ability. Comparing the optical density of the *S. aureus*-cultured TC4 with those of the different composite coatings, it was observed that the optical density decreased significantly with increasing $CuSO_4$ concentration. The optical densities of the four *S. aureus*-cultured composite coatings were lower than that of the *S. aureus*-cultured TC4. As Cu content in MAO-Cu/Cu-(HEA)N composite coatings increased, the precipitation of Cu ions increased, and the sterilization performance of the MAO-Cu/Cu-(HEA)N composite coating were also enhanced. The optical density of the composite coating containing 5 g/L $CuSO_4$ was 0.19, and this presented the most favorable condition for bacterial deactivation of all the composite coatings.

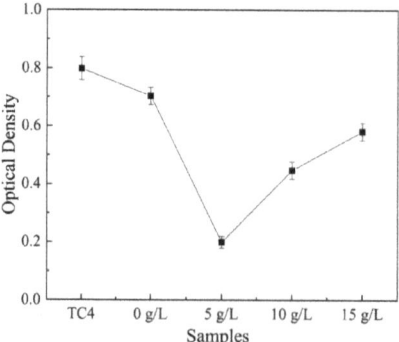

Figure 8. Optical density of *S. aureus* cultured by TC4 and MAO-Cu/Cu-(HEA)N composite coatings with different CuSO$_4$ concentrations.

Figure 9 represents the microcosmic surface adhesion of *S. aureus* on TC4 and the MAO-Cu/Cu-(HEA)N composite coatings. A large number of *S. aureus* were attached to the surface of the TC4 sample, and the TC4 provided favorable living conditions for the growth and reproduction of the *S. aureus*, as depicted in Figure 9a. The adhesion rates of the *S. aureus*-cultured composite coatings containing different concentrations of CuSO$_4$ in basic electrolyte were comparatively lower than that of the *S. aureus*-cultured TC4. Some of the *S. aureus* agglomerated to form a grape-like structure which adhered to the surface of the MAO-Cu/Cu-(HEA)N composite coatings, as shown in Figure 9c. In addition, there were almost no agglomerations of *S. aureus* on the surface of the MAO-Cu/Cu-(HEA)N composite coatings containing 10 g/L and 15 g/L CuSO$_4$ (Figure 9d,e).

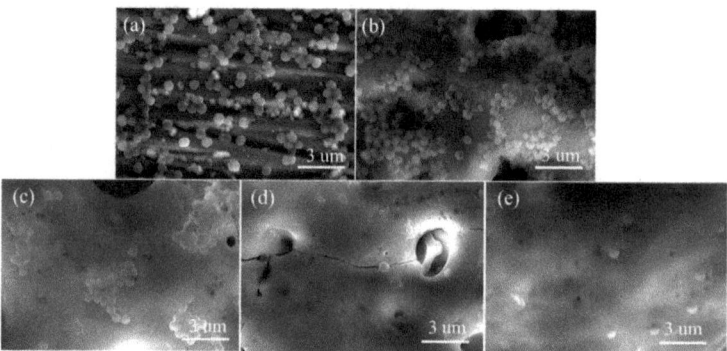

Figure 9. The microcosmic adhesion of *S. aureus* on TC4 and MAO-Cu/Cu-(HEA)N composite coatings surface with different CuSO$_4$ concentrations attached: (**a**) TC4; (**b**) 0 g/L; (**c**) 5 g/L; (**d**) 10 g/L; and (**e**) 15 g/L.

Cu$_2$O has been proven to have good antibacterial effects [50,52–56]. Studies have shown that reactive oxygen species (OH, H$_2$O$_2$ and O$_2^-$) produced by Cu$_2$O may interact with bacterial cell membranes and promote bacterial permeation [57]. Disturbance of bacterial cell membranes by Cu$_2$O can cause a number of functional disorders that impede the growth of bacteria and may eventually lead to the growth of bacterial species and possibly eventually their death [57,58]. The small size (nanoscale) of Cu$_2$O makes it easy for Cu$_2$O to penetrate the cell membrane compared to the pore size (microns) of bacteria [59]. The XPS result displayed in Figure 4 shows that after the MAO process and sputtering, TiO$_2$ and Cu$_2$O were the main components on the surface of the coatings. Based on the above observation, it is reasonable to conclude that Cu$_2$O was generated after the addition of CuSO$_4$ to the electrolyte, and it played an important role in the antibacterial behavior of the coatings. As shown in Figure 10a, there was no Cu_2O in the pores on the MAO layer without $CuSO_4$ addition. Here, the antibacterial effect was mainly dependent on the Cu_2O on the surface of the Cu-(HEA)N films. This means that the incompletely killed bacteria continued to grow and reproduce in the pores. The increasing concentration of $CuSO_4$ changed the structure and Cu_2O content of the MAO-Cu/Cu-(HEA)N composite coatings. From Figure 10b, the Cu_2O in the pores of the MAO layer interacted with the active bacteria on the surface of the MAO-Cu/Cu-(HEA)N film and continued to kill them, hence bacterial inactivation continued in the pores [60]. Moreover, the rough structure of the coatings increased their surface areas or contact interface areas, thereby improving their antibacterial performances.

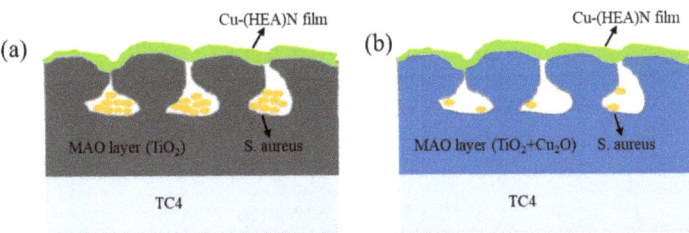

Figure 10. Mechanism of the effect of $CuSO_4$ on the antibacterial ability of MAO-Cu/Cu-(HEA)N composite coating: (**a**) without $CuSO_4$; and (**b**) with $CuSO_4$.

4. Conclusions

MAO-Cu/Cu-(HEA)N composite coatings were successfully prepared on titanium alloy by MAO and MS. With the addition of $CuSO_4$ to the electrolyte, the surface of the composite coating became flat and smooth with the existence of dense and uniform pores. Cu mainly existed as Cu_2O in the composite coatings, and the introduction of $CuSO_4$ promoted both the growth rates and thicknesses of the coatings. The wear resistance and antibacterial property of the composite coating was improved in simulated seawater with increments in $CuSO_4$ concentration. The friction coefficient of the composite coating with 5 g/L $CuSO_4$ was reduced to about 0.2, and the wear marks were shallow and narrow. Compared with the TC4 material, the optical density of the composite coatings decreased to 0.19, and the adhesion of *S. aureus* on the surface of the composite coatings were significantly weakened. MAO-Cu/Cu-(HEA)N composite coatings are important for surface modification of titanium alloys, but the wear and antimicrobial mechanisms of the composite coatings need to be investigated in more depth. In addition, the composite coatings research will focus on corrosion and erosion resistance in the marine environment.

Author Contributions: Conceptualization, N.L. and Z.W.; methodology, Z.W.; software, Y.Z.; validation, Y.Z., Z.W. and W.D.; formal analysis, Z.W.; investigation, Y.Z.; resources, W.D.; data curation, Z.W. and Y.Z.; writing—original draft preparation, Z.W.; writing—review and editing, N.L.; visualization, N.L; supervision, N.L; project administration, Z.W; funding acquisition, Z.W. All authors have read and agreed to the published version of the manuscript.

Funding: The authors gratefully acknowledge financial support of National Natural Science Foundation of China (No. 52071252) and Key research and development plan of Shaanxi province industrial project (2021GY-208, 2022GY-407 and 2021ZDLSF03-11).

Institutional Review Board Statement: Not applicable.

Informed Consent Statement: Not applicable.

Data Availability Statement: Not applicable.

Conflicts of Interest: The authors declare no conflict of interest.

References

1. Bai, H.; Zhong, L.; Kang, L.; Liu, J.; Zhuang, W.; Lv, Z.; Xu, Y. A review on wear-resistant coating with high hardness and high toughness on the surface of titanium alloy. *J. Alloy. Compd.* **2021**, *882*, 160645. [CrossRef]
2. Bai, X.; Tu, B. Ellipsoid non-probabilistic reliability analysis of the crack growth fatigue of a new titanium alloy used in deep-sea manned cabin. *Theor. Appl. Fract. Mech.* **2021**, *115*, 103041. [CrossRef]
3. Pan, X.; He, W.; Cai, Z.; Wang, X.; Liu, P.; Luo, S.; Zhou, L. Investigations on femtosecond laser-induced surface modification and periodic micropatterning with anti-friction properties on Ti6Al4V titanium alloy. *Chin. J. Aeronaut.* **2021**, *35*, 521–537. [CrossRef]
4. Dhanda, M.; Haldar, B.; Saha, P. Development and Characterization of Hard and Wear Resistant MMC Coating on Ti-6Al-4V Substrate by Laser Cladding. *Procedia Mater. Sci.* **2014**, *6*, 1226–1232. [CrossRef]
5. Chowdhury, M.A.; Hossain, N.; Shahid, M.A.; Alam, M.J.; Hossain, S.M.; Uddin, M.I.; Rana, M.M. Development of SiC–TiO_2-Graphene neem extracted antimicrobial nano membrane for enhancement of multiphysical properties and future prospect in dental implant applications. *Heliyon* **2022**, *8*, e10603. [CrossRef]
6. Kujala, S.; Ryhänen, J.; Danilov, A.; Tuukkanen, J. Effect of porosity on the osteointegration and bone ingrowth of a weight-bearing nickel–titanium bone graft substitute. *Biomaterials* **2003**, *24*, 4691–4697. [CrossRef]
7. Wang, X.; Li, Y.; Xiong, J.; Hodgson, P.D.; Wen, C. Porous TiNbZr alloy scaffolds for biomedical applications. *Acta Biomater.* **2009**, *5*, 3616–3624. [CrossRef]
8. Li, Y.; Ding, Y.; Munir, K.; Lin, J.; Brandt, M.; Atrens, A.; Xiao, Y.; Kanwar, J.R.; Wen, C. Novel β-Ti35Zr28Nb alloy scaffolds manufactured using selective laser melting for bone implant applications. *Acta Biomater.* **2019**, *87*, 273–284. [CrossRef]
9. Muhaffel, F.; Kaba, M.; Cempura, G.; Derin, B.; Kruk, A.; Atar, E.; Cimenoglu, H. Influence of alumina and zirconia incorporations on the structure and wear resistance of titania-based MAO coatings. *Surf. Coat. Technol.* **2019**, *377*, 124900. [CrossRef]
10. Dong, H. Tribological properties of titanium-based alloys. In *Surface Engineering of Light Alloys*; Dong, H., Ed.; Elsevier: Amsterdam, The Netherlands, 2010; pp. 58–80.
11. He, R.; Wang, B.; Xiang, J.; Pan, T. Effect of copper additive on microstructure and anti-corrosion performance of black MAO films grown on AZ91 alloy and coloration mechanism. *J. Alloy. Compd.* **2021**, *889*, 161501. [CrossRef]
12. Chen, J.; Xu, J.L.; Huang, J.; Dai, L.; Xue, M.S.; Luo, J.M. Corrosion resistance of T-ZnOw/PDMS-MAO composite coating on the sintered NdFeB magnet. *J. Magn. Magn. Mater.* **2021**, *534*, 168049. [CrossRef]
13. Lan, N.; Yang, W.; Gao, W.; Guo, P.; Zhao, C.; Chen, J. Characterization of ta-C film on micro arc oxidation coated titanium alloy in simulated seawater. *Diam. Relat. Mater.* **2021**, *117*, 108483. [CrossRef]
14. Zuo, Y.; Li, T.; Yu, P.; Zhao, Z.; Chen, X.; Zhang, Y.; Chen, F. Effect of graphene oxide additive on tribocorrosion behavior of MAO coatings prepared on Ti6Al4V alloy. *Appl. Surf. Sci.* **2019**, *480*, 26–34. [CrossRef]
15. Parichehr, R.; Dehghanian, C.; Nikbakht, A. Preparation of PEO/silane composite coating on AZ31 magnesium alloy and in-vestigation of its properties. *J. Alloys Compd.* **2021**, *876*, 159995. [CrossRef]
16. Sopchenski, L.; Robert, J.; Touzin, M.; Tricoteaux, A.; Olivier, M.-G. Improvement of wear and corrosion protection of PEO on AA2024 via sol-gel sealing. *Surf. Coat. Technol.* **2021**, *417*, 127195. [CrossRef]
17. Rizwan, M.; Alias, R.; Zaidi, U.Z.; Mahmoodian, R.; Hamdi, M. Surface modification of valve metals using plasma electrolytic oxidation for antibacterial applications: A review. *J. Biomed. Mater. Res. Part A* **2017**, *106*, 590–605. [CrossRef]
18. Sun, W.; Liu, Y.; Li, T.; Cui, S.; Chen, S.; Yu, Q.; Wang, D. Anti-corrosion of amphoteric metal enhanced by MAO/corrosion inhibitor composite in acid, alkaline and salt solutions. *J. Colloid Interface Sci.* **2019**, *554*, 488–499. [CrossRef]
19. Mashtalyar, D.V.; Sinebryukhov, S.L.; Imshinetskiy, I.M.; Gnedenkov, A.S.; Nadaraia, K.V.; Ustinov, A.Y.; Gnedenkov, S.V. Hard wearproof PEO-coatings formed on Mg alloy using TiN nanoparticles. *Appl. Surf. Sci.* **2019**, *503*, 144062. [CrossRef]
20. Li, X.; Dong, C.; Zhao, Q.; Pang, Y.; Cheng, F.; Wang, S. Characterization of Microstructure and Wear Resistance of PEO Coatings Containing Various Microparticles on Ti6Al4V Alloy. *J. Mater. Eng. Perform.* **2018**, *27*, 1642–1653. [CrossRef]
21. Jiang, B.L.; Wang, Y.M. Plasma electrolytic oxidation treatment of aluminium and titanium alloys. In *Surface Engineering of Light Alloys*; Elsevier: Amsterdam, The Netherlands, 2010; pp. 110–154. [CrossRef]
22. Yao, Z.; Xu, Y.; Jiang, Z.; Wang, F. Effects of cathode pulse at low frequency on the structure and composition of plasma elec-trolytic oxidation ceramic coatings. *J. Alloys Compd.* **2009**, *488*, 273–278. [CrossRef]
23. Rafieerad, A.; Ashra, M.; Mahmoodian, R.; Bushroa, A. Surface characterization and corrosion behavior of calcium phosphate-base composite layer on titanium and its alloys via plasma electrolytic oxidation: A review paper. *Mater. Sci. Eng. C* **2015**, *57*, 397–413. [CrossRef] [PubMed]
24. Gowtham, S.; Hariprasad, S.; Arunnellaiappan, T.; Rameshbabu, N. An investigation on ZrO_2 nanoparticle incorporation, surface properties and electrochemical corrosion behaviour of PEO coating formed on Cp-Ti. *Surf. Coat. Technol.* **2017**, *313*, 263–273.
25. Cerchier, P.; Pezzato, L.; Brunelli, K.; Dolcet, P.; Bartolozzi, A.; Bertani, R.; Dabalà, M. Antibacterial effect of PEO coating with silver on AA7075. *Mater. Sci. Eng. C* **2017**, *75*, 554–564. [CrossRef] [PubMed]
26. Xia, Z.; Min, J.; Zhou, S.; Ma, H.; Zhang, B.; Tang, X. Photocatalytic performance and antibacterial mechanism of Cu/Ag-molybdate powder material. *Ceram. Int.* **2021**, *47*, 12667–12679. [CrossRef]
27. Lu, M.; Zhang, Z.; Zhang, J.; Wang, X.; Qin, G.; Zhang, E. Enhanced antibacterial activity of Ti-Cu alloy by selective acid etching. *Surf. Coat. Technol.* **2021**, *421*, 127478. [CrossRef]
28. Athinarayanan, J.; Periasamy, V.S.; Krishnamoorthy, R.; Alshatwi, A.A. Evaluation of antibacterial and cytotoxic properties of green synthesized Cu_2O/Graphene nanosheets. *Mater. Sci. Eng. C* **2018**, *93*, 242–253. [CrossRef]

29. Stranak, V.; Wulff, H.; Rebl, H.; Zietz, C.; Arndt, K.; Bogdanowicz, R.; Nebe, B.; Bader, R.; Podbielski, A.; Hubicka, Z.; et al. Deposition of thin titanium–copper films with antimicrobial effect by advanced magnetron sputtering methods. *Mater. Sci. Eng. C* **2011**, *31*, 1512–1519. [CrossRef]
30. Zhang, X.; Wu, Y.; Wang, J.; Xia, X.; Lv, Y.; Cai, G.; Liu, H.; Xiao, J.; Liu, B.; Dong, Z. Microstructure, formation mechanism and antifouling property of multi-layered Cu-incorporated Al2O3 coating fabricated through plasma electrolytic oxidation. *Ceram. Int.* **2019**, *46*, 2901–2909. [CrossRef]
31. Li, X.; Guo, P.; Sun, L.; Wang, A.; Ke, P. Ab Initio Investigation on Cu/Cr Codoped Amorphous Carbon Nanocomposite Films with Giant Residual Stress Reduction. *ACS Appl. Mater. Interfaces* **2015**, *7*, 27878–27884. [CrossRef]
32. Wu, Y.; Zhou, S.; Zhao, W.; Ouyang, L. Comparative corrosion resistance properties between (Cu, Ce)-DLC and Ti co-doped (Cu, Ce)/Ti-DLC films prepared via magnetron sputtering method. *Chem. Phys. Lett.* **2018**, *705*, 50–58. [CrossRef]
33. Cui, P.; Li, W.; Liu, P.; Zhang, K.; Ma, F.; Chen, X.; Feng, R.; Liaw, P.K. Effects of nitrogen content on microstructures and mechanical properties of (AlCrTiZrHf)N high-entropy alloy nitride films. *J. Alloys Compd.* **2020**, *834*, 155063. [CrossRef]
34. Xu, Y.; Li, G.; Li, G.; Gao, F.; Xia, Y. Effect of bias voltage on the growth of super-hard (AlCrTiVZr)N high-entropy alloy nitride films synthesized by high power impulse magnetron sputtering. *Appl. Surf. Sci.* **2021**, *564*, 150417. [CrossRef]
35. Yang, W.; Ke, P.; Fang, Y.; Zheng, H.; Wang, A. Microstructure and properties of duplex (Ti:N)-DLC/MAO coating on magnesium alloy. *Appl. Surf. Sci.* **2013**, *270*, 519–525. [CrossRef]
36. Mazinani, A.; Nine, M.J.; Chiesa, R.; Candiani, G.; Tarsini, P.; Tung, T.T.; Losic, D. Graphene oxide (GO) decorated on multi-structured porous titania fabricated by plasma electrolytic oxidation (PEO) for enhanced antibacterial performance. *Mater. Des.* **2021**, *200*, 109443. [CrossRef]
37. Sunilraj, S.; Blessto, B.; Sivaprasad, K.; Muthupandi, V. Microstructural and corrosion behavior of MAO coated 5052 aluminum alloy. *Mater. Today Proc.* **2020**, *41*, 1120–1124. [CrossRef]
38. Poulston, S.; Parlett, P.M.; Stone, P.; Bowker, M. Surface oxidation and reduction of CuO and Cu2O studied using XPS and XAES. *Surf. Interface Anal.* **1996**, *24*, 811–820. [CrossRef]
39. Cheng, Y.; Zhu, Z.; Zhang, Q.; Zhuang, X.; Cheng, Y. Plasma electrolytic oxidation of brass. *Surf. Coat. Technol.* **2020**, *385*, 125366. [CrossRef]
40. Shaikh, J.S.; Pawar, R.C.; Moholkar, A.V.; Kim, J.H.; Patil, P.S. CuO-PAA hybrid films: Chemical synthesis and supercapacitor behavior. *Appl. Surf. Sci.* **2011**, *257*, 4389–4397. [CrossRef]
41. Park, J.; Lam, S.S.; Park, Y.K.; Kim, B.J.; An, K.H.; Jung, S.C. Fabrication of Ni/TiO2 visible light responsive photocatalyst for de-composition of oxytetracycline. *Environ. Res.* **2023**, *216*, 114657. [CrossRef]
42. Tang, Y.; Hu, X.; Liu, Y.; Chen, Y.; Zhao, F.; Zeng, B. An antifouling electrochemiluminescence sensor based on mesoporous CuO2@SiO2/luminol nanocomposite and co-reactant of ionic liquid functionalized boron nitride quantum dots for ultrasensitive NSE detection. *Biosens. Bioelectron.* **2022**, *214*, 114492. [CrossRef]
43. Xie, R.; Lin, N.; Zhou, P.; Zou, J.; Han, P.; Wang, Z.; Tang, B. Surface damage mitigation of TC4 alloy via micro arc oxidation for oil and gas exploitation application: Characterizations of microstructure and evaluations on surface performance. *Appl. Surf. Sci.* **2018**, *436*, 467–476. [CrossRef]
44. Yang, X.; Xiong, J.; Guo, Z.; Wu, B.; You, Q.; Liu, J.; Deng, C.; Fang, D.; Gou, S.; Yu, Z.; et al. Effects of CrMnFeCoNi additions on microstructure, mechanical properties and wear resistance of Ti (C, N)-based cermets. *J. Mater. Res. Technol.* **2022**, *17*, 2480–2494. [CrossRef]
45. Santaella-González, J.B.; Hernández-Torres, J.; Morales-Hernández, J.; Flores-Ramírez, N.; Ferreira-Palma, C.; Rodríguez-Jiménez, R.C.; García-González, L. Effect of the number of bilayers in Ti/TiN coatings on AISI 316L deposited by sputtering on their hardness, adhesion, and wear. *Mater. Lett.* **2022**, *316*, 132037. [CrossRef]
46. Chowdhury, M.A.; Hossain, N.; Al Masum, A.; Islam, M.S.; Shahin, M.; Hossain, M.I.; Nandee, M.R. Surface coatings analysis and their effects on reduction of tribological properties of coated aluminum under motion with ML approach. *Mater. Res. Express* **2021**, *8*, 086508. [CrossRef]
47. Wang, S.Q.; Wang, Y.M.; Zou, Y.C.; Chen, G.L.; Wang, Z.; Ouyang, J.H.; Jia, D.C.; Zhou, Y. Generation, Tailoring and Functional Ap-plications of Micro-Nano Pores in Microarc Oxidation Coating: A Critical Review. *Surf. Technol.* **2021**, *250*, 1–22.
48. Dan, M.; Tong, H.H.; Shen, L.R. The effect of sodium hexametaphosphate on the structure and corrosion resistance of LD7 aluminum alloy micro-arc oxidation ceramic film. *Mater. Prot.* **2011**, *44*, 11–13.
49. Li, J.Z.; Shao, Z.C.; Tian, Y.W. The action of the forms of P element on the process of the micro arc oxidation. *J. Chin. Soc. Corros. Prot.* **2004**, *24*, 222–225.
50. Zhang, W.; Zhang, S.; Liu, H.; Ren, L.; Wang, Q.; Zhang, Y. Effects of surface roughening on antibacterial and osteogenic properties of Ti-Cu alloys with different Cu contents. *J. Mater. Sci. Technol.* **2021**, *88*, 158–167. [CrossRef]
51. Meyers, A.; Furtmann, C.; Jose, J. Direct optical density determination of bacterial cultures in microplates for high-throughput screening applications. *Enzym. Microb. Technol.* **2018**, *118*, 1–5. [CrossRef]
52. Lee, Y.-J.; Kim, S.; Park, S.-H.; Park, H.; Huh, Y.-D. Morphology-dependent antibacterial activities of Cu2O. *Mater. Lett.* **2011**, *65*, 818–820. [CrossRef]
53. Zhang, J.; Liu, J.; Peng, Q.; Wang, A.X.; Li, Y. Nearly Monodisperse Cu2O and CuO Nanospheres: Preparation and Applications for Sensitive Gas Sensors. *Chem. Mater.* **2006**, *18*, 867–871. [CrossRef]

54. Pang, H.; Gao, F.; Lu, Q. Morphology effect on antibacterial activity of cuprous oxide. *Chem. Commun.* **2009**, *9*, 1076–1078. [CrossRef]
55. Ren, J.; Wang, W.; Sun, S.; Zhang, L.; Wang, L.; Chang, J. Crystallography Facet-Dependent Antibacterial Activity: The Case of Cu_2O. *Ind. Eng. Chem. Res.* **2011**, *50*, 10366–10369. [CrossRef]
56. Jung, H.Y.; Seo, Y.; Park, H.; Huh, Y.D. Morphology-controlled synthesis of octahedral-to-rhombic dodecahedral Cu_2O microcrystals and shape-dependent antibacterial activities. *Bull. Kor. Chem. Soc.* **2015**, *36*, 1828–1833. [CrossRef]
57. Kabir, M.H.; Ibrahim, H.; Ayon, S.A.; Billah, M.; Neaz, S. Structural, nonlinear optical and antimicrobial properties of sol-gel derived, Fe-doped CuO thin films. *Heliyon* **2022**, *8*, e10609. [CrossRef]
58. Ayon, S.A.; Billah, M.M.; Nishat, S.S.; Kabir, A. Enhanced photocatalytic activity of Ho3þ doped ZnO NPs synthesized by modified sol-gel method: An experimental and theoretical investigation. *J. Alloys Compd.* **2021**, *856*, 158217. [CrossRef]
59. Das, D.; Nath, B.C.; Phukon, P.; Dolui, S.K. Synthesis and evaluation of antioxidant and antibacterial behavior of CuO nanoparticles. *Colloids Surf. B Biointerfaces* **2013**, *101*, 430–433. [CrossRef]
60. Liu, S.; Zhang, Z.; Zhang, J.; Qin, G.; Zhang, E. Construction of a TiO_2/Cu_2O multifunctional coating on Ti-Cu alloy and its influence on the cell compatibility and antibacterial properties. *Surf. Coat. Technol.* **2021**, *421*, 127438. [CrossRef]

Article

Thermal Stability of the Copper and the AZO Layer on Textured Silicon

Ping-Hang Chen, Wen-Jauh Chen * and Jiun-Yi Tseng

Graduate School of Materials Science, National Yunlin University of Science and Technology, 123 University Road, Section 3, Douliou 64002, Taiwan; M108470121@yuntech.edu.tw (P.-H.C.); jytseng@yuntech.edu.tw (J.-Y.T.)
* Correspondence: chenwjau@yuntech.edu.tw; Tel.: +886-5-534-2601-3069

Abstract: Transparent conductive oxide (TCO) film is the most widely used front electrode in silicon heterojunction (SHJ) solar cells. A copper metallization scheme can be applied to the SHJ process. The abundance of zinc in the earth's crust makes aluminum-doped zinc oxide (AZO) an attractive low-cost substitute for indium-based TCOs. No work has focused on the properties of the copper and AZO layers on the textured silicon for solar cells. This work deposited an aluminum-doped zinc oxide layer and copper metal layer on textured (001) silicon by a sputtering to form Cu/AZO/Si stacks. The structures of Cu/AZO/Si are characterized by scanning electron microscope (SEM), scanning transmission electron microscope (STEM), and energy-dispersive X-ray spectrometer (EDS). The results show that the copper thin film detached from AZO in the valley of the textured silicon substrate at a temperature of 400 °C. Additionally, the gap between the copper and AZO layers increases as temperature increases, and the 65 nm thickness AZO layer was found to be preserved up to 800 °C.

Keywords: textured silicon; aluminum-doped zinc oxide; solar cells; diffusion barrier; thermal stability

1. Introduction

In silicon heterojunction (SHJ) solar cells, the hydrogenated amorphous silicon (a-Si:H) thin layer is too low in transverse conductivity to collect charge carriers horizontally over the metal electrodes effectively. Additional transparent conductive oxide layers such as Sn-doped In_2O_3 (ITO) are deposited at the top. In addition to charge collection, another essential function of the front-end transparent conductive oxide (TCO) layer is to act as an anti-reflective layer. Because ITO has excellent performance with low resistivity and high transparency, ITO is currently the most widely used transparent conductive oxide [1]. However, In_2O_3 has the disadvantage of high refining and production costs and the relative scarcity of indium ore. Sn is also susceptible to price fluctuations (compared to other elements (e.g., Zn). Finding a replacement for ITO could reduce the costs of the production of SHJ solar cells. Aluminum-doped zinc oxide (AZO) is based on zinc oxide. Zinc oxide (ZnO) is a potential candidate in TCO, known as ZnO, a natural type of semiconductor with an energy gap of 3.4 eV (intrinsic n-type semiconductor) [2]. Compared with ITO, AZO has high thermal stability and a low price, and is mineral-rich, relatively non-toxic, and has an increased transparency of the visible light range, as well as other advantages [3–6].

As mentioned above, reducing costs and improving conversion efficiency have always been the solar cell industry's theme. With the continuous progress of solar cell industry technology and policy promotion, the public's attention gradually shifted to the cost of electricity, and high-efficiency batteries have attracted attention. The three high costs of SHJ cells are for the silicon chip, conductive silver paste, and target material. Because of these three high-cost components, various cost reductions will help improve SHJ cells' competitiveness, including reducing the consumption of raw materials and introducing new technologies.

In SHJ technology, the low-temperature silver paste is used. Still, the conductivity of low-temperature silver paste is significantly lower than that of bulk silver, so the conductivity of silver electrodes in SHJ solar cells is lower than that of silver electrodes in c-Si solar cells. Therefore, the use of silver paste in a typical 3–5 bus configuration limits efficiency and increases the cost of SHJ solar cells, estimated to cost 30% due to the use of a large number of expensive low-temperature net-printed silver paste processes. The electroplating technology can effectively reduce the cost of electrodes. The Cu metallization process introduces to SHJ solar cells, and it is crucial to improve the SHJ solar cell industry's competitiveness. At present, there are some reports on the application of copper plating technology to SHJ solar cells [7–10].

Li et al. used a rotating coating method to coat the resin (photo-resistance) on the n-type 156 mm SHJ surface [10]. Then pattern the resin by the inkjet printer (Dimatix DMP 2800). Copper is then plated to the p-i type a-Si: H's ITO surface using field-induced power generation plating (FIP). The efficiency of the SHJ solar cells produced by Li is 18.8% (Voc s 717 mV, Jsc s 35.4 mA/cm^2, and F.F. s 74%). Dabirian et al. introduced the double-mask layer strategy for the metallization of SHJ solar cells. They combined the nanosecond laser patterning and Ni-Cu electroplating. The results showed that the Ni-Cu metallization qualities of their process are comparable to Ag screen-printing and lithography-based Cu plating [11]. Meza et al. compared ZnO:Al films instead of the usual ITO as the front TCO in rear emitter SHJ solar cells. An indium-free mono facial cell achieved η = 22.5% showing that the replacement of ITO/Ag by ZnO:Al/Ag as a back contact produces cells with practically the same efficiency [12].

In SHJ technology, the metallic contacts are generally deposited on a transparent conductive oxide (TCO). Contact metallization on TCOs is typically performed by screen-printing of low-temperature Ag pastes. Electroplating of copper is becoming more attractive to reduce precious Ag consumption. However, copper has a high diffusion coefficient and high solubility in silicon and formation copper silicide at low temperature. TCO can also serve as a barrier to Cu migration. Additionally, the Cu directly electroplated on TCO is usually tricky. Kang et al. also reported the electroplated copper films fail to adhere to Si during rapid heating and cooling [13]. Thus, an additional seed layer is usually used. The plating process sequence involves seed layers usually deposited by physical vapor deposition (PVD). Additionally, copper electroplating by the D.C. power supplier is the most commonly adopted method.

Li et al. and Dabirian et al. confirmed that the plated copper process could be applied to SHJ solar cells. However, their reports focused on the electroplating Ni-Cu (or Cu) on ITO [10,11]. Meza et al. showed that the ITO/Ag could be replaced by ZnO:Al/Ag [12]. There have been no reports about the electroplating Cu on AZO till now. Additionally, about the role of seed copper layer on AZO. In this work, we studied the thermal stability of the copper seed layer and the AZO layer on the textured silicon.

2. Materials and Methods

As substrate, commercially available single crystal phosphorus-doped (0 0 1) oriented silicon wafers with textured roughness around 3–5 μm were used. The acetone and H_2SO_4/H_2O_2 solution were applied to clean the textured silicon substrate. To remove the native oxide of the silicon, the substrate dip into hydrogen fluoride solution before loading into the vacuum chamber. The AZO and copper films were sputter-deposited onto textured silicon substrates in a direct current/radio-frequency (dc/rf) magnetron sputtering system. The AZO films were sputtered from an AZO target with an rf power supply in an Ar ambient of 99.999% purity. The base pressure of the vacuum chamber was 2×10^{-7} Torr. The rf power was held at 60 W during deposition. Additionally, a fixed Ar flow rate and the operation pressure were 50 sccm and 6×10^{-3} Torr, respectively. A fixed Ar flow rate of 25 sccm and the dc power held at 30 W during the copper film deposition.

The AZO and copper films were deposited onto a textured silicon substrate at 25 °C during the sputtering process. The sputtered AZO and copper films' thicknesses were 65 nm and 220 nm, respectively. AZO films and copper films were subsequently sputter-deposited without breaking the vacuum. The samples are designed for Cu/AZO/Si.

To evaluate the copper seed layer and the barrier property of the AZO layer, an annealing temperature from 300 to 800 °C was selected. The as-deposited samples were annealed in the furnace at 300–800 °C for 10 min in an Ar/H_2 atmosphere. The structure was obtained using a scanning electron microscope (SEM, JEOL Ltd., Tokyo, Japan) and a scanning transmission electron microscope (STEM, JEOL Ltd. L, Tokyo, Japan). SEM and STEM were performed on JSM-6360 and a JEM-ARM200, respectively. A STEM that was equipped with an energy dispersive X-ray spectrometer (EDS, Oxford Instruments, Abingdon, UK) was used to determine the chemical composition and STEM-EDS compositional maps of all samples. A focus ion beam (FIB) resembled a scanning electron microscope (SEM) operating at 20 kV and was used for a cross-sectional view of SEM examination. The sample's surface is protected with an about 2 μm Pt layer formed by an electron beam within the FIB chamber. The sample was then ion-milled using a focused beam of gallium ions to give a cross-section view. The tape and peel-off tests were conducted on the as-deposited sample. For the tape and peel-off tests, about 15 μm thick copper layer was deposited on the Cu/AZO/Si by electroplating. Tape tests were conducted by peeling the 3 M Scotch tape off at 180°. The peel-off tests were conducted at an angle of 180° with a constant speed of 30 mm/min.

3. Results and Discussion

SEM images for the Cu/AZO/Si samples annealed at 300, 500, 600, and 800 °C are presented in Figure 1. The pyramid shapes with uniform Cu and AZO thin layers exist on the surface of the samples after annealing at 300 °C (Figure 1a). The surface is smooth for the annealing of the samples at temperatures 300 °C. The surface of the as-deposited sample is the same as the sample annealing at 300 °C for 10 min. The surface became rough when the annealing temperature was higher than 500 °C, as shown in Figure 1b–d.

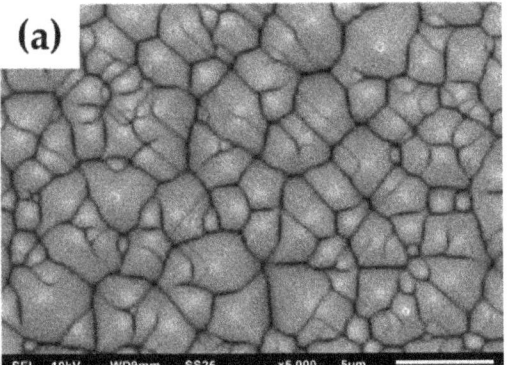 

Figure 1. *Cont.*

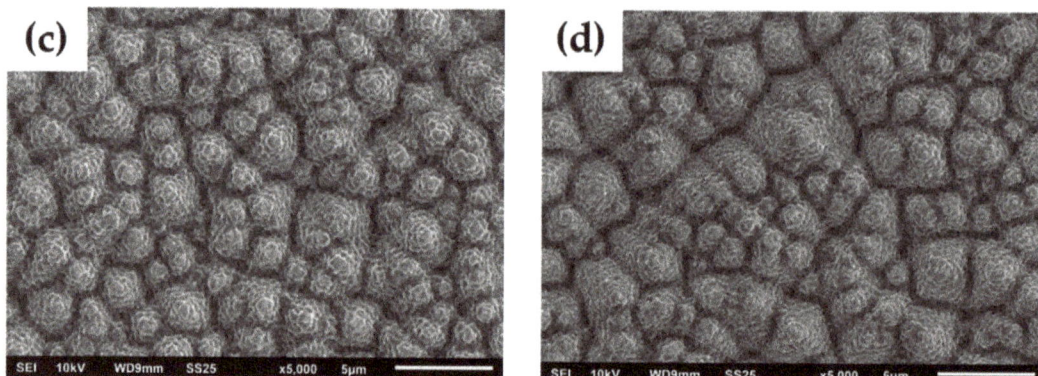

Figure 1. SEM images for the Cu/AZO/Si samples annealed at (**a**) 300, (**b**) 500, (**c**) 600, and (**d**) 800 °C.

The cross-sectional views of SEM for the Cu/AZO/Si samples annealed at 400, 600, and 800 °C are shown in Figure 2. The copper thin films were continuous for all samples. However, the thin films (Cu/AZO or Cu) were detached from the substrate (silicon or AZO/Si substrate) in some areas for the samples after 400 °C annealing. It was hard to know the composition of the detached thin film (Cu/AZO or Cu) from FIB-SEM. Additionally, the separation of thin films and silicon substrate (or AZO/Si) became severe at temperatures of 600 and 800 °C. It was also hard to confirm the gap formed at the Cu and AZO/Si or Cu/AZO and silicon interfaces.

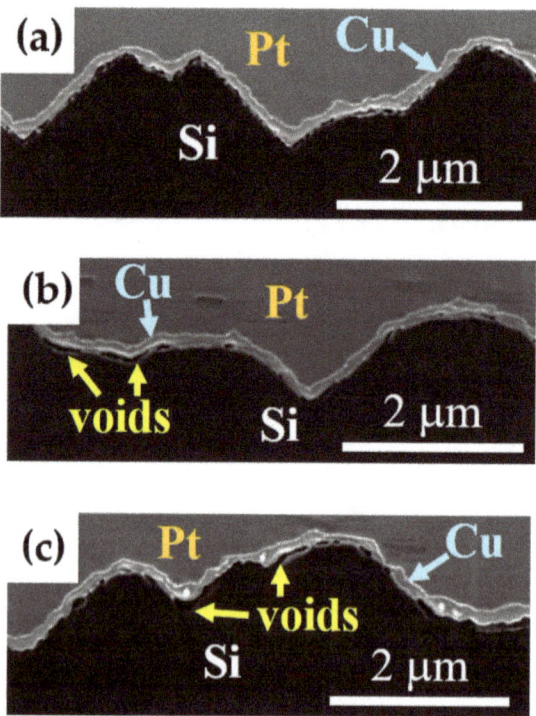

Figure 2. The cross-sectional views of SEM for the Cu/AZO/Si samples were annealed at (**a**) 400, (**b**) 600, and (**c**) 800 °C.

The low and high magnification TEM micrographs of Cu/AZO/Si sample annealed at 300 °C are shown in Figure 3a,b, respectively. The multilayer is preserved after annealing at 300 °C for 10 min. The structure of the as-deposited sample is the same as the sample annealing at 300 °C for 10 min. Figure 3c–f shows the EDS maps of Si, Cu, Zn, and O, respectively. The Zn and O elements are overlapped and located between the copper and silicon elements. The elemental maps reveal that the Cu/AZO/Si stack structure is very stable after annealing at 300 °C for 10 min. The line scan across the Cu/AZO/Si using STEM-EDS is shown in Figure 4a. Figure 4b shows the intensity signal of the Si, Zn, O, Al, and Cu elements along the yellow line, shown in Figure 4a. The signal of the aluminum element is almost at the noise level due to the small aluminum amount in the AZO layer. The spot analysis of the AZO layer is shown in Figure 4c. The amount of aluminum is only about 1.3 at.%. The distribution of zinc and oxygen in the Cu/AZO/Si stack is located at a scale between 180 and 245 nm, which reveals that the thickness of the AZO layer is near 65 nm. There were interfacial layers at the scale of 165–190 nm and 235–260 nm (Figure 4b). The width of the interface layer was caused by the situation where the interface plane was not perpendicular to the figure shown in Figure 4a. The thickness of the AZO thin film is also near 65 nm, from Figure 3b–f. The thickness of the copper thin film is about 220 nm, as shown in Figure 3b,d. Only a silicon signal emerges at a scale below 160 nm. Additionally, the copper signal was presented at a scale almost greater than 260 nm. The line scan shows that Cu/AZO/Si structure is maintained after annealing at 300 °C. The result of the line scan (Figure 4) agrees with the work of the TEM and EDS map (Figure 3).

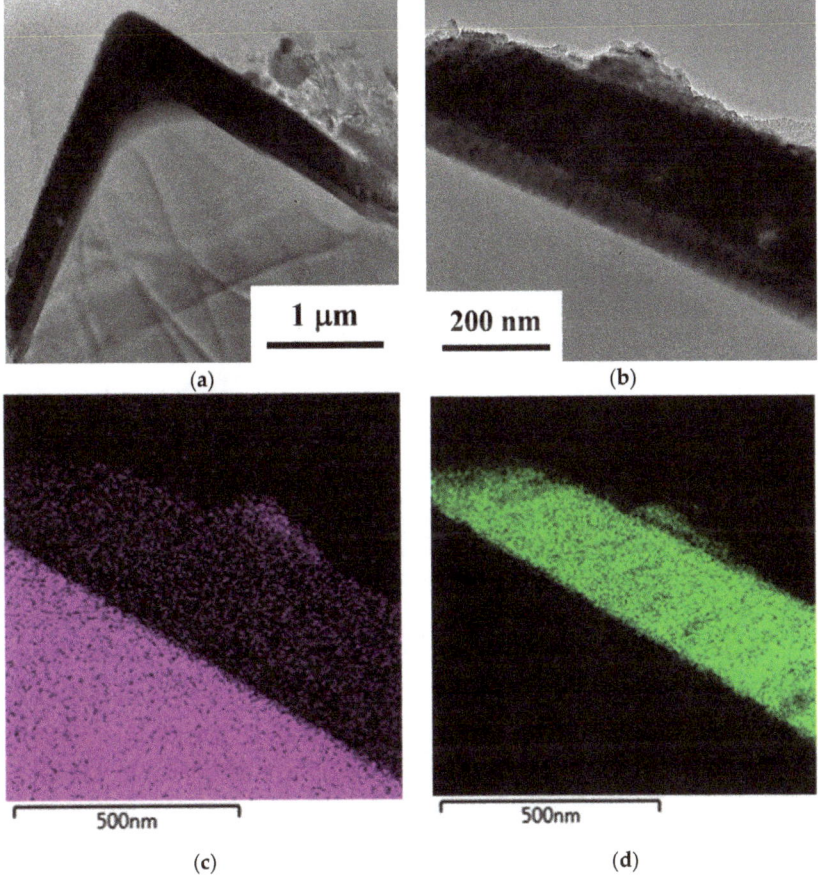

Figure 3. Cont.

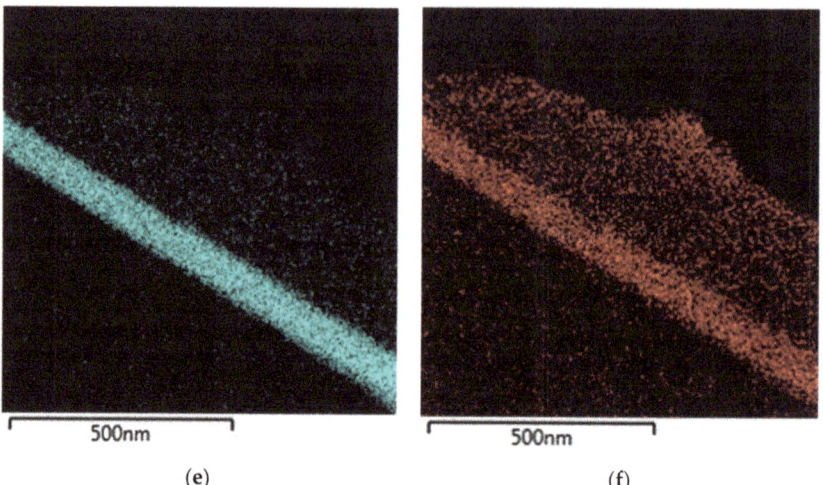

Figure 3. (a) Low and (b) high magnification TEM micrographs of Cu/AZO/Si sample annealed at 300 °C, respectively. EDS map of (c) Si, (d) Cu, (e) Zn, and (f) O elements.

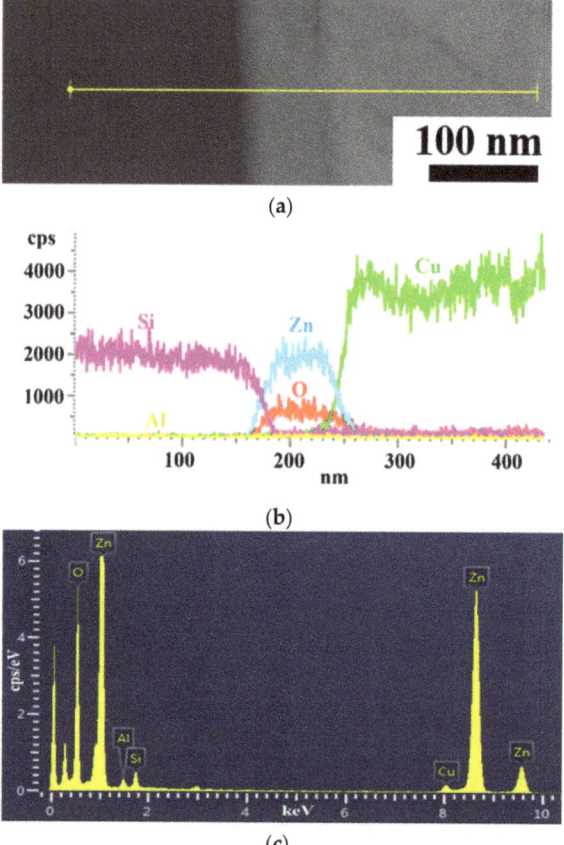

Figure 4. STEM-EDS line scan across the Cu/AZO/Si sample annealed at 300 °C. (a) STEM image and (b) intensity signal of Si, Zn, O, Al, and Cu elements along the yellow line. (c) EDS spectra of AZO layer.

For the annealing of the samples at temperatures of 500, 600, and 700 °C, STEM images also confirm that the copper thin film detached from AZO/Si in the valley of the textured silicon substrate after annealing. Additionally, the gap between the copper and AZO layers increases as temperature increases. For the sample annealed at 800 °C, the TEM and STEM micrographs of Cu/AZO/Si are presented in Figure 5a,b, respectively. Additionally, the EDS map of Si, Cu, Zn, and O is illustrated in Figure 5c–f, respectively. The copper layer starts to agglomerate and almost leaves the textured structure's valley area from Figure 5. The AZO layer seems stable even after annealing at 800 °C, and the copper element stays above the AZO layer. The copper layer also presents the oxygen signal from Figures 3f and 5f. The origin of oxygen migration to the copper layer needs further identification. The STEM-EDS evidence that copper does not diffuse through the AZO layer to the silicon substrate. Therefore, the AZO thin film is also an excellent diffusion barrier layer to prevent copper diffuse into the silicon substrate.

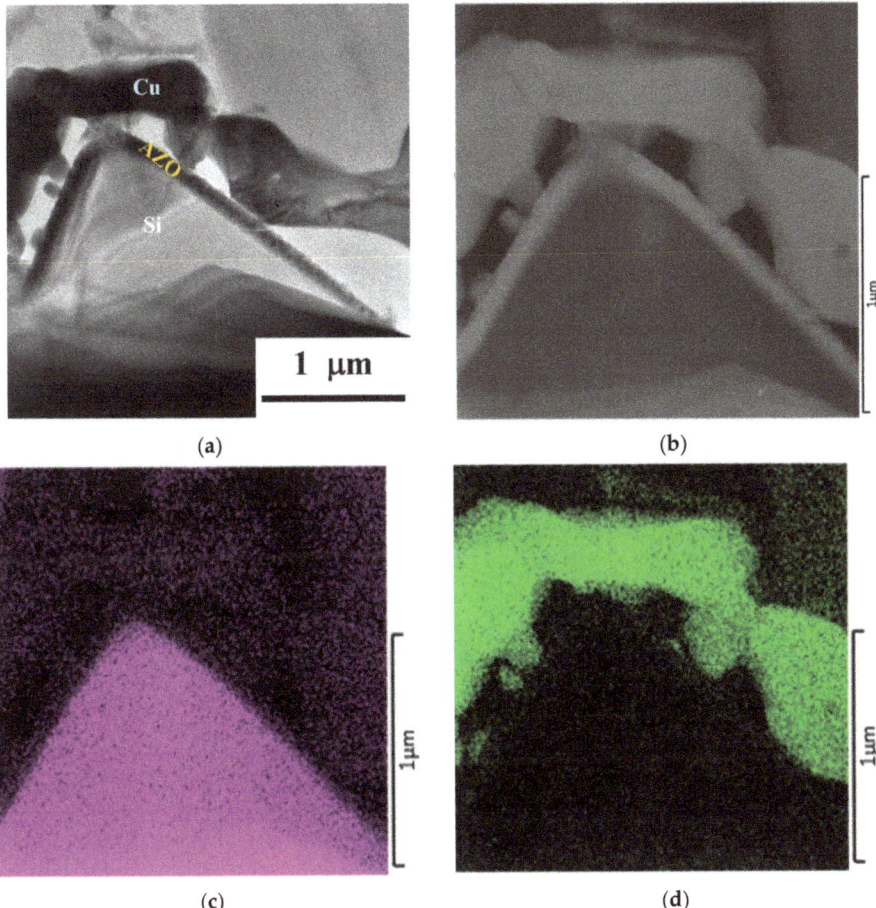

Figure 5. Cont.

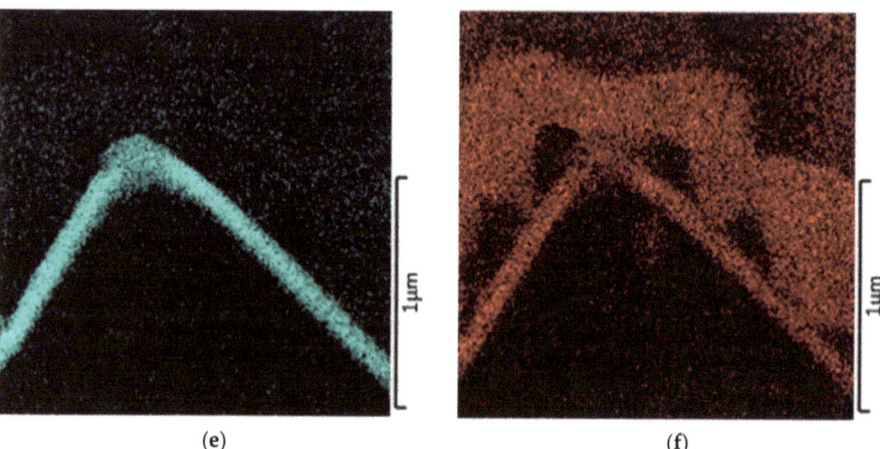

Figure 5. (**a**) TEM micrograph and (**b**) STEM micrograph for the sample annealed at 800 °C for 10 min. Additionally, EDS maps of (**c**) Si, (**d**) Cu, (**e**) Zn, and (**f**) O elements.

To understand the formation of the gap at the copper and AZO/Si interface, the tape and peel-off tests were used to evaluate copper's adhesion on the AZO/Si substrate. About 15 μm thick copper layer deposited on the as-deposited Cu/AZO/Si by electroplating for tape and peel-off tests. The sample is designed as E-Cu/Cu/AZO/Si. No copper remained on the AZO/Si after the 180° tape test. Figure 6 shows the peel force diagram for the E-Cu/Cu/AZO/Si sample. The maximum and average peel force values were 0.39 and 0.2 N/mm, respectively. The results indicated that the adhesion between copper and AZO/Si is weak. It can imply that the adhesion strength will decrease as annealed temperature increases due to the voids emerging when the temperature is higher than 400 °C. Lee also reported a copper seed layer deposited on an indium tin oxide (ITO) by electron-beam evaporation [14]. After the copper seed layers were deposited, copper/silver (Cu/Ag) metal stacks were plated on the sample using the light-induced plating (LIP) technique. Additionally, adhesive contact between copper and ITO was evaluated by the tape test [14]. The results show that pure copper was mostly detached from the ITO surface after the tape test. Their results also indicated that the adhesive force between copper and ITO/Si is very weak. Kang deposited the (Ti(30 nm)/Cu(100 nm)) on p-type (1 0 0) Si as a seed layer. Then thick copper was electroplated on the seed layer [13]. They found that rapid thermal annealing (RTA) caused the electroplated films to fail to adhere to the Si. The adhesion strength of copper and ITO or Si was deficient from Lee's and Kang's reports. Our result shows that the adhesion between copper and AZO/Si is very weak, from the tape test. Additionally, the gap formed at the copper and AZO/Si interface easily after high-temperature annealing. It can imply that the copper detached from the AZO/textured silicon after annealing due to the weak adhesion of copper thin films on the AZO/Si. Additionally, the separation of copper and AZO/Si became serious at high-temperature treatments.

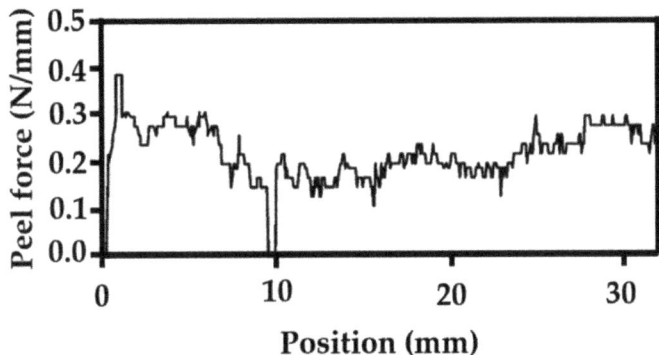

Figure 6. The peel-force measurement of E-Cu/Cu/AZO/Si sample.

4. Conclusions

This paper reports the thermal stability of the copper layer and the AZO layer on textured silicon. The Cu/AZO/Si stack seems preserved after annealing at temperatures of 300–800 °C. Although the copper thin film detached from the AZO/Si substrate, the sputtering AZO layer with a thickness of 65 nm was a sound diffusion barrier against Cu up to 800 °C. The copper thin film detached from AZO in the valley of the textured silicon substrate at 400 °C. Additionally, the gap between the copper and AZO layers increases as temperature increases. It can imply that the copper detachment from AZO occurs due to the weak adhesive strength of copper thin films on the AZO.

Author Contributions: Performing the experiments and data collection, P.-H.C.; conceptualization, funding acquisition, writing—original draft preparation, reviewing and editing, supervision, project administration, W.-J.C.; writing—review & editing, supervision, J.-Y.T. All authors have read and agreed to the published version of the manuscript.

Funding: This research was funded by the Ministry of Science and Technology of Taiwan, grant number MOST 109-2221-E-224-034.

Institutional Review Board Statement: Not applicable.

Informed Consent Statement: Not applicable.

Data Availability Statement: Not applicable.

Acknowledgments: This research was funded by the Ministry of Science and Technology of Taiwan, grant number MOST 109-2221-E-224-034. The authors would like to thank Tseng's laboratory at the Graduate School of Materials Science, National Yunlin University of Science and Technology, for the support.

Conflicts of Interest: The authors declare no conflict of interest. The funders had no role in the study's design, in the collection, analyses, or interpretation of data, in the writing of the manuscript, or in the decision to publish the results.

References

1. Tohsophon, T.; Dabirian, A.; Wolf, S.D.; Morales-Masis, M.; Ballif, C. Environmental stability of high-mobility indium-oxide based transparent electrodes. *APL Mater.* **2015**, *3*, 116105. [CrossRef]
2. Sernelius, B.E.; Berggren, K.F.; Jin, Z.C.; Hamberg, I.; Granqvist, G. Band-gap tailing of ZnO by means of heavy Al doping. *Phys. Rev. B* **1988**, *37*, 10244–10248. [CrossRef] [PubMed]
3. Cruz, A.; Wang, E.C.; Morales-Vilches, A.B.; Meza, D.; Neubert, S.; Szyszka, B.; Schlatmann, R.; Stannowski, B. Effect of front TCO on the performance of rear-junction silicon heterojunction solar cells: Insights from simulations and experiments. *Sol. Energy Mater. Sol. Cells* **2019**, *195*, 339–345. [CrossRef]
4. Madani Ghahfarokhi, O.; Chakanga, K.; Geissendoerfer, S.; Sergeev, O.; Maydell, K.V.; Agert, C. DC-sputtered ZnO:Al as transparent conductive oxide for silicon heterojunction solar cells with μc-Si:H emitter. *Prog. Photovolt Res. Appl.* **2015**, *23*, 1340. [CrossRef]

5. Senaud, L.L.; Christmann, G.; Descoeudres, A.; Geissbuhler, J.; Barraud, L.; Badel, N.; Allebe, C.; Nicolay, S.; Despeisse, M.; Paviet-Salomon, B.; et al. Aluminium-Doped Zinc Oxide Rear Reflectors for High-Efficiency Silicon Heterojunction Solar Cells. *IEEE J. Photovolt.* **2019**, *9*, 1217. [CrossRef]
6. Wu, Z.P.; Duan, W.Y.; Lambertz, A.; Qiu, D.P.; Pomaska, M.; Yao, Z.R.; Rau, U.; Zhang, L.P.; Liu, Z.X.; Ding, K.N. Low-resistivity p-type a-Si:H/AZO hole contact in high-efficiency silicon heterojunction solar cells. *Appl. Surf. Sci.* **2021**, *542*, 148749. [CrossRef]
7. Geissbühler, J.; De Wolf, S.; Faes, A.; Badel, N.; Jeangros, Q.; Tomasi, A.; Barraud, L.; Descoeudres, A.; Despeisse, M.; Ballif, C. Silicon heterojunction solar cells with copper-plated grid electrodes: Status and comparison with silver thick-film techniques. *IEEE J. Photovolt.* **2014**, *4*, 1055. [CrossRef]
8. Yu, J.; Bian, J.; Duan, W.; Liu, Y.; Shi, J.; Meng, F.; Liu, Z. Tungsten doped indium oxide film: Ready for bifacial copper metallization of silicon heterojunction solar cell. *Sol. Energy Mater Sol. Cells* **2016**, *144*, 359. [CrossRef]
9. Khanna, A.; Ritzau, K.-U.; Kamp, M.; Filipovic, A.; Schmiga, C.; Glatthaar, M.; Aberle, A.G.; Mueller, T. Screen-printed masking of transparent conductive oxide layers for copper plating of silicon heterojunction cells. *Appl. Surf. Sci.* **2015**, *349*, 880. [CrossRef]
10. Li, Z.T.; Hsiao, P.-C.; Zhang, W.; Chen, R.; Yao, Y.; Papet, P.; Lennon, A. Patterning for plated heterojunction cells. *Energy Procedia* **2015**, *67*, 76. [CrossRef]
11. Dabirian, A.; Lachowicz, A.; Schüttauf, J.-W.; Paviet-Salomon, B.; Morales-Masis, M.; Hessler-Wyser, A.; Despeisse, M.; Ballif, C. Metallization of Si heterojunction solar cells by nanosecond laser ablation and Ni-Cu plating. *Sol. Energy Mater. Sol. Cells* **2017**, *159*, 243. [CrossRef]
12. Meza, D.; Cruz, A.; Morales-Vilches, A.B.; Korte, L.; Stannowski, B. Aluminum-Doped Zinc Oxide as Front Electrode for Rear Emitter Silicon Heterojunction Solar Cells with High Efficiency. *Appl. Sci.* **2019**, *9*, 862. [CrossRef]
13. Kang, J.; You, J.S.; Kang, C.S.; Pak, J.J.; Kim, D.W. Investigation of Cu metallization for Si solar cells. *Solar Energy Mater. Sol. Cells* **2002**, *74*, 91–96. [CrossRef]
14. Lee, S.H.; Lee, D.W.; Kim, H.J.; Lee, A.R.; Lee, S.H.; Lim, K.J.; Shin, W.S. Study of Cu-X alloy seed layer on ITO for copper-plated silicon heterojunction solar cells. *Mater. Sci. Semicon. Proc.* **2018**, *87*, 19. [CrossRef]

MDPI
St. Alban-Anlage 66
4052 Basel
Switzerland
www.mdpi.com

Coatings Editorial Office
E-mail: coatings@mdpi.com
www.mdpi.com/journal/coatings

Disclaimer/Publisher's Note: The statements, opinions and data contained in all publications are solely those of the individual author(s) and contributor(s) and not of MDPI and/or the editor(s). MDPI and/or the editor(s) disclaim responsibility for any injury to people or property resulting from any ideas, methods, instructions or products referred to in the content.

www.ingramcontent.com/pod-product-compliance
Lightning Source LLC
LaVergne TN
LVHW070702100526
838202LV00013B/1012